The International Library of Environmental, Agricultural and Food Ethics

Volume 22

More information about this series at http://www.springer.com/series/6215

Bernhard Freyer • Jim Bingen
Editors

Re-Thinking Organic Food and Farming in a Changing World

Springer

Editors
Bernhard Freyer
Department of Sustainable
Agriculture Systems
Division of Organic Farming
University of Natural Resources
and Life Sciences (BOKU)
Vienna, Austria

Jim Bingen
Department of Community Sustainability
Michigan State University
East Lansing, MI, USA

ISSN 1570-3010
ISBN 978-94-017-9189-2 ISBN 978-94-017-9190-8 (eBook)
DOI 10.1007/978-94-017-9190-8
Springer Dordrecht Heidelberg New York London

Library of Congress Control Number: 2014952291

Printed on acid-free paper

Springer is part of Springer Science+Business Media (www.springer.com)

Foreword

I met Bernhard Freyer at a coffee shop on Michigan Avenue in Lansing about 6 years ago. He came with his friend and my Michigan State University colleague Jim Bingen (who I have known for a long time—too long to recall). Bernhard and I hit it off at once, engaging in a very stimulating philosophical conversation that probed the ontological conditions that give rise to that phenomenon we call "ethics." I would like to think that our conversation was (in some way or another) the impetus for this fine collection of essays on ethics and organic farming, though I am confident that my influence was too insignificant to deserve recognition as a contributing cause. Organic farming has almost always been conceptualized in light of ethics and the philosophy of agriculture. Some view it as an outcrop of Rudolf Steiner's metaphysical reconstruction of human rationality, while others see it as the fruit of Sir Albert Howard's ecological understanding of the relationship between soil health, on the one hand, and the health of plants grown in that soil and of animals that eat the plants, on the other. Still others have an ethic grounded in place-based social relationships that emerge among small, diversified farms and the localities of merchants, tradesmen, and professionals that they support—a vision now promulgated by the American poet Wendell Berry. Perhaps the main thing that all such advocates of organic farming would agree upon is that the version of organic food derived from the U.S. Department of Agriculture's organic standard and its various global equivalents fails precisely because of its lack of responsiveness to the ethical commitments articulated in any of these founding statements of the organic creed.

But I would also say that proponents of organic farming have themselves failed to engage in the kind of critical and reflective conversation about ethics that is necessary to give life to ideas. I understand this. Farmers are, in part, venerated by moral philosophers precisely because they are few in words but rich in deeds. Their ethics are embodied in their actions. Yet, there needs to be an ongoing conversation—an ethical discourse—that subjects the implied norms or moral commitments of organic farming to periodic reevaluation and assessment. It is not enough to simply presume that one's own vision is the ethical one or that the practitioners of industrialized farming are venal and lacking in ethics of any kind. Not only will visions and practices of organic farming flourish in response to dialogue and exchange of ideas among the various philosophical strands that have contributed to the organic ethos, it will also challenge the more utilitarian, efficiency-based ethics that continue to rationalize the development of technologically intensive methods of agricultural production. Such conversations and debates are the venue in which ethics are constructed and performed, as much or possibly more than in organic farming practices.

I commend the essays in this volume to a new generation of farmers, eaters, and readers. They exemplify the conversation all of us should be having.

W.K. Kellogg Professor of Agricultural, Paul B. Thompson
Food and Community Ethics, Michigan State University,
East Lansing, MI, USA

Acknowledgment

We are grateful to the organizers of the 2011 Agriculture, Food and Human Values Society Conference in Missoula, Montana, for the opportunity to bring diverse researchers from around the world together in a workshop on "Rethinking Organic." The presentations and discussions gave us the idea to create this book. We thank the contributing authors for their great input and patience with the publication process. In addition, we thank the two anonymous reviewers who gave us profound and constructive feedback that helped us to improve the text and the structure of the book. We especially want to thank Valentin Fiala for his editing assistance. Finally, we thank Dr. Helene Murray and the committee of the Minnesota Institute of Sustainable Agriculture, University of Minnesota, that offered the financial and ideal background for this transatlantic collaboration. We also want to express our greatest appreciation to the University of Natural Resources and Life Sciences, Vienna for supporting the sabbatical of Bernhard Freyer and to the Fulbright Austrian-American Education Commission for the fellowship to Jim Bingen that allowed us to launch our collaboration in Vienna.

Contents

Editors and Contributors

Editors

Dr. Jim Bingen is a Professor Emeritus of Community, Food and Agriculture. He works on a range of food, farming, and rural development issues in Michigan, Western Europe, and French-speaking Africa. He has collaborated with many colleagues across the social and agricultural sciences on numerous applied research studies of farmers, markets, and farmer-market vendors in Michigan; the transition to organic farming by Michigan fruit and vegetable growers; farmer access to organic markets; and, more recently, the contribution of origin- or place-based and quality food to the development in Michigan. He is the Associated Codirector of the Working Group for Transdisciplinary Systems Research at BOKU, the national agricultural school in Vienna, Austria, and holds the *Chevalier d'Ordre du Mérite Agricole* (Order of Agricultural Merit) awarded by the Government of France. From October 2009 to January 2010, he was a Fulbright Distinguished Chair at the University of Natural Resources and Applied Life Sciences in Vienna.

Department of Community Sustainability, Michigan State University, East Lansing, MI, USA

Dr. Bernhard Freyer is a Professor at the University of Natural Resources and Life Sciences, Vienna (Austria), Head of the Division of Organic Farming, Head of the Working Group for Transdisciplinary Systems Research, and Senior Fellow at the University of Minnesota (USA). He leads several research projects in the fields of organic agriculture, sustainable development, and societal transformation and organic agriculture in tropical and subtropical environments. His current research focus is on ethical and philosophical approaches to societal transformation processes and organic agriculture.

Department of Sustainable Agriculture Systems, Division of Organic Farming, University of Natural Resources and Life Sciences (BOKU), Vienna, Austria

Contributors

Dr. Ginevra Adamoli received her PhD in Communication from Florida State University. Her research focuses on social media, food communication, and social movements. She is currently a social media consultant for clients across Europe and the United States. She likes to keep different blogs where she talks about food, social media, and fashion.

School of Communication, Florida State University, Tallahassee, FL, USA

Bradford Barham is a Professor in the Agricultural and Applied Economics Department at the University of Wisconsin-Madison. Brad's research and teaching areas include patterns and effects of agricultural technology/management system adoption and innovation (e.g., organic, fair trade, biotechnology) in Wisconsin and Latin America; the economic logic of specialization, diversification, and structural change in agriculture, and how these patterns relate to productivity and sustainability outcomes; and the impact of land, education, and credit policies and programs on rural poverty and the environment in Latin America.

Agricultural and Applied Economics, University of Wisconsin, Madison, WI, USA

Bridget K. Behe is currently a Professor of Horticulture at Michigan State University. She teaches classes and conducts research on marketing horticultural products, both edible and ornamental. Each year, she teaches courses on marketing and management for horticulture majors at Michigan State. Bridget has conducted over 75 consumer and market research projects, written more than 500 publications in the trade press, and peer-reviewed journals.

Department of Horticulture, Michigan State University, East Lansing, MI, USA

Caroline Brock is an Assistant Professor in the Rural Sociology Department at the University of Missouri-Columbia. Caroline's research interests include value-based farm decision making within a bounded rationality framework among Amish, organic, graziers and conventional dairy farmers in Wisconsin. Caroline's main professional activities include teaching several undergraduate classes about the Amish community and science, technology, and society. She is researching in her course and other similar courses around the country what college students can learn from the Amish as they explore questions of values, technology adoption, and community and relate these questions to their own lives. In addition, she is currently working on creating informational support networks for agricultural and conservation professionals working with Amish and Mennonite farmers in the Midwest.

Rural Sociology, University of Missouri, Columbia, MO, USA

Hugh Campell Over the past decade and a half, Hugh Campbell's research has mainly focused on the social dynamics involved in sustainable agriculture. He has been one of the research leaders of the $13.5 m Agricultural Research Group on

Sustainability (ARGOS) program, a joint venture between the Agribusiness Group, Lincoln University, and the University of Otago.

Department of Sociology, Gender and Social Work, University of Otago, Dunedin, New Zealand

Dr. Jin Young Choi is an Associate Professor in the Department of Sociology at Sam Houston State University. Her research focuses on health, quality of life, and farming in rural communities.

Department of Sociology, Sam Houston State University, Huntsville, TX, USA

David S. Conner is an Assistant Professor in the Department of Community Development and Applied Economics at the University of Vermont. His teaching and research interests are centered on food systems economics and contributions to sustainable community development. His topics of interest include consumer demand, business management and entrepreneurship, and alternative supply chains. He holds a PhD in Agricultural Economics from Cornell University.

Community Development and Applied Economics, University of Vermont, Burlington, VT, USA

Dr. Douglas H. Constance is Professor of Sociology at Sam Houston State University. His research focuses on the community impacts of the industrialization and globalization of the agrifood system and alternative agrifood systems. He is Past President of the Southern Rural Sociological Association and the Agriculture, Food and Human Values Society.

Damian Lara is a master's candidate in the Department of Sociology at Sam Houston State University. His research interests include organizational practices of health-care actors and institutions and the resulting clinical outcomes.

Department of Sociology, Sam Houston State University, Huntsville, TX, USA

John Fairweather made a major contribution to the management of the ARGOS program and provided research results to sector organizations in order to better chart pathways to sustainability. His research work included a specific focus on the social aspects of different farming systems and a general focus on integrating the results by way of transdisciplinary research.

Agribusiness and Economics Research Unit, Lincoln University, Lincoln, New Zealand

Kristin L. Getter is an outreach specialist at Michigan State University. Her professional interests include educating and performing applied research to benefit Michigan producers.

Department of Horticulture, Michigan State University, East Lansing, MI, USA

Rafi Grosglik is a PhD candidate in the Department of Sociology and Anthropology, Ben-Gurion University of the Negev. He holds a master's degree in Sociology and Anthropology. Since 2008, he has been researching the cultural field of organic food in Israel. His dissertation deals with the cultural globalization and sociology of Israeli culinary culture.

Department of Sociology and Anthropology, Ben-Gurion University of the Negev, Beer-Sheva, Israel

Dr. Maki Hatanaka is an Assistant Professor of Sociology at Sam Houston State University. Her recent research examines global social movements that seek to make food production more environmentally sustainable and socially just. In particular, she is interested in issues of certification, labeling, and standards and relations between consumers in the global North and producers in the South.

Department of Sociology, Sam Houston State University, Huntsville, TX, USA

Philip H. Howard is Associate Professor of Community, Food and Agriculture at Michigan State University. He received his PhD in Rural Sociology from the University of Missouri. His scholarly interests focus on characterizing food system changes, particularly industry consolidation, and assisting communities to respond positively to these changes.

Department of Community Sustainability, Michigan State University, East Lansing, MI, USA

Lesley Hunt is a sociologist of work who is employed as a Senior Research Officer in the Agribusiness and Economics Research Unit at Lincoln University in New Zealand. She has had what could be called a varied career! It started with lecturing in statistics at the University of Otago, including teaching in schools in London and New Zealand, lecturing in research methods at the University of Canterbury and Lincoln University, teaching the piano, being a biometrician, and now, after her PhD study, working as a researcher in the Agriculture Research Group on Sustainability (ARGOS – www.argos.org.nz). Her PhD, an ethnographic study of scientists working in one of New Zealand's Crown Research Institutes, was titled "Compliance at work: protecting identity and science practice under corporatization." With her skills and experience, she brings to the fore the views of farmers and growers on sustainability and how people working in conventional and organic farming systems make meaning of their work.

Agribusiness and Economics Research Unit, Lincoln University, Lincoln, New Zealand

Dr. Rebecca Jones is a Postdoctoral Fellow at the Centre for Environmental History and School of History at the Australian National University. Her research interests include agricultural sustainability and organic agriculture, rural social and emotional health and well-being, and environmental history. For many years, her research has addressed both history and the social aspects of health. Rebecca completed her PhD at Monash University, Department of Rural and Indigenous Health, exploring the history of Australian organic growing. She is the author of *Green Harvest: A History of Organic Farming and Gardening in*

Australia, published by CSIRO publishing. Rebecca is currently working on "Slow Catastrophes: drought resilience amongst farmers and agricultural communities in Australia, 1880s–2000s," an ARC Discovery Early Career Research Award (DECRA)-funded project, exploring the way people in Australia respond to and survive drought.

School of History, College of Arts and Social Sciences, Australian National University, Canberra, ACT, Australia

Milena Klimek is a PhD candidate at the University of Natural Resources and Life Sciences in the Division of Organic Farming and a participant of the Working Group for Transdisciplinary Systems Research. Topics that she has explored or focuses on currently are food and farming issues concerning organic farming and direct marketing in Austria and Minnesota, farmer and consumer partnerships, and more specifically the role of values in farmers' markets.

Department of Sustainable Agriculture Systems, Division of Organic Farming, University of Natural Resources and Life Sciences (BOKU), Vienna, Austria

Damian Lara is a master's of arts student in the Department of Sociology at Sam Houston State University. His research interests include organizational practices of health-care actors and institutions.

Department of Sociology, Sam Houston State University, Huntsville, TX, USA

Dr. Allison Loconto is a researcher at the French National Institute for Agricultural Research (INRA), Science in Society Laboratory (INRA-SenS), and the French Institute for Research and Innovation in Society (IFRIS) at the University of Paris-Est Marne-la-Vallée. She holds a PhD in Sociology from Michigan State University and an MA in International Affairs and Development from the American University in Washington, DC. Allison Loconto's research examines the socioeconomic questions raised by the development, implementation, and evaluation of sustainability standards in global agrifood systems. Her recent research has been published in the *Review of International Political Economy, Regulation and Governance* and the *International Journal of Sociology of Agriculture and Food.* Allison Loconto is currently a Visiting Expert at FAO, where she is working on voluntary standards and institutional innovations that link sustainable agricultural practices with markets for sustainable goods.

Institut National de la Recherche Agronomique, Sciences en Société (INRA-SenS), Institut Francilien Recherche Innovation et Société, Université Paris-Est Marne-La-Vallée, Champs-sur-Marne, France

Rebecca Paxton is a PhD candidate at the University of Natural Resources and Life Sciences in the Division of Organic Farming and is a research assistant in the Working Group for Transdisciplinary Systems Research. Rebecca's current research examines social representations of health and health practices within the

Austrian organic agrifood system, in relation to different roles and distributions of responsibility for public health provision.

Department of Sustainable Agriculture Systems, Division of Organic Farming, University of Natural Resources and Life Sciences (BOKU), Vienna, Austria

Dr. Chris Rosin is Senior Research Fellow with the Centre for Sustainability: Agriculture, Food, Energy, Environment (CSAFE) at the University of Otago, New Zealand. His research includes land managers' responses to economic change and its impacts on human–environment relationships. As part of his involvement in the ARGOS project, he examines the social negotiation of sustainability in agriculture, including the processes through which concerns about and claimed achievements in social and environmental impacts of agricultural practice are communicated and legitimized within global food systems.

Centre for Sustainability: Agriculture, Food, Energy, Environment (CSAFE), University of Otago, Dunedin, New Zealand

Lia M. Spaniolo received an MS from the Department of Community, Agriculture, Recreation and Resource Studies at Michigan State University. Her thesis explored the role of new ecolabels in political consumerism efforts in the United States.

Department of Community Sustainability, Michigan State University, East Lansing, MI, USA

Dr. Maarten Van der Kamp is Managing Director of Value in Enterprise Limited, a consulting firm helping businesses and social enterprises build capacity around sustainability. He received his PhD in Management from Lancaster University, where his research examined the enactment of standards for organic agriculture in the UK through the practices of farmers, certification bodies, policy makers, and market actors. Grounded in science and technology studies, his research takes a practice-based approach to studying standardization processes. He is particularly interested in how the notions of "sustainability" and "responsibility" are enacted in everyday life through the production and consumption of food.

Value in Enterprise Limited, Cambridge, UK

Acronyms and Abbreviations

ACB	IFOAM Accredited Certification Body
ACO	Australian Certified Organic
AFHVS	Agriculture, Food and Human Values Society
Agrexco	Agricultural Export Company Ltd. (trading as Carmel *Agrexco*)
AOFGS	Australian Organic Farming and Gardening Society
AQIS	Commonwealth Government Australian Quarantine and Inspection Service
ARGOS	Agricultural Research Group on Sustainability
ASFS	Association for the Study of Food and Society
BOKU	(Universität für) Bodenkultur
CFP	Corporate Financial Performance
CIW	Coalition of Immokalee Workers
CLA	Conjugated linoleic acid
CoMoRe	Corporate Moral Responsibility
COROS	Common Objectives and Requirements of Organic Standards
CSA	Community Supported Agriculture
CSP	Corporate Social Performance
CSR	Corporate Social Responsibility
EC	European Commission
ELLS-NA	Euro-League for Life Sciences and North America
EPOPA	Export Promotion of Organic Products from Africa
FAO	Food Agriculture Organization
FDA	U.S. Food and Drug Administration
FLO	Fairtrade Labeling Organizations International
GMO	Genetically modified organisms
IAR	IFOAM Accreditation Requirements
ICO	Indiana Certified Organic
ICS	International Control System
IFOAM	International Federation of Organic Agriculture Movements
IMO	Institute of Market ecology

IOAS	International Organic Accreditation Service
IPM	Integrated pest management
IS	IFOAM Standard
ISOFAR	International Society of Organic Agriculture Research
ITC	International Trade Center
MAF	Ministry of Agriculture and Forestry
MRL	Maximum Residue Level
MSC	Marine Stewardship Council
NASAA	National Association for Sustainable Agriculture Australia
NGO	Non-governmental organization
NIS	New Israelian Shekels
NOP	National Organic Program
PA	*Perlindungan Alam*
PGS	Participatory Guarantee Systems
RA	Rainforest Alliance
rBST	recombinant Bovine Somatotropin
SAFN	Study of Food and Nutrition social movement organizations (SMOs)
SN	Sustainable Network
SOAAN	Sustainable Organic Agriculture Action Network
STS	Science and Technology Studies
TOAM	Tanzanian organic agriculture movement
TPC	Australian Trade Practices Commission
USDA	U.S. Department of Agriculture
WTO	World Trade Organization

List of Figures

List of Tables

Part I
Overview and Foundations

Chapter 1
Introduction

Bernhard Freyer and Jim Bingen

1.1 Foreword

As we witness the continuing growth in organic food production and markets around the world, we join with many others in our concern that "organic has lost its way" or, lost sight of its first or fundamental philosophical principles and assumptions.

The discussions that led to this volume started in June 2007 during a short visit by Jim Bingen to the Division of Organic Agriculture, University of Natural Resources and Life Sciences (BOKU), Vienna, Austria in order to explore of teaching and research exchanges between Michigan State University and BOKU under the auspices of ELLS-NA.[1] Based on shared concerns discussed during this visit, we organized an open roundtable to explore these concerns with other researchers at the Second Scientific Conference of ISOFAR[2] in Modena, Italy from 18 to 20 June 2008. At this meeting, Bernhard Freyer invited Jim Bingen to apply for a Fulbright Distinguished Chair position at BOKU. During this incredibly rich and stimulating 4-months (October 2009–January 2010), we jointly offered a seminar on Organic

[1] ELLS-NA, the Euro- League for Life Sciences and North America; June 2007.

[2] International Society of Organic Agriculture Research.

B. Freyer (✉)
Department of Sustainable Agriculture Systems, Division of Organic Farming,
University of Natural Resources and Life Sciences (BOKU), Vienna, Austria
e-mail: Bernhard.Freyer@boku.ac.at

J. Bingen
Department of Community Sustainability, Michigan State University, East Lansing, MI, USA
e-mail: Bingen@msu.edu

B. Freyer, J. Bingen (eds.), *Re-Thinking Organic Food and Farming in a Changing World*, The International Library of Environmental, Agricultural and Food Ethics 22, DOI 10.1007/978-94-017-9190-8_1

Agriculture, Politics and Society, co-advised students, and prepared several conference presentations that explored trans-disciplinary approaches to organic issues.

Bernhard Freyer's appointment as Senior Fellow at the Minnesota Institute for Sustainable Agriculture, University of Minnesota from late 2010 to June 2011 offered us a special opportunity to continue our Vienna conversations largely by Skype. We proposed a panel discussion on the principles and fundamental assumptions of organic practices for the 2011 annual meetings of AFHVS, ASFS, and SAFN Societies.[3] In response, the conference program committee invited us to organize five sets of panel discussions to addressed organic issues.[4]

In our call for papers for these meetings, we asked for empirically grounded discussions that focused on core principles and practices of organic food and farming. We specifically sought papers that drew upon clearly articulated and well-defined conceptual frameworks that might offer new insights into organic practices. Given the quality of the submitted papers, we decided to select some of them to be used for our proposal to Springer for this collection of original papers that offer several different perspectives on, and issues in the worldwide organic movement.

1.2 The Broader Idea of This Book

The explosive growth of organic food and farming in recent decades has raised numerous challenges throughout the organic sector (Willer and Kilcher 2011). Some of these include: a new generation of young organic farmers and others converting to organic; government subsidies for organic; the development of numerous new processed organic products; the emergence of diversified market opportunities that range from grass root food coops, to several types of farmer-consumer collaborative arrangements, farmers markets as well as other local markets, and organic supermarkets. Several revisions in the organic regulatory framework e.g., in Europe and in North America have accompanied this expansion and diversification of markets. Finally, several private and public sector groups and agencies have been established to deal with organic certification and control, to provide advisory services to farmers and processors, to carry-out organic research, and to publicize activities in the organic sector.[5]

All of these changes have been occurring within the context of a wide and diverse number of macro-level changes in food and farming in Europe and North

[3]Joint annual meetings of the Agriculture, Food and Human Values Society, the Association for the Study of Food and Society, and the Society for the Study of Food and Nutrition, "Food and Ag Under the Big Sky" held in Missoula, Montana.

[4]Challenges of Standardization; Between Diversification and Holism; Social and Gender Dimensions; Differentiations in Products, Markets and Consumers; and, Values and Ethics.

[5]http://www.isofar.org/

America. Some of these that affect both organic and non-organic farmers (albeit somewhat differently), include: increasing land prices; increasing extreme weather events; energy costs; international trade regulations; genetically modified crops; the power of large retailers; and, the creation of consumer demand for new food products.

In response, the International Federation of Organic Agriculture Movements (IFOAM 2012), representing over 100 country member organizations around the world, undertook far reaching revisions to the IFOAM Norms. These norms provide the foundation for the organic movement, and they include: the IFOAM Principles, the Standards and guidelines for certification agencies and firms.

The IFOAM Principles (Health, Ecology, Fairness and Care) represent the cornerstone of the organic movement (Luttikholt 2007). They offer the normative, ethical framework on which all the IFOAM regulatory instruments practice along the organic agrofood chain.

However, over the last two decades many in the organic movement have started to raise critical questions and concerns about the apparent loss of influence, and popular awareness of the IFOAM Principles on organic practices (IFOAM 2009). Instead of discussions related to the ethical ideas of the IFOAM Principles, more popular discussions of organic tend to focus on a different set of largely sociological or business-related features of organic such as: "professionalization, industrialization, conventionalization, bifurcation, competition, internationalization" and other socio-organizational features of organic. At the same time, many "pragmatic" organic actors, driven by ideas of a healthy and environmental friendly future, seek to keep alive ethically oriented organic practices through numerous forms of localisms and farmer-consumer relationships that embody the IFOAM Principles.

Confronted by these developments, we felt that it would be important to rethink the role of the IFOAM Principles and their potential to orient the future development of the organic idea. Three guiding questions helped us to select and organize the chapters in this volume: What was, is and could be the future role of an ethically based organic agriculture? What are the challenges of an ethically driven organic future? And what are the conditions for establishing an ethically driven organic agriculture for the future?

Our underlying premise is: solving the social, environmental and economic challenges of the future cannot be served only with a techno-economic approach or procedural, technical or social regulations. The absence of a more ethical orientation in future decision making, for food, feed, fiber and energy production and consumption will jeopardize future sustainable and resilient problem solving.

We are aware that this volume only addresses a selected number of the issues raised above. Nevertheless, we hope that we have assembled thoughtful and insightful reflections on several of the critical issues and ethical concerns in the organic movement. As such, we hope this volume offers a solid starting point for a reinvigorated debate on the development of an ethically centered organic future.

1.3 Structure of the Book

In Chap. 2, we introduce the IFOAM Principles and their ethical foundation. We provide an overview on the history of ethics in the organic movement starting in the 1920s (Part I, Chap. 2) and discuss the diverse streams of organic ethics that arose until the 1990s.

The following ten invited chapters discuss and explore different dimensions of organic ethics. These papers are presented in three parts. Each part includes a summary discussion that provides overview comments and highlights the ethical issues raised in the chapters. The parts are: Part II Standards and Certification (Chaps. 3, 4, and 5); Part III Markets and Consumers (Chaps. 6, 7, and 8); Part IV The Interplay of Conventional and Organic (Chaps. 9, 10, 11, and 12).

Our final Part V "Framework for Re-thinking Ethics in the Organic Movement," first raises questions about the current role of ethics based on a typology that is sensitive to the different ethics that currently characterize the organic movement (Chap. 13). In doing so, we seek to shed light on the environment of organic that influences the organic future and to review the challenges for an ethically driven organic future. Chapter 14 summarizes and concludes our discussion and offers our reflections on "Positioning Organic Ethics".

The following provides a brief overview of contributed chapters.

Part I Overview and Foundations

Chapter 2: The history of organic is not a homogeneous one, but is build on a colorful development based on different thinkers and societal movements over now approx. 100 years. Beginning with the current IFOAM Principles and their ethical foundations, Bernhard Freyer, Jim Bingen and Milena Klimek summarize and discuss the ethical foundations of the organic approach from their origins in the 1920s and follow the development of the different ethically/value driven streams until the 1990s.

The IFOAM Principles Health, Ecology, Fairness and Care, developed in a participatory process and with scientific support, serve as an ethical framework for the IFOAM Standards and certification issues. They are based on a normative ethic, grounded on an ecocentric/holistic approach, a moderate deontology and a virtue approach to human-nature relationships. Current ethical thinking about organic can trace its European origins to primarily in German-speaking countries and was heavily influenced during the 1920s by the life reform movement and by and the Philosopher Rudolf Steiner. Both introduced notions of a "circular agricultural economy" and nineteenth century Christianity that were at the foundation of many organic discussions. Since this time, organic has been characterized by diverse trends, largely in the UK, the US, New Zealand, Australia before becoming more "mainstream" through the establishment of IFOAM and its acceptance in more than 100 countries (see for details Chaps. 10 and 11). However, over this time, organic has been characterized by a remarkably consistent set of values that were embodied in the first set of IFOAM Principles and Standards in 1972.

Part II Standards and Certification

The maturing of the organic movement in recent years has brought significant pluralism to organic. Do government or other (e.g., IFOAM) regulations define organic, or are these simply official guidelines? What does organic certification mean, and what political and ethical principles does it embody? The three chapters in this part offer different, but complementary, responses to these questions and on the meaning of certification.

Chapter 3: Using a longitudinal study of a sustainable shrimp project in Indonesia, Maki Hatanaka examines how organic certification affects the principles, practices, and goals of the project. Her findings indicate that the emphasis on objectivity, calculability, and expert knowledge that characterize certification constrained both farmer and consumer participation in the governance of the sustainable shrimp project. Building on the case study, Hatanaka argues that certification may be producing alternative agrifood initiatives that are highly rationalized and embody shallow forms of social justice and environmental sustainability.

Chapter 4: Allison Loconto and Maarten Van der Kamp draw upon their separate research in Tanzania and in the UK to explore how context influences the way in which 'organic' is defined through its practices. Using the notion of performativity, they examine the organic standard as a calculative device that defines how organic tea is grown in Tanzania and how organic cereals are grown in the UK. They argue that while organic certification renders these products 'singular' in the UK market, the products embody 'multiple' production practices. They conclude that despite the use of standards to create a singular organic market, the practice of organic farming remains diverse. This tension between singularity and multiplicity is necessary for organic markets to develop and be maintained.

Chapter 5: Instead of being preoccupied with the mechanistic supply and demand formulation of world food security issues, Bernhard Freyer, Jim Bingen, Milena Klimek and Rebecca Paxton ask, what would happen if we started to focus on the idea of ethical values in the agrofood system? Their main thesis is: strengthening the discussion of the ethical values in the agrofood system should play a key role in our assessments of world food supply and demand in the future. This paper offers some preliminary reflections on selected ethical values raised by concerns with food supply, food demand and food access. The paper does not offer calculations of food and human nutrition, but instead discuss the critical relationships between ethical values and their quantitative influence on food security. The authors argue that a value-centered discussion is essential to explicating many of the issues related to organic practices and the question of how "to feed the world." More specifically, they suggest that the ethics embodied in the IFOAM Principles offer a framework for identifying how organic agrofood systems might contribute to a sustainable food future. This has also consequences for standard and certification procedures.

Part III Markets and Consumers

For years, many academics have critically discussed how organic is advertised, and how consumers react to diverse marketing strategies. The three chapters in this part offer insights into market and consumer issues raised by organics. They discuss the broad range of farmer-retailer-consumer chains and how consumers perceive, react and are affected by market strategies.

Chapter 6: Ginevra Adamoli explores the concept of organic marketing in terms of compliance marketing and green selling. In light of the growing organic products and the rise of the hybrid citizen-consumer, defined by Johnston (2008) as a concept that implies a social practice, she studies the connection between consumer's agency and organic packages of food products. She examines the concept of organic marketing based on King (1984) and Peattie and Crane's (2005) discussions of compliance marketing and green selling. Through a qualitative textual and visual analysis of the organic egg carton by *Esselunga*, one of the largest Italian chain supermarkets, she argues that these forms of textual and visual communication are consistent with practices of marketing, rather than serving to empower the buyer through valuable information. She concludes with a discussion of the implications of her findings for the citizen-consumer.

Chapter 7: Kristin Getter, Bridget Behe, Philip H. Howard, David Conner and Lia M. Spaniolo present the findings of their innovative research on the attributes and images that consumers prefer with respect to pasture-based dairy. They discuss the result of their research that created and tested promotional messages with potential consumers, while also investigating current consumer perceptions and attitudes about milk and its attributes. Their research asked participants in six focus groups around Michigan to create hand-drawn milk labels that represented attributes they sought when purchasing milk. Research participants also answered oral questions about their milk purchases and evaluated images designed by a graphic artist. The most common themes to emerge in the hand-drawn images included cows in pasture, blue skies, and sunshine. Words appearing on these same drawings indicated that participants generally wanted organic, local, and grass-fed milk products. By looking at the themes and concerns of milk drinkers, the authors identify the need for effective communication to consumers in order to improve profitable sales of pastured milk.

Chapter 8: Ralf Grosglik raises important questions that will command increasing analytic and policy attention as organic production and marketing spreads beyond North America and Western Europe. Based on his research in Israel, he examines the appearance of politicized organic practices in tandem with the globalization that Israel underwent. Grosglik points that the emergence of Israeli organic food is essentially a part of the economic and cultural globalization in Israel. Furthermore, he describes how a variety of production and distribution methods have responded to the increased demand for organic food by Israeli consumers. These new methods embody a variety of symbolic and materialistic aspects of globalization and anti-globalization. As these new activities emerge, they help us consider the political dimensions of a range of alternative organic practices.

Part IV The Interplay of Conventional and Organic

The four chapters in this part allow us to gain fresh perspectives on the enduring questions related to the "conventionalization" of organic and to debates over the "co-existence" of conventional and organic production.

Chapter 9: Following an insightful review of the evolution of organic policy in the US, including a thorough review of the debates over, and perspectives on several political issues related to the "conventionalization" of organic farming, Doug Constance, Jin Young Choi, and Damian Lara draw on empirical evidence from Texas to reflect on the transformative potential of organic. The conventionalization of organic production has attracted substantial attention in the agrofood literature in recent years. Some authors note a dilution from deep-organics to organic lite; others see little evidence of such dilution and critique the concept of conventionalization as being over-stated and not taking into account national differences. Within the US, recent changes in USDA policies regarding organics indicate an increase in support for organic conversion in the face of expanding consumption but lagging domestic production. Constance et al. use results from research on a representative sample of Texas commodity producers to engage the conventionalization thesis. More specifically, they investigate pragmatic conventional producers, those producers who identify themselves as conventional but have an interest in organic production, across a set of structural and attitudinal variables related to organic agriculture. The authors argue that the dynamics of worldwide organics offers valuable insights into the contours of a new food regime that is being contested by social movements, corporate interests, nation-states, and supra-national organizations. The paper concludes with a discussion of the implications of organic entry by pragmatic conventional producers for the conventionalization debate.

Chapter 10: Lesley Hunt, Chris Rosin, Hugh Campbell and John Fairweather draw on research undertaken by the transdisciplinary ARGOS program that compares the sustainability of organic, integrated and conventional farming/orcharding in the dairy, sheep/beef and kiwifruit sectors of New Zealand. They observe that organic farmers can contribute to the resilience of the organic sector more fully when they are recognized as "good farmers" in their local communities, as well as by others in the sector. The authors examine three themes. First, studying organic farming in isolation limits the understanding of its broader contribution. By comparing different management systems within and across sectors (sheep/beef, dairy, kiwifruit) one can see the contributions that organics can make to the primary sector as a whole. Second, organic farmers and their farms contribute to the resilience of the food system. Resilience implies adaptability and redundancy at many levels – farm/farmer, community, national and global. They suggest that organic farms and farmers perform a useful function as part of a resilient farming system because they introduce alternative practices, provide alternative possibilities to non-organic farmers, and expand the diversity of products produced by an agriculture sector. Third, the extent to which organic farmers are likely to influence the resilience of a given supply network is related to their relative level of acceptance as good farmers. The good farming literature suggests that the status of 'good farmer' is awarded by farmers to other farmers who follow cultural rules established over time by the farming community and government policies/regulations. Based on ARGOS studies in three agricultural sectors the authors demonstrate that in the kiwifruit sector, organic management is one of many models of 'good orcharding' while organic practices challenge the precepts of 'good farming' in the other sectors.

Chapter 11: Drawing on a larger study of the history of organic farming and gardening in Australia, Rebecca Jones examines whether the values that inspired the foundation of organic agriculture remain relevant in contemporary society. Following a discussion of the principles upon which Australian organic agricultural societies were founded in the 1940s and 1950s, she examines whether these principles still resonate with, and are relevant to, Australian organic agriculture today. She concludes that the original key principles remain fundamental to organic agriculture today: the production of humus-rich fertile soil, chemical free production and the maintenance and development of ecological wellbeing. While Australian organic farmers grapple with many contemporary concerns such as genetic modification, climate change, and a plethora of commercially manufactured organic products and certification, she concludes that indeed a founder of the early organic societies would still recognize the key principles of organic farming and gardening in contemporary Australia. But similar to other countries, there are also tendencies toward conventionalization.

Chapter 12: Caroline Brock and Bradford Barham use an Oikonomia-bounded rationality framework to explore how values shape farming choices among two Old Order Amish communities when full and complete information is not always available and/or utilized according to the standard assumptions of microeconomic decision-making models. Despite economic, legal, and organizational challenges, Amish farmers in these Wisconsin communities use many farming methods and technologies similar to those used decades ago by non-Amish farmers. Based on interviews with Amish farmers in these settlements, Brock and Barham discuss the complex interplay of sustaining traditional religious values, bounded rationality, and decisions to adopt "new" systems–based practices such as organic dairy farming methods. They find diversity across the settlements with respect to the likelihood of going organic which reflects how values and bounded rationality may shape farming choices not only for the Amish but also for other farmers facing diverse management and technological choices.

Part V Re-thinking Ethics in the Organic Movement

In this part, the authors refer to current and future perspectives of an ethically driven organic movement. The central questions are: What are the ethics in the organic practice in the last two decades and today? How are these ethics framed and accompanied by organizations and institutions? And what are steps forward an ethical driven organic movement?

Chapter 13: Based on current scientific discussions and the findings in the different parts of this volume, Bernhard Freyer, Jim Bingen and Milena Klimek look at the role of ethics in the organic movement in more recent decades covered by the chapters in Part II–IV, and offer additional insights on the broad range of international research done over the last two decades. Consistent with the objective of this volume to stimulate a rethinking of the role of ethics in the organic agrofood chain, this chapter uses a value-based typology to discuss the contemporary values of farmers, consumers and other actors along the organic supply chain. The authors introduce a broad range of ethical contributions and challenges that are critical for an ethically oriented development of organic in the future. They describe an ethically based philosophical perspective of a regulatory framework to reflect critically on the ethical features of standards, as well as the control and certification processes in the organic agrofood chain. The authors look at alternative approaches to assuring control and trust in new forms of certification and farmer-consumer collaboration. Since consumers are free from control and certification processes of their behavior, the authors review specific ethical challenges that consumer behavior raises. They also examine the ethical questions initiated by the activities of organic advisory services, higher education and research, as well as what could improve the use of organic values[6] in these activities. Finally, the authors look at three challenges of more value-based organic practice: the influence of political and industrial power on the organic movement; the status of social and economic justice and social coalitions and communication strategies; and, some key factors that are relevant if the organic movement is to continue and strengthen a value based community.

Chapter 14: In the final chapter, Bernhard Freyer and Jim Bingen review the main findings on organic ethics in this volume. They emphasize and deepen key findings from Chap. 13, asking what concrete actions might be necessary to bring ethics more into the center of the organic movement and related institutions. The authors close with some philosophical reflections on how realistic it might be to establish the idea of an organic ethically driven agrofood chain.

References

De Wit, J., and H. Verhoog. 2007. Organic values and the conventionalization of organic agriculture. *NJAS-Wageningen Journal of Life Sciences* 54(4): 449–462.

IFOAM. 2009. *The principles of organic agriculture*. Retrieved from http://www.ifoam.org/about_ifoam/principles/index.html. Accessed 1 Sept 2011.

IFOAM. 2012. *The IFOAM norms for organic production and processing*. Retrieved from http://www.ifoam.org/en/ifoam-norms. Accessed 1 Jan 2013.

Johnston, J. 2008. The citizen-consumer hybrid: Ideological tensions and the case of Whole Foods Market. *Theory and Society* 37(3): 229–270.

King, S. 1984. Has marketing failed, or was it never really tried? *Journal of Marketing Management* 1: 1–19.

Luttikholt, L.W.M. 2007. Principles of organic agriculture as formulated by the International Federation of Organic Agriculture Movements. *NJAS-Wageningen Journal of Life Sciences* 54(4): 347–360.

Peattie, K., and A. Crane. 2005. Green marketing: Legend, myth, farce or prophesy? *Qualitative Market Research: An International Journal* 8(4): 357–370.

Willer, H., and L. Kilcher. 2011. *The world of organic agriculture: Statistics and emerging trends 2011*. Bonn/Frick: IFOAM/FIBL.

[6]The meaning of value in most sociological research on organic "is almost identical to the motives of different stakeholders in the organic sector" (De Wit and Verhoog 2007, p. 455).

Chapter 2
Ethics in the Organic Movement

Bernhard Freyer, Jim Bingen, and Milena Klimek

In the last two decades, several publications have addressed the ethical foundations of the organic movement.[1] Papers on the ethics in, and of the organic movement have been especially sensitive to human-nature interrelationships, and the equality of all living things. Many of these discussions of ethics are integrated into, or arise in the latest IFOAM Principles.[2] Other publications review the historical development of organic agriculture (e.g., Conford 1995; Vogt 2000; Heckman 2006; Lockeretz 2007).

Our aim in this chapter is to offer a review that lays the groundwork for the ethical discussions and development of ethics in the organic movement that are illustrated in the contributions to this volume. We begin with the development and introduction of the most recent version of the IFOAM Principles (IFOAM 2009). Then we discuss the relationship between ethical concepts and their significance in the IFOAM Principles. A comprehensive historical overview of the ethical roots

[1](e.g., Browne et al. 2000; Goodman 2000; McEachern and Mcclean 2002; Lund 2006; Padel and Gössinger 2008, p. 7).

[2](e.g., Benbrook and Kirschenmann 1997, p. 1; Thompson and Nardone 1999, p. 112; DARCOF 2000, p. 12; Lund and Röcklinsberg 2001, pp. 391, 402; Alrøe and Kristensen 2002, p. 1; Taylor 2003, p. 75; Verhoog et al. 2003, p. 44, Verhoog et al. 2007; Alrøe et al. 2006; Padel et al. 2007; Padel and Gössinger 2008, p. 6).

B. Freyer (✉) • M. Klimek
Department of Sustainable Agriculture Systems, Division of Organic Farming,
University of Natural Resources and Life Sciences (BOKU), Vienna, Austria
e-mail: Bernhard.Freyer@boku.ac.at; Milena.Klimek@boku.ac.at

J. Bingen
Department of Community Sustainability, Michigan State University, East Lansing, MI, USA
e-mail: Bingen@msu.edu

B. Freyer, J. Bingen (eds.), *Re-Thinking Organic Food and Farming in a Changing World*, The International Library of Environmental, Agricultural and Food Ethics 22, DOI 10.1007/978-94-017-9190-8_2

of the organic movement from the 1920s until the 1990s provides insights into the relevance of different ethical concepts that have arisen in the organic movement over the last century.[3] Our review does not claim to be exhaustive. Instead, we seek to integrate our understanding of the socio-political, cultural and agricultural background of the organic movement with the development of its values and ethics.

2.1 Implementation of the IFOAM Principles

In 1980, IFOAM formulated the first series of principles to serve as the ethical guidelines for organic practices (e.g., Lockeretz 2007, p. 117). The latest IFOAM Principles arose from an international consultative process established by the IFOAM World Board[4] from 2003 to 2005 (Luttikholt and Vijayalaksmi 2004). This worldwide stakeholder-based discourse embodied IFOAM's specific ethical commitment to develop the principles through a bottom-up participatory process.

The initiative to reformulate the principles stemmed from the decline of IFOAM's role in the organic movement and the need to assure the application of its Basic Standards in an increasingly globalized organic market. "The [consultative] process aimed to bridge the values from the pioneers of organic agriculture to the present time of globalization and to extend growth of the organic sector" (Luttikholt 2007, p. 347).

As a result of this process, the four IFOAM Principles—Health, Ecology, Fairness and Care—were established as the pillars of the ethical framework to support and guide organic agriculture and the global organic movement (long version see annex) (IFOAM 2012):

- **Principle of Health:** Organic Agriculture should sustain and enhance the health of soil, plant, animal and human as one and indivisible.
- **Principle of Ecology:** Organic Agriculture should be based on living ecological systems and cycles, work with them, emulate them and help sustain them.
- **Principle of Fairness:** Organic Agriculture should build on relationships that ensure fairness with regard to the common environment and life opportunities.
- **Principle of Care:** Organic Agriculture should be managed in a precautionary and responsible manner to protect the health and well-being of current and future generations and the environment.

As the IFOAM Norms state, these principles are the foundation for writing separate national standards and regulations (IFOAM 2012).[5] The Principles:

---

[3] An overview of the latest development is discussed in the Chap. 13.

[4] http://www.ifoam.org/about_ifoam/principles/history_of_principles.html

[5] The IFOAM Norms (IFOAM 2012) are composed of three documents, which are the Common Objectives and Requirements of Organic Standards (COROS) – IFOAM Standards Requirements; the IFOAM Standard for Organic Production and Processing; and the IFOAM Accreditation Requirements for Bodies Certifying Organic Production and Processing.

- express the potential contribution that organic agriculture can make in the world and they inspire a vision for improving all dimensions of the organic agrofood chain in a global context (IFOAM 2012);
- are meant to be universal and are proposed for those outside of, as well as those within the organic movement (Alrøe and Kristensen 2004);
- are intended to be applied *in toto* and with attention to their interdependence (Luttikholt 2007; IFOAM 2009); and,
- combine a strong focus on nature, and the human-nature relationship through the coexistence of humans and the well-being of humans and communities (Freyer 2008, p. 395; IFOAM 2009). They can be broadly applied, addressing social relations as well as those with other living beings. They stress that organic agriculture should maintain and conduct these relationships in a manner that ensures health, ecology, fairness and care, and that includes equity, respect, and stewardship for future generations and the environment as a whole.

Given these features, the Principles offer a basis for examining broader ecological, social, economic and political relationships (Table 2.1). Compared to earlier versions, they are more sensitive to social issues such as the idea of justice, mutual respect, quality of life, fair salaries and prices etc. (Kristiansen and Merfield 2006, p. 16; IFOAM 2009).

The application of these principles to decision-making processes and to guiding further innovations in the organic system is quite controversial (Schmid and Lockeretz 2007, p. 167). At least two positions on this issue are obvious in the

Table 2.1 Five common human-nature relationships and how they may be practiced in agriculture

Views of Human-Nature Relationship	
Anthropocentric	Humans take hierarchical precedence over nature; the value of nature is mainly instrumental (Nash 1989; Elmore 1996)
Theocentric	Nature, and all that lies within, is God's creation, and humans are to act as stewards of those creations (Schaefer 2009; Gudorf 2012)
Pathocentric	All living things can experience suffering (Bentham 1978; Birnbacher cited in Fenner 2010)
Biocentric	Non-human value of nature is recognized in plants, animals and soils (Nash 1989; Kirchmann and Thorvaldsson 2000)
Ecocentric/holistic	A non-anthropocentric belief, which blends ecocentrism and holism in which the non-human value of nature is recognized as holistic systems i.e. ecosystems and the benefits for soils and minerals have in such systems, as well as the intrinsic value of nature (Nash 1989; Sterba 2003; Hay 2010)
Practices of Human-Nature Relationship in Farming (examples)	
Anthropocentric	Health, benefits of subsidies and price premiums
Theocentric	Farmers need to manage God's creation, to use but not abuse
Pathocentric	Animal welfare and pesticide free plant production
Biocentric	Closing the cycle by producing own inputs, care for soil fertility, animals and plants
Ecocentric/Holistic	Closed-cycle; farm seen as an organism; recognizing the importance of soils and diversity and integrating that in practices; using nature as a model to grow polyculture crops and healthy animals

discussions in this volume: (1) the principles provide an ethical orientation, to follow them is a question of individual choice (e.g., Steiner 1892; Piaget 1965); (2) the principles are worthless since they are not a binding part of a regulation and certification system.

With the continuing growth of the movement that leads to increasing anonymity among the partners along the agrofood chain, and thus the potential for a loss of mutual commitment, voluntary ethical acting might be one of the key challenges of the contemporary organic movement. Thus, it becomes critical to bring to the foreground a discussion of the ethics in the IFOAM Principles.

2.2 The Ethical Foundation of the IFOAM Principles

This section identifies and discusses the embodied worldviews and normative ethics of the IFOAM Principles in order to help gain fresh insights into current challenges confronting, and issues raised by organic worldwide.

2.2.1 Worldviews and IFOAM Principles

We believe that it is useful to identify and discuss five different worldviews reflected by the Principles (Table 2.2). These worldviews are commonly not explicit or acknowledged by most actors in their different modes of organic practice. Nevertheless, we suggest that these worldviews, or elements of them, are embodied in most of the issues and controversies surrounding organic. Most actors hold a "core" worldview, but commonly draw upon elements of others to "round out" the grounds upon which they see the world, and more specifically think about organic. Making these worldviews explicit could help to: (1) improve the conversations and debates about organic; and, (2) help bring an explicit articulation of values back into the center of organic discourse.

2.2.1.1 The Anthropocentric View

It is no exaggeration to state that this worldview makes up the core of what we all think and believe. It is, by definition, human-centered, but covers a wide range of instrumentalist interpretations of the ways in which nature is subordinate to human interests and needs (Kirchhoff 2011). In this worldview, humans are "above" nature and are not accountable to "nature" or ecosystems for their actions. Here, nature does not have intrinsic value, and humans hold a privileged position in comparison to other species that have no more than instrumental importance (Daly et al. 1995).

Anthropocentrism underlies most utilitarian and instrumentalist perspectives on, as well as mainstream economic approaches to, organic (Peet 1996). With this view, being or practicing organic is a pragmatic matter of simply following sets of stipulated technical practices defined as organic.[6] Perhaps the most egregious illustration of this instrumentalist organic, is seen with farmers who convert in order to capture high-premiums and government subsidies, as well as those consumers who purchase organic predominantly for personal health reasons (Magnusson et al. 2003; Gilg et al. 2005). The "conventionalization" of organic, and the seemingly endless controversy over specific cropping or animal husbandry practices—which are covered by the guidelines, yet copy conventional approaches—is a variant of instrumentalist organic (Best 2008).

Farmers, processors or traders who are interested in organic for purely instrumentalist (and economic) reasons are rarely, if ever, interested in discussions about organic and health, ecology, fairness or concerns for future generations. Furthermore, an Anthropocentric perspective overlooks the complex consequences of internationalization, commodification and industrialization of organic food and farming, including its critical social and environmental impacts (Raynolds 2004).[7]

Equally instrumentalist are actions or policies based on the notion of duty to future human generations or those based on valuing nature as a way to avoid ecological disaster, thus conserving for the future need of people. Many organic marketing strategies illustrate this approach in promoting a "feel good" rationale for buying organic without reference to specific crop production or livestock practices that would be important in terms of the IFOAM Principles (cf. Thompson 1998; Tanner and Wölfing Kast 2003; Ginsberg and Bloom 2004; Hamm and Gronefeld 2004).

2.2.1.2 The Theocentric View

This view of human-nature relationships was historically the foundation of organic thinking (see sections below on the Müllers and Lady Balfour), but is much less prominent in contemporary discussions (see Chap. 13). In this view, nature is seen as God's creation and thus, is sacred. All human and non-human life, including the land and the earth, have value (Carruthers 2009, p. 302), and humans hold a moral responsibility to God to be faithful stewards for all of God's creation

[6]The discussion about (pragmatic) anthropocentrism in environmental ethics reminds us that there are multiple understandings of these ethics (cf. Minteer and Manning 2005; Katz 2008; P. Thompson 2008), but also controversies about how they contrast with other ethical concepts (Jacob 1994; Johnson 1996; Norton 2008).

[7]This largely utilitarian position is in conflict with what we call "daily morals" (Alltagsmoral) (Daly et al. 1995); individuals who use this approach cannot prevent endangering individuals or minorities as they pursue their goals. In contrast, daily morals based on Christian traditions guarantee through a set of rules that the weakest in society enjoy legal and public protection (Fischer 2003).

(Kirchhoff 2011).[8] Human actors are responsible for balancing their actions between what is necessary for survival and the consequences for others.[9] Lochbühler argues that this exceptional relationship between human and the rest of creation divine reflects a "moderate anthropocentric in a theocentric context" (Lochbühler 1996, p. 117). For White, the Judaeo-Christian monotheism at the core of this anthropocentric perspective is the ideological source of the modern environmental crisis (White 1967). The IFOAM Principles represent an ethic that is independent of any religious association. While a relationship to a spiritual dimension is not excluded, it is not specifically indicated.

2.2.1.3 The Pathocentric View

Pathocentrism is a perspective that is fundamental in the IFOAM Principles. From this perspective, all living things, except plants (cf. Willemsen 2008), can suffer or feel pain (e.g., Bentham 1978; Fenner 2010). As Singer (1993) argues, since animals have both the capacity to suffer and an interest in avoiding pain, we have a moral obligation to respect this interest (see also Fenner 2010; Vaarst and Alrøe 2012). Having distinguished a difference between animals and plants, however, it is worth nothing that techniques advanced by Kirlian technology suggest that plants may also experience pain.[10] This adds a new element to this view that has not yet been thoroughly explored, yet may play an important role in the future.

2.2.1.4 The Biocentric View

From this perspective, all living organisms without distinction between humans, animals and plants, have intrinsic value (e.g., Taylor 1989; Schweitzer et al. 1999). "Being alive" (*lebendig sein*) is of value in and of itself, and it includes an interest in staying alive (Schweitzer 1976; Jonas 1979, 2004). In contrast, non-living things do not have moral value; they have instrumental value. This is at the core of concerns with living soils and efforts to nurture and protect soil fertility (Abaidoo and Dickinson 2002). The IFOAM Principles support the idea of intrinsic value, but there are different perspectives on the instrumental value of non-living things, that are rooted in individual belief patterns.

---

[8]The relationship to nature in the tradition of Catholic Christianity, establishes a divine order of man and nature that is apart from human egoism and intentions (Hoffman and Sandelands 2005). It is God who is in charge of nature and the role of humans is that of faithful stewards of creation, a service for God, mandated through God (Elmore 1996).

[9]Vorster (2005, p. 882) argues that "thus stewardship implies kinship over and against kingship because a theocentric approach renders any anthropocentrism null and void. Real theocentrism can lead to only one attitude: responsibility to God that will be expressed in humankinds care for its creation." This rather positive picture of a nature-sensitive theocentrism is also under critique.

[10]http://www.thesynergycompany.com/v/superfood_article10.html

2.2.1.5 The Ecocentric and Holistic View

An ecocentric/holistic perspective is central in the IFOAM Principles.[11] From this perspective, all non-living and living things have moral rights (Foster and Burkett 2000; Gilg et al. 2005; Schlüns and Voget 2008; Kings and Ilbery 2010) and human have a responsibility for all things animate and inanimate (Meyer-Abich 2006). In the definition of ecocentrism, humans and nature are connected, but ecosystems and nature have precedence over human interests (Hay 2010), a statement which might be discussed controversially in the organic movement. This perspective offers an individual and a socially oriented ethic for action by individual, organizational and institutional, including corporate and other types of collective organizations (cf. Schroth 2009). Within this context, "the Principles concern the way people interact with living landscapes, relate to one another and shape the legacy of future generations" (IFOAM 2012, p. 9).[12]

The IFOAM Principles are close to what can be called pluralistic holism. They assume a holistic and systemic understanding of the farm as an organism in which all objects (organic, non-organic) deserve respect and enjoy the right to exist, while serving a purpose on the farm. Humans neither dominate nor control nature, but are integral to it so that human activities are as much as part of "nature".

The concept of conviviality (*Gastlichkeit*) offers an additional idea for interpreting human-nature-relationship in an organic context. This idea specifies that human action should avoid bringing about suffering, damage or destruction to other living things (Littig and Grießler 2004). As Illich (1973) suggests, conviviality, defined as an understanding of friendship or playfulness in interpersonal relationships, replaces a technological value with an ethical value based on the realization of individual freedom among different actors in the production process.[13]

---

[11](Lund and Röcklinsberg 2001; Lammerts Van Bueren et al. 2003; Verhoog et al. 2003, Verhoog et al. 2007; Padel and Gössinger 2008); It should be noted that the concept ecocentric/ecocentrism is not named in the IFOAM Principles. Holistic/holism is only mentioned directly within the context of breeding (IFOAM 2012, p. 9).

[12]The more holistic interpretation draws from the idea that ecosystems co-evolved and that their existence is a result of reciprocal dependency, similar to the organs of an individual organism (Kirchhoff 2011, p. 18). All things have moral value and rights: humans have responsibility for all living and non-living things including soils, minerals and other natural resources (Callicott 1989). Some also criticize that humans are seen as part of the problem (Tokar 1990). American author, farmer and forester Aldo Leopold's land ethic highlights an aesthetic dimension of ecocentrism (Callicott 2008). For Leopold, "A thing is right when it tends to preserve the integrity, stability, and beauty of the biotic community. It is wrong when it tends otherwise" (Leopold 1949, pp. 224, 225; Wenz 2003).

See also Gorke (2006) who incorporates individuals into a holistic perspective by arguing that individuals and superior entities have both the same intrinsic moral status. With Norton (1987 in Gorke 2006, p. 261) this ethical approach is categorized as a pluralistic holism.

[13]The concept of conviviality includes also a technological perspective: A convivial technology is oriented towards cooperation and not domination of technology so long as technology meets three requirements: (1) an increase in personal capabilities; (2) situations free of either slave or master relationships; and, (3) creates economic benefits without destroying personal autonomy (Illich 1975, p. 32).

2.2.1.6 Final Observations on Worldviews and the IFOAM Principles

As expressed in the IFOAM Principles, these diverse views appeal to actors in the organic movement to weigh their actions in terms of their lives, but also for human generations and nature now and in the future. They focus on ecological dimensions and invite respect for the needs of all others, without regard to religious or spiritual persuasion. Consistent with this, and even though not specifically acknowledged, they do not preclude more radical positions that subordinate the individual to ecosystems (Stenmark 2004, p. 104) such as deep ecology (Naess 1973, 1989), eco-feminism, new age movements (Krebs 1977, p. 362), or cosmological interpretations (Siep cited in Fenner 2010, p. 169).

The ethical message of the Principles that captures an ecocentric/holistic perspective is best expressed with: "Fairness is characterized by equity, respect, justice and stewardship of the shared world, both among people and in their relations to other living beings" (IFOAM 2012, p. 10). In this way, the IFOAM Principles represent what Alrøe and Kristensen (2003) call a "systemic ethic". This perspective offers considerable freedom in designing alternative/organic agrofood systems.[14] Moreover, the IFOAM Principles offer an ethically based structure for evaluating the organic agrofood chain as a whole (Comstock 1995).[15]

2.2.2 *Beyond Organic as a Moral Obligation*

In addition to reflecting these worldviews, the IFOAM Principles stand as a moral guide for all actors along the organic agrofood chain (IFOAM 2009). They offer a deontological ethic (duty, obligation) that stipulates what is "right" and what is "wrong" (Mepham 2001; Barnett et al. 2005; Padel and Gössinger 2008; Padel et al. 2009) in different environments. Understanding the Principles exclusively as deontological would not offer much room for a flexible interpretation of standards and certifications schemes, or for addressing unforeseen situations in daily practice.

In this section, we therefore broaden the ethical perspective through a discussion of four relevant ethical positions (deontology, consequentialism, moderate deontology and virtue ethics) (Carruthers 2009, p. 296) in organic.[16] This will help us to deepen our understanding of the IFOAM Principles and how they might be applied in organic decision making processes.

[14]The ecocentric view is founded on the belief that 'our deepest moral guidance comes from understanding nature and our 'natural' place in it' (Armstrong and Botzler 1993, p. 54, cited in Carruthers 2009, p. 297).

[15]In contrast, the definition of sustainable development in the Brundtland report (1987) is explicitly anthropocentric (Shearman 1990; Rennings and Wiggering 1997), while others also identify partly biocentric characteristics (Weinschenck 1994).

[16]see also Alrøe and Kristensen's approach toward a systemic ethic (Alrøe and Kristensen 2003).

2.2.2.1 The Deontological Perspective of Organic

IFOAM offers "principles to guide behavior in order that such behavior becomes 'normal'" (Vardi and Grosch 1999, p. 109). They call for a personal obligation and serve as moral norms to be respected (cf. Pettit 1993; Barnett et al. 2005, p. 1). In the ethical tradition of deontology, an act is moral, if the acting is done based on a morally oriented decision (cf. McNaughton and Rawling 2007) and conforms to a moral norm or rule. It asks about the inner nature of an act, and the rightness of the act (cf. Barnett et al. 2005, p. 5). It orients decisions along "what one should do" / (*"Was man tun soll"*) (Fenner 2010, p. 34), independent of their consequences; it does not automatically include a moral future oriented decision. Fulfilling the norms is fundamental for those who hold the organic ecocentric/holistic approach—the IFOAM Principles express this ideal.[17]

2.2.2.2 The Consequentialism Perspective of Organic

The consequentialist or teleological perspective argues, "the purpose sanctifies the means" (*"Der Zweck heiligt die Mittel"*) (Schroth 2009). Consequentialism declares the rightness of the outcome or the good result of the action as the moral instance of their acting (Barnett et al. 2005, p. 5). It is also clear that a growing number of actors apply the IFOAM Standards because they are mandatory and not from a sense of duty or moral obligation. This utilitarian orientation favors the individual and consumer perspective rather than a community- and citizen based approach (Carruthers 2009, p. 299). Such approaches arise in organic practices that are followed primarily to maximize profit, with little or no attention to social and economic justice or ecological concerns. This position essentially disregards the IFOAM Principles.

2.2.2.3 Moderate Deontology as a Step Forward

Both approaches, the deontological and the consequential (teleological) alone are critical for several reasons (Alrøe and Kristensen 2003, pp. 62, 63; Clarke et al. 2008, p. 221). First, a pure deontological approach ignores that in practice we often do not know what might be a deontological-based decision. This is because farming is always a process-oriented decision between short-term and long-term perspectives in a complex environment. Second, the teleological oriented approach contradicts the IFOAM Principles, by focusing only on the result but not on the process, and might ignore minorities (Fenner 2010, p. 34).

---

[17] As introduced above, ecocentrism can also be seen as a form of deontological contractarianism (Carruthers 2009, 299). The IFOAM Principles are not of a legal contract, but they do offer a type of voluntary social contract (cf. Clark 2012) that "provides a rationale for individuals to act morally and for governments to create and maintain a just and ordered society" (cf. Carruthers 2009, p. 297).

A third concept that is described with the term "weak" or moderate deontology combines both perspectives (Hunt and Vasquez-Parraga 1993, pp. 80, 81, 87; cf. Harel and Sharon 2008). The right acting depends not only but also on their consequences.[18] This duty includes how to act now and with attention to the long-term effects of decisions. This is, for example, an important perspective in organic breeding (Alrøe et al. 2001, p. 12; Lammerts Van Bueren and Struik 2005, p. 484). The Principles claim both a responsibility to act in an ethical way now and with respect to the consequences for future generations. The IFOAM Principle of Care embodies this position: "Organic agriculture is managed in a precautionary and responsible manner to protect the health and well-being of current and future generations and the environment" (IFOAM 2012). In other words: in organic, we must address both current and future needs.

2.2.2.4 Organic as Virtue

This type of normative ethics brings in a perspective that puts the actor in the center (Borchers 2001). While deontological ethics focuses on the kind of activity and instrumental values (first order moral), virtue ethics is about the inner disposition and attitude of a person or non-instrumental values (second order moral) (cf. Alrøe and Kristensen 2003; Carruthers 2009, p. 299).

Virtue ethics specifically defines a moral life in the context of social relationships, community, traditions, socially established co-operative human activities, a strong relation in places and on the land and human flourishing (Carruthers 2009, p. 299). This has been illustrated by many, including: Aldo Leopold's land ethics (Hull 2005; Minteer 2006; Frasz 2008; Shaw 2008); Rachel Carson's critique on the environmental destruction; Henry David Thoreau's lifestyle (Sandler 2007; Cafaro 2008); and, Arne Naess's deep ecology (Hursthouse 2007). Virtue ethics have several commonalities with the IFOAM Principles of Fairness and Care. The holistic ethic of the Principles is sensitive to the ecological and human needs beyond a purely economic perspective.

2.2.2.5 Thoughts on Re-conceptualizing Organic Ethics

The ethical perspective of IFOAM Principles can be described with an ecocentric/holistic ethic, the concept of conviviality and in the tradition of normative ethics as a moderate type of deontology, as well as with virtue ethics. It embodies a culture of life similar to what Wendell Berry and others have referred to as agrarian stewardship that is diametrically opposed to industrial approaches, or thinking about

[18]The more deontological perspective is also evident in an organic economy that is in contrast to a consequentialist approach, integrating values such as autonomy, basic liberties, truth-telling, and promise-keeping over the promotion of good outcomes (Zamir and Medina 2010).

the "farm as a factory" (Wirzba 2003). Similar to the concept of agrarian stewardship (cf. Thompson and Hilde 2000; Thompson 2001, 2007; Wirzba 2002), the IFOAM Principles also refer to humanness, community, place and stewardship.

However, whether from a perspective of systemic or agrarian ethics (see Alrøe et al. 2001), we suggest that most interpretations of the IFOAM Principles are too narrow, and are not sensitive enough to the organic agrofood chain as a whole. Several discussions in agricultural ethics address the importance of reconnecting food producers and consumers (e.g., Carruthers 2009, pp. 303, 304). To honor the ethical promise of the Principles, the perspectives and conditions of the consumers must be strengthened to become relevant in the daily lives of consumers. Moreover, there is a need for more critical ethical discussion of corporate commercial activities and approaches (Williams and Murphy 1990). It is time to address issues of compassion, fairness, loyalty and openness. All are emphasized by the IFOAM Principles of Fairness and Care, but are not addressed in the context of commercial marketing. Finally, both the agrarian stewardship and the IFOAM Principles need to be extended toward "urban" stewardship that integrates the urban realities of citizens and markets. The challenge is then how to think about an "urban ethic" in an industrialized and globalized society.

2.3 Diversification of Ethics in the Organic Movement – A Historical Perspective

What has been the role and development of diverse ethics in the history of the organic movement (see also Constance et al., Hunt et al. and Jones, Chaps. 9, 10 and 11)? In this section we discuss the history behind the organic ethics in order to make explicit that the ethical foundation has changed over time, but include several contributions that continue to be of relevance in the IFOAM Principles.

We start our journey with the second half of the eighteenth century, long before organic was established as a movement and before there was a differentiation between organic and conventional. We conclude in the late 1980s, when organic had become progressively well-known and well-established in society. Later, in Chap. 13 we discuss the significance of these ethics for different organic farming groups.

2.3.1 Framing the Organic Roots: Agriculture in the Nineteenth and Early Twentieth Century – A European Perspective

In the eighteenth century, theocentric perspectives about agriculture were central. There was a strong belief in a "God" who understood that even weeds served some higher purpose (see Becker 1788/1980 in Dirlinger et al. 1998).

Early theological literature draws a picture of a good world in which one finds harmony between nature and society and in which all parts intertwine. This is very much a bourgeois, aesthetic perception of nature removed from the notion of the natural world as hostile (Dirlinger et al. 1998, p. 28). As such, this world was not to be destroyed by human beings; its harmony would be guaranteed through "Devine" balance (cf. Sieferle 1990, p. 53, cited in Dirlinger et al. 1998, p. 29). Trust in the everlasting use of nature, with its fertility and its imperishability is given by providential precaution (ibid). From this theocentric perspective, agricultural practices needed to reduce risks to nature and to God's creations. The IFOAM Principles still express elements of this type of theological thinking and an holistic ethic, however without religious beliefs (cf. Mohr cited in Vieth 2008, p. 166).

Starting in the nineteenth century, animal manure and crop rotation became central tools for increasing production, and in doing so expressed the strong relationship between soil fertility and animal husbandry/livestock production. Interestingly, the idea of crop rotation during this time was described as a type of "division of labor." Similarly, leguminous fodder plants were used to fertilize the soil and recycle minerals, while other plants served as subsistence food or as cash crops (Thaer, cited in Dirlinger et al. 1998, pp. 32–34). These farming systems were built on the idea of continuous humus production through fodder legumes, which also offered nitrogen to the whole cropping system and provided the basis for protein fodder for animal feed. In this relationship, humus was both the result of life and the condition for it. The human-nature-relationship was "place-based". These production practices relied on limited inputs, a principle that would be later incorporated into the IFOAM Principle of Ecology.

With the ideas of Albrecht Thaer, a techno-economic oriented agricultural understanding was set into place that marked the turning point to an industrialized form of agriculture, and the phasing out of an organic based agriculture. Nineteenth-century industrialization and early agricultural research led to a new profit-driven paradigm. This era marked a transition from a fairly closed agricultural system, including animals and the recycling of nutrients, to a system that was increasingly open or reliant on external inputs of nutrients and feedstuff. Maximizing profits became more important than maximizing production (Dirlinger et al. 1998, pp. 32, 33). This conceptualization of agriculture was limited to capital intensive and market-oriented farms. The majority of family farms participated in the use of only some elements of modernized agriculture, and continued to rely largely on organically driven crop rotation.

At the turn of the twentieth century, Fritz Haber created the basis for the development of industrially produced nitrogen and thereby tremendously changing agriculture and the world[19] (Smil 2004). From this point on, the legumes and animal manure became markedly less important as a source of nitrogen and other minerals. Farming had taken the first steps toward becoming an industry-produced nitrogen driven system.

[19]Fritz Haber's invention did not only change agriculture fundamentally, it also became the driving factor for war activities—enabling the destruction of both nature and societies (Dirlinger et al. 1998, p. 29).

The change from a humus based cropping system fundamentally endangered soils. Because fodder legumes were no longer needed for crop rotation, the other contributions of legumes to the soil disappeared, including the production of humus through high amounts of carbon from root biomass, and the capacity to protect and to store nutrients and water. Instead of being fed through a diversity of plants and microorganisms, as well as green manure and compost from animals, feeding the soil became dependent on industrial produced nitrogen.

The human-nature relationship was moving towards a new paradigm. This marked the beginning of thinking about farming with more industrial-type language and images: ideas related increasingly to economic oriented input-output regimes (cf. Ropohl 1978) and a shift in thinking about the relationship between the farm and its environment.

The extensive production and dissemination of industrially produced mineral fertilizer started after World War II (Charles 2005). These industrialization processes also came to symbolize the societal division of labor between farms and industries. This phase marked a turning point from a holistic religious-oriented understanding of agriculture and nature toward an anthropocentric and egocentric business-oriented organization of farming.

To summarize: before the beginning of the use of industrially produced fertilizer, farmers applied a fairly holistic-theocentric, closed, site-specific practice. This new type tend to ignore the need to invest in the sustainable production of soil fertility. More broadly, through this process, the farmer became less responsible for the sustainable production of food from the farm's internal resources. From an ethical perspective, it might be said that more anthropocentric oriented values replaced a holistic and religious orientation. Some nineteenth century elements and ethics of agriculture survived until the 1950s, for example, as documented in Jean Gionos novel "Harvest" (Giono 1978). This type of farming is still found among small farms in mountainous or abandoned regions in Eastern European countries and in religious movements all around the world.

2.3.2 Pioneers of the Organic Movement

The organic movement developed in the early 1900s during a period of politically oriented counter-movements in both the German and English speaking world. At the beginning of the twentieth century, the German Ewald Könemanns (1899–1976) established the natural farming (*Natürlicher Landbau*) or "*back to nature*" movement that emerged from the life-reform movement. The educated middle class, laborers and artists shared in the counter-movement and partly supported these new lifestyles.[20]

This back to nature movement was composed mainly of vegetarians who believed in agriculture without animals and a self-subsistence form of gardening

[20](see Krabbe 1974; Linse 1983, 1986; Rothschuh 1983; Huerkamp 1986; von Loesch 1986; Rollins 1997; Hotaka 2000).

focused primarily on fruit and vegetable production. They refused to use mineral fertilizers and synthetic pesticides, just as today's organic farmers. They recycled and composted organic materials (including composting human and urban waste), established green manure, mulch and used minimal tillage methods, as well as low soluble mineral fertilizers and stone meals. In general, those following the back to nature approach established a low-input system, based on the continual recycling of organic materials found on the farm (Vogt and Lockeretz 2007).

These organic practices were accompanied by an ethically oriented lifestyle. For example, in the life-reform movement, participants refer to an understanding and respect of animals (Biocentrism), and of ecosystems (ecocentrism). Specifically, the movement political origins led the life reformers to focus on fairness and care. Clearly, some of the IFOAM Principles have their roots in this movement.

In the early 1920s, the introduction of larger machinery and the use of mineral fertilizers on large farms led to soil compaction, the loss of soil fertility and a serious decline in yields. In response, a delegation of big landowners in Germany asked the philosopher Rudolf Steiner (1861–1925), who was neither a farmer nor an agricultural scientist, to advise on how to reduce the negative effects of these changes on the soil (Vogt 2000; Patzel 2009; Patzel and Lindenthal 2009). Steiner, founder of anthroposophy, emphasized the humanistic fundamentals of and biodynamic farming (Steiner 1984), which included ideas about the role of the individual and society instead of the dominate natural science perspective (Steiner 1984, pp. 48, 76). "It is infinitely important that agriculture should be closely related to the social life" (Steiner 1973, p. 249). Steiner's intention was not to offer a complete description of agricultural practices, but to present a perspective that farmers could put into practice. Steiner's idea for a biodynamic agriculture was the "*Verlebendigung der Erde*" (roughly the 'vitalization of the earth') through organic fertilizers. According to Steiner, living soils, animals and compost are key to the system. In Steiner's biodynamic approach, each farm is seen as an organism, and field practices must account for cosmic forces (Steiner 1984, p. 169). For Steiner, bio-dynamic farmers need to develop their individual farm identity that accounts for the evolution of nature, society, humankind and the cosmos.

Steiner's perspective does not fundamentally contradict the ethical standpoints of Könemanns's back to nature approach, yet Steiner's ethics and spiritual individualism goes much further. For Steiner (1892) " [...] ethical human life, in a real sense, only begins where justification by utilitarian principles ends. ([...] *das im eigentlichen Sinne ethische Leben des Menschen fängt aber da erst an, wo diese auf Nützlichkeit begründeten Gesetze aufhören*") (p. 170). Steiner held that rules, principles or norms alone do not make an ethical society. To act ethically in a deeper sense, he argues, is what an individual has to arrange with him/herself (Steiner 1892, p. 172). Steiner also put the individual's role in society more in the foreground. Steiner's contributions to the ethical roots of organic were profound and comprehensive. We find them today in the IFOAM Principles, however less emphatic than he did, e.g., in seeing the farm as an organism as a whole.

In the 1950s, the British agriculturalist working in India, Sir Albert Howard (1873–1947), and the British farmer and educator Lady Eve Balfour (1898–1990), engaged into organic farming with contributions that are relevant until today. Both pioneers underlined the significance of the soil. To them, the increase and maintenance of humus, or organic matter, was essential for assuring overall soil health and soil fertility. Howard focused on a more scientific argument and the composition of composting: cow dung (known as Indore compost); urban waste; the addition of micro-organisms to the soils and compost; and, the integration of leguminous fodder crops and green manure (Howard 1946). Balfour concentrated on the close relation between soil, animal and human health and the organic cycle (Balfour 1948, pp. 16, 23). Following Steiner, both understood the necessary contribution of animals to the farming system (Howard 1946). It should be noted that both Howard and Balfour were familiar with, and were seeking to adapt long-standing agricultural practices from China, East Asia and India, as documented by King (1911), to the definition of organic farming in Europe.

Balfour and Howard shared an ecocentric ethic that was sensitive to living and non-living organisms. In addition, Howard is recognized for bringing ecological and health related values into the organic movement, while Balfour valued Christian social ethics as a foundation for agriculture (Balfour 1948, pp. 184–190). She focused on the social fairness and care, and she was highly critical of capitalist economic relations and materialism that she saw as responsible for the economic exploitation of nature. Balfour specifically underlined Christian values and social justice in agriculture and food systems that reflected combined ecocentric – theocentric ethics. She argued that, "we cannot escape from the ethical and spiritual values of life for they are part of wholeness" (Patzel 2010, p. 271). Overall, she demanded the holistic necessity of "service to God, service to our soil, service to each other, and, through each other, to the community and the world" (Balfour 1948, p. 188; Patzel 2010). Clearly, her holistic environmental and social philosophy is at the core of the IFOAM Principles.

In 1949, the Swiss couple Hans (1891–1988) and Maria Müller (1894–1969) established an approach called organic-biologic agriculture (Moser 1994 cited in Patzel and Lindenthal 2009, p. 6). During the 1930s, Hans Müller was engaged in public policy concerning soil and economy. He argued that since the soil was essential for farmers, it should not be considered as a commodity. Based on extensive scientific reading and their own experiences, the Müllers created the 'organic-biologic agriculture and gardening method.' This method drew upon Sir Albert Howard's ideas as well as those from Rudolf Steiner and his biodynamic attention to humus, soil organisms, composting and the cycling of systems. The Müller's approach to compost, in contrast to Howard's use of the Indore method, focused on promoting a system of surface composting and mulching.

The Müllers also sought to secure independence for farmers from trade and agricultural inputs. They were committed to the economic survival of small family farms, respecting tradition, assuring fair prices, producing high quality healthy food, knowledgeable consumers and close relationships between farmers and consumers through regional and cooperative markets (Vogt 2001). Hans Müller's journal, entitled "Culture and Policy (Kultur und Politik)," publicized his political convictions, and he led the Swiss farmer-home-movement that was founded on Christian-based responsibility for nature and consumers (Vogt 2001). In summary, the Müllers contributed to the ecological and health related ethics in the current organic movement, as well as to the initial thinking about the value of fairness and care in agriculture.

The natural scientist and medical doctor, Hans Peter Rusch (1906–1977), provided the scientific evidence for the Müller's organic-biological agriculture approach. He highlighted the use of organisms for their contribution to the cycle of living substances, including soil and compost organisms, and the microbiological activity of soils (Rusch 1968). Rusch understood biology as a holistic guiding science, expressing what he called the "biological reason" or "the result of processes which include all of which comprises our entire being; not only logic, but our cognitive, mental and bodily being, our character, as well as instinct and intuition" (Patzel 2010, p. 271). In short, Rusch provided the ethical background for the ecological and health related principles of today's organic movement. From an ecological perspective, he comment that, "life is a unified whole, where every part is of equal value and given equal rights, regardless if it is a simple organisms or humans" (Rusch 1968, p. 34) illustrating his contribution to the IFOAM Principle of Fairness.

Aldo Leopold (1884–1948), a US ecologist and forester, introduced the concept of a land ethic to capture the idea that all beings are interdependent parts of "ecological communities" (Callicott 1992). He highlighted the community, not the individual, as a value of the common good.[21] To realize such a society, Leopold identified the importance of a rational and sensitive process based on changing our "land ethic" from one of conqueror, to one of being a mutually respectful community member and citizen (Thompson 1988). Leopold's land ethic enlarges the boundaries of community to include soil, water, plants and animals, or collectively – the land. To Leopold, land was a community and the basic concept of ecology. The land ethic was to love and respect the land (Leopold 1949).

A land ethic in Leopold's eyes could not be forced. Society needed to realize that it possesses a land ethic. According to Leopold, this ethic should reflect the existence of an ecological conscience that, in turn would reflect a conviction of individual responsibility for the health of the land (Leopold 1949). Such an ethic could be promoted through higher education and governmental support.

The land ethic has been predominantly embraced by the environmental movement, and is tied logically to agrarianism.[22] Leopold's ecological perspective contributed to understanding the human-nature relationship. His ethic went beyond

[21]This contrasts with Steiner's view of individuality as a precondition for community.

[22]"Agrarianism—the celebration of agriculture and rural life for the positive impact thereof on the individual and society..." (Danbom 1991, p.1) (see also: Oren 1973; Flinn and Johnson 1974; Chase 1988; Montmarquet 1989; Dalecki and Coughenour 1992).

the borders of a farm to the landscape (Principle of Ecology) and a community (Principle of Fairness and Care). For Leopold, the land is a common good and not the property of an individual (cf. Foster 1995; Freyfogle 2003). It is important to note that Leopold sheds light on the relevance of individual freedom and responsibility while identifying the need for a place-based education within a governmental framework (see Knapp 2005).

Organic farmer and author, Jerome Irving Rodale (1898–1970), was first inspired by Albert Howard's experiments and observations (Jerome Irving Rodale 1971). In 1930 he established his own experimental farm for organic agriculture in Pennsylvania. His book "The Organic Front" (1948), presents his thinking about organic agriculture, soils and health, including the interaction of organic fertilizer, soil health and human health. Rodale's approach was very techno-ecological oriented, and he advocated innovation and a more modernized type of organic agriculture. As such, he therefore contributed a strong ecological and health oriented perspective to the current organic movement.

Each of these organic pioneers influenced the development of ethics and value that undergird organic production and consumption today. Although they were very aware that food production inevitably required human intervention with nature, each one understood that this intervention must respect nature. In summary, each pioneer contributed to promoting the need for ethics or values which are documented today in the IFOAM Principles of Health, Ecology, Fairness and Care.

2.3.3 Organic and Environmentalism

In the early 1960s, a time in which diverse groups of ranging values and objectives already comprised the organic movement, Rachel Carson's (Carson 2002) critique of pesticide use launched a new phase for the organic movement. Even though the British Soil Association had already given environmental issues attention in the '50s (Conford and Dimbleby 2001), Carson was able to broaden attention to the organic movement and to organic as an environmental friendly agricultural practice. This coincided with the rise of the counter-cultural movements in which many started organic farming (back to the land movement) as a protest against the 'industrial-military complex' (Sligh and Cierpka 2007, p. 33). The discourse that economic profits and self-interest should be less important than more holistic oriented values was important during this time.

In addition to the traditional organic practitioners, who followed many of the pioneers and the back to the land movement in the 1920s, many family farmers wanting to avoid the trend of 'get big or get out' turned to organic agriculture. During this time, there was also a growing movement of food cooperatives that promoted the value of healthy, local, and locally processed food (Allen et al. 2003). These diverse motives and values illustrated the ways in which the organic movement had broadened and had become much more than an agricultural movement, but also a societal movement of 'organic activism' (Sligh and Cierpka 2007, p. 34).

In the 1990ties, in the US organic has been called a counter-culture movement (Tovey 1997, 2002; Reed 2002; Allen et al. 2003, pp. 63, 65). With the increase of farmers' markets, community supported agriculture, urban farming and food cooperatives, today many such social activities for producing and procuring food have taken a 'stand' against corporate organics, and have actively included values associated with such innovations (Tovey 2002), closely linked to the IFOAM Principles. In North America, these activities have been popularized by well known farmers, authors and activists today such as Wendell Berry (2002), Wes Jackson (1980), Joe Salatin (2013), Barbara Kingsolver (Kingsolver et al. 2009) and Michael Pollan (2007). In Europe, on the other hand, organic simply became part of its long tradition of farmers' markets, and introduced in supermarkets, while food coops played a minor role, and community supported agriculture and urban farming were unheard of (Seyfang 2006).

2.3.4 Organic Principles and Standards

Between the 1970s and 1980s there was a fundamental shift that profoundly affected the reflection and practices of ethics in the organic movement – the "institutionalization of organic farming" (Michelsen et al. 2001). In 1972, the International Federation of Organic Agriculture Movements (IFOAM) was created to improve communication and trade for organic. In 1980, IFOAM formulated the first set of principles to serve as the ethical guidelines for organic and then codified as the Basic Standards (Schmid and Lockeretz 2007, p. 154). During the early stages of organic, rules, norms or standards were formulated largely on a private and informal basis. IFOAM was the first organization that initiated a worldwide formalization process of organic.

From an ethical standpoint, this was a remarkable change. While in the pioneer days, the "control" of production was part of a personal relationship between farmers and consumers, among farmers and within farmer associations, the IFOAM Standards transferred the "verification of organic" to a separate certification and inspection system. Additionally, the focus on what was relevant for organic farming changed, for example, processing and animal welfare began to be of interest (Schmid and Lockeretz 2007, p. 152). Originally, "in the pioneer phase the standards brought organic farmers together, whereas later the standards seemed to divide them" (ibid 2007, 158). Additionally, there became increased competition between different rules and labels. While this stimulated the development of organic, it contributed to confusion about what constituted an organic product.

The popularization of the organic movement has also led to a measure of detachment or weakening of former values and their substitution by a new understanding of organic agriculture (Vogt 2001; Patzel 2009, p. 9). The earlier organic values were shaped by small farmers and Christian attitudes about life yet the environmental movement, and often intellectuals from outside agriculture, drove this newest redefinition of organic. However, this redefinition has also opened the door for new coalitions within society and more specifically to the movement's political acceptance (see Patzel 2009, p. 10). These new coalitions also include technical, ecological and market oriented practices with a rising environmental-political consciousness. Perhaps this new approach may be a starting point for several societal changes: rural development (Libery and Kneafsey 1998; Darnhofer 2005), the emergence of "agri"-culture (Parrott et al. 2002) and (in an urban context) new lifestyles (Gilg et al. 2005; Pellegrini and Farinello 2009).

Without question, the first set of IFOAM Principles was obviously necessary for a growing movement as a foundation for coming together and for communicating a common understanding of the meaning of organic agriculture. In contrast to the time of the organic pioneers, urban organic consumers have brought new perspectives into the organic movement. Now, a wide range of views—from anti-militarist, to back to the land, to alternative energy movements, to development work, and the European green parties—have introduced many important issues related to social justice, fair trade and relationships between farmers and consumers. All of these influence the continuing development of the organic movement and its ethical foundations and contribute to the ethical differentiation of the organic approach.

Annex

Table 2.2 IFOAM Principles

IFOAM Principles (2009)
Background
The IFOAM Principles arose from an international consultation process from 2003 to 2005; of a task force and a consulting group formed by the IFOAM World Board. They consist of ethical principles organized to inspire action in all dimensions of the organic agro-food chain. This worldwide stakeholder-based consultation process embodied a specific ethical position to develop ethical guidelines through a participatory process. The four Principles – Health, Ecology, Fairness and Care – "serve to inspire the organic movement in its full diversity. They guide IFOAM's development of positions, programs and standards" (IFOAM 2009)
The Principle of Health
Organic Agriculture should sustain and enhance the health of soil, plant, animal, human and planet as one and indivisible. This principle points out that the health of individuals and communities cannot be separated from the health of ecosystems – healthy soils produce healthy crops that foster the health of animals and people
Health is the wholeness and integrity of living systems. It is not simply the absence of illness, but the maintenance of physical, mental, social and ecological well-being. Immunity, resilience and regeneration are key characteristics of health
The role of organic agriculture, whether in farming, processing, distribution, or consumption, is to sustain and enhance the health of ecosystems and organisms from the smallest in the soil to human beings. In particular, organic agriculture is intended to produce high quality, nutritious food that contributes to preventive health care and well-being. In view of this it should avoid the use of fertilizers, pesticides, animal drugs and food additives that may have adverse health effects

(continued)

Table 2.2 (continued)

IFOAM Principles (2009)
The Principle of Ecology
Organic Agriculture should be based on living ecological systems and cycles, work with them, emulate them and help sustain them
This principle roots organic agriculture within living ecological systems. It states that production is to be based on ecological processes, and recycling. Nourishment and well-being are achieved through the ecology of the specific production environment. For example, in the case of crops this is the living soil; for animals it is the farm ecosystem; for fish and marine organisms, the aquatic environment
Organic farming, pastoral and wild harvest systems should fit the cycles and ecological balances in nature. These cycles are universal but their operation is site- specific. Organic management must be adapted to local conditions, ecology, culture and scale. Inputs should be reduced by reuse, recycling and efficient management of materials and energy in order to maintain and improve environmental quality and conserve resources
Organic agriculture should attain ecological balance through the design of farming systems, establishment of habitats and maintenance of genetic and agricultural diversity. Those who produce, process, trade, or consume organic products should protect and benefit the common environment including landscapes, climate, habitats, biodiversity, air and water
The Principle of Fairness
Organic Agriculture should build on relationships that ensure fairness with regard to the common environment and life opportunities
Fairness is characterized by equity, respect, justice and stewardship of the shared world, both among people and in their relations to other living beings
This principle emphasizes that those involved in organic agriculture should conduct human relationships in a manner that ensures fairness at all levels and to all parties – farmers, workers, processors, distributors, traders and consumers. Organic agriculture should provide everyone involved with a good quality of life, and contribute to food sovereignty and reduction of poverty. It aims to produce a sufficient supply of good quality food and other products
This principle insists that animals should be provided with the conditions and opportunities of life that accord with their physiology, natural behavior and well-being.
The Principle of Care
Organic Agriculture should be managed in a precautionary and responsible manner to protect the health and well-being of current and future generations and the environment
Organic agriculture is a living and dynamic system that responds to internal and external demands and conditions. Practitioners of organic agriculture can enhance efficiency and increase productivity, but this should not be at the risk of jeopardizing health and well-being. Consequently, new technologies need to be assessed and existing methods reviewed. Given the incomplete understanding of ecosystems and agriculture, care must be taken
This principle states that precaution and responsibility are the key concerns in management, development and technology choices in organic agriculture. Science is necessary to ensure that organic agriculture is healthy, safe and ecologically sound. However, scientific knowledge alone is not sufficient. Practical experience, accumulated wisdom and traditional and indigenous knowledge offer valid solutions, tested by time. Organic agriculture should prevent significant risks by adopting appropriate technologies and rejecting unpredictable ones, such as genetic engineering. Decisions should reflect the values and needs of all who might be affected, through transparent and participatory processes

Source: IFOAM (2009)
http://www.ifoam.org/about_ifoam/principles/history_of_principles.html

References

Abaidoo, S., and H. Dickinson. 2002. Alternative and conventional agricultural paradigms: Evidence from farming in Southwest Saskatchewan*. *Rural Sociology* 67(1): 114–131.

Allen, P., M. FitzSimmons, M. Goodman, and K. Warner. 2003. Shifting plates in the agrifood landscape: The tectonics of alternative agrifood initiatives in California. *Journal of Rural Studies* 19(1): 61–75.

Alrøe, H.F., and E.S. Kristensen. 2002. Towards a systemic research methodology in agriculture: Rethinking the role of values in science. *Agriculture and Human Values* 19(1): 3–23.

Alrøe, H.F., and E.S. Kristensen. 2003. Toward a systemic ethic. *Environmental Ethics* 25(1): 59–78.

Alrøe, H.F., and E.S. Kristensen. 2004. Why have basic principles for organic agriculture?... and what kind of principles should they be? *Ecology & Farming* 36: 27–30.

Alrøe, H.F., M. Vaarst, and E.S. Kristensen. 2001. Does organic farming face distinctive livestock welfare issues? A conceptual analysis. *Journal of Agricultural and environmental Ethics* 14(3): 275–299.

Alrøe, H.F., J. Byrne, and L. Glover. 2006. Organic agriculture and ecological justice: Ethics and practice. In *Global development of organic agriculture: Challenges and prospects*, ed. N. Halberg, H.F. Alroe, and M. Trydeman Knudsen, 75–112. Wallingford: CAB International.

Armstrong, S.J., and R.G. Botzler (eds.). 1993. *Environmental ethics. Divergence and convergence*. London: McGraw-Hill.

Balfour, E. 1948. *The living soil. 8 «edition»*. Lodon: Faber & Faber.

Barnett, C., P. Cloke, N. Clarke, and A. Malpass. 2005. Consuming ethics: Articulating the subjects and spaces of ethical consumption. *Antipode* 37(1): 23–45.

Becker, R. Z. 1788/1980. *Noth- und Hülfsbüchlein für Bauersleute*. Nachdruck der Erstaus-gabe von 1788. Herausgegeben und mit einem Nachwort versehen von Reinhard Siegert (1980) Dortmund.

Benbrook, C.M., and F. Kirschenmann. 1997. Proposed principles for evaluating the organic 'rule' *DARCOF Discussions*.

Bentham, J. 1978. *An introduction to the principles of morals and legislation*. London: Routledge.

Berry, W. 2002. *The art of the common-place: The agrarian essays of Wendell Berry*. Berkeley: Counterpoint Press.

Best, H. 2008. Organic agriculture and the conventionalization hypothesis: A case study from West Germany. *Agriculture and Human Values* 25(1): 95–106.

Borchers, D. 2001. *Die neue Tugendethik: Schritt zurück im Zorn?* Münster: Mentis Verlag.

Browne, A.W., P.J. Harris, A.H. Hofny-Collins, N. Pasiecznik, and R.R. Wallace. 2000. Organic production and ethical trade: Definition, practice and links. *Food Policy* 25(1): 69–89.

Cafaro, P. 2008. Thoreau, Leopold, and Carson. *Environmental Ethics* 23(1): 3–17.

Callicott, J.B. 1989. *In defense of the land ethic: Essays in environmental philosophy*. Albany: Suny Press.

Callicott, J.B. 1992. Aldo Leopold's metaphor. In *Ecosystem health: New goals for environmental management*, ed. B.G. Norton and B. Haskell, 42. Washington, DC: Island Press.

Callicott, J.B. 2008. Do deconstructive ecology and sociobiology undermine Leopold's land ethic? *Environmental Ethics* 18(4): 353–372.

Carruthers, P. 2009. The land debate–'doing the right thing': Ethical approaches to land-use decision-making. In *What is land for? The food, fuel and climate change debate*, 293–318. London: Earthscan.

Carson, R. 2002. *Silent spring – 40th anniversary edition*. New York: Houghton Mifflin Harcourt.

Charles, D. 2005. *Master mind: The rise and fall of Fritz Haber, the Nobel laureate who launched the age of chemical warfare*. New York: HarperCollins.

Chase, M. 1988. *"The people's farm": English radical agrarianism, 1775–1840*. Oxford: Clarendon Press.

Clark, K.J. 2012. Why be moral? Social contract theory versus Kantian-Christian morality. *Journal of Markets & Morality* 6(1): 81–98.
Clarke, N., P. Cloke, C. Barnett, and A. Malpass. 2008. The spaces and ethics of organic food. *Journal of Rural Studies* 24(3): 219–230.
Comstock, G.L. 1995. Do agriculturalists need a new, an ecocentric, ethic? 1994 Presidential address to the agriculture, food, and human values society. *Agriculture and Human Values* 12(1): 2–16.
Conford, P. 1995. The alchemy of waste: The impact of Asian farming on the British organic movement. *Rural History* 6(01): 103–114.
Conford, P., and J. Dimbleby. 2001. *The origins of the organic movement*, vol. 15. Edinburgh: Floris Books.
Dalecki, M.G., and C.M. Coughenour. 1992. Agrarianism in American society. *Rural Sociology* 57(1): 48–64.
Daly, H., M. Jacobs, and H. Skolimowski. 1995. Discussion of Beckerman's critique of sustainable development. *Environmental Values* 4(1): 49–70.
Danbom, D.B. 1991. Romantic agrarianism in twentieth-century America. *Agricultural History* 65(4): 1–12.
DARCOF. 2000. *Principles of organic farming: Discussion document prepared for the DARCOF Users Committee*. Tjele: Danish Research Centre for Organic Farming.
Darnhofer, I. 2005. Organic farming and rural development: Some evidence from Austria. *Sociologia Ruralis* 45(4): 308–323.
Dirlinger, H., M. Fliegenschnee, F. Krausmann, G. Liska, and M.A. Schmidt (eds.). 1998. *Bodenfruchtbarkeit und Schädlinge im Kontext von Agrargesellschaften*, vol. 51. Wien: Iff Social Ecology.
Elmore, R.W. 1996. Our relationship with the ecosystem and its impact on sustainable agriculture. *Journal of Production Agriculture* 9(1): 42–45.
Fenner, D. 2010. *Einführung in die angewandte Ethik*. Tübingen: Francke UTB.
Fischer, P. 2003. *Einführung in die Ethik*. München: W. Fink.
Flinn, W.L., and D.E. Johnson. 1974. Agrarianism among Wisconsin farmers. *Rural Sociology* 39: 187–204.
Foster, J.B. 1995. Global ecology and the common good. *Monthly Review* 46(9): 1–10.
Foster, J.B., and P. Burkett. 2000. The dialectic of organic/inorganic relations Marx and the Hegelian Philosophy of Nature. *Organization & Environment* 13(4): 403–425.
Frasz, G.B. 2008. Environmental virtue ethics. *Environmental Ethics* 15(3): 259–274.
Freyer, B. 2008. *The differentiation process in organic agriculture (OA), between Capitalistic Market System and IFOAM Principles.* Paper presented at the Second Scientific Conference of the International Society of Organic Agriculture Research (ISOFAR), Modena.
Freyfogle, E. 2003. Private land made (too) simple. *Environmental Law Reporter* 33: 10.
Gilg, A., S. Barr, and N. Ford. 2005. Green consumption or sustainable lifestyles? Identifying the sustainable consumer. *Futures* 37(6): 481–504.
Ginsberg, J.M., and P.N. Bloom. 2004. Choosing the right green marketing strategy. *MIT Sloan Management Review* 46(1): 79–84.
Giono, J. 1978. *Ernte*. Frankfurt: S. Fischer.
Goodman, D. 2000. Organic and conventional agriculture: Materializing discourse and agro-ecological managerialism. *Agriculture and Human Values* 17(3): 215–219.
Gorke, M. 2006. Albert Schweitzers Ethik der Ehrfurcht vor dem Leben als Wegbereiterin einerholistischen Umweltethik – Gemeinsamkeiten und Unterschiede. In *Ethik des Lebens. Albert Schweitzer als Philosoph*, ed. M. Hauskeller, 259–277. Baden-Baden: Die Graue Edition.
Gudorf, C.E. 2012. Ecologies of grace: Environmental ethics and Christian theology. By Willis Jenkins. *Journal of the American Academy of Religion* 80(1): 248–252.
Hamm, U., and F. Gronefeld. 2004. *The European market for organic food: Revised and updated analysis*. Wallingford: CAB International.

Harel, A., and A. Sharon. 2008. What is really wrong with torture? *Journal of International Criminal Justice* 6(2): 241–259.
Hay, R. 2010. The relevance of ecocentrism, personal development and transformational leadership to sustainability and identity. *Sustainable Development* 18(3): 163–171.
Heckman, J. 2006. A history of organic farming: Transitions from Sir Albert Howard's War in the Soil to USDA National Organic Program. *Renewable Agriculture and Food Systems* 21(03): 143–150.
Hoffman, A.J., and L.E. Sandelands. 2005. Getting right with nature anthropocentrism, ecocentrism, and theocentrism. *Organization & Environment* 18(2): 141–162.
Hotaka, T. 2000. Contact with nature as urban culture in the modern age: The gardening movement in the second imperial age in Germany. In *New directions in urban history. Aspects of European art, health, tourism and leisure since the enlightenment*, ed. P. Borsay and G. Hirschfelder, 127–146. Münster: Waxman.
Howard, S.A. 1946. *Soil and health*. New York: Devin-Adair.
Huerkamp, C. 1986. Medizinische Lebensreform im späten 19. Jahrhundert. Die Naturheilbewegung in Deutschland als Protest gegen die naturwissenschaftliche Universitätsmedizin. *VSWG: Vierteljahrschrift für Sozial-und Wirtschaftsgeschichte* 73(2): 158–182.
Hull, R. 2005. All about eve: A report on environmental virtue ethics today. *Ethics & The Environment* 10(1): 89–110.
Hunt, S.D., and A.Z. Vasquez-Parraga. 1993. Organizational consequences, marketing ethics, and salesforce supervision. *Journal of Marketing Research* 30: 78–90.
Hursthouse, R. 2007. Environmental virtue ethics. In *Working virtue: Virtue ethics and contemporary moral problems*, 155–172. Oxford: Clarendon Press.
IFOAM. 2009. *The principles of organic agriculture*. Retrieved from http://www.ifoam.org/about_ifoam/principles/index.html. Accessed 1 Sept 2011.
IFOAM. 2012. *The IFOAM norms for organic production and processing*. Retrieved from http://www.ifoam.org/en/ifoam-norms. Accessed 1 Jan 2013.
Illich, I. 1973. *Tools for conviviality*. New York: Harper & Row.
Illich, I. 1975. *Selbstbegrenzung. Eine politische Kritik der Technik (German translation of "Tools for Conviviality")*. Hamburg: Reinbek.
Jackson, W. 1980. *New roots for agriculture*. Lincoln: University of Nebraska Press.
Jacob, M. 1994. Sustainable development and deep ecology: An analysis of competing traditions. *Environmental Management* 18(4): 477–488.
Johnson, P.C. 1996. Development of an ecological conscience: Is ecocentrism a prerequisite? *The Academy of Management Review* 21(3): 607–611.
Jonas, H. 1979. *Das Prinzip Verantwortung: Versuch einer Ethik für die technologische Zivilisation*, 1 Auflth ed. Insel-Verlag: Frankfurt am Main.
Jonas, H. 2004. *Leben, Wissenschaft, Verantwortung: ausgewählte Texte*. Stuttgart: Reclam.
Katz, E. 2008. A pragmatic reconsideration of anthropocentrism. *Environmental Ethics* 21(4): 377–390.
King, F.H. 1911. *Farmers of forty centuries; or, permanent agriculture in China, Korea and Japan*. Madison: Mrs. FH King.
Kings, D., and B. Ilbery. 2010. The environmental belief systems of organic and conventional farmers: Evidence from central-southern England. *Journal of Rural Studies* 26(4): 437–448.
Kingsolver, B., C. Kingsolver, and S.L. Hopp. 2009. *Animal, vegetable, miracle*. New York: HarperCollins.
Kirchhoff, T. 2011. *Naturethik, Ökologie und Naturästhetik – eine kritische Analyse des Ökozentrismus und seiner kulturellen Basis*, 39–47. Retrieved from http://www.wzw.tum.de/loek/publikationen/abstracts/592.pdf. Accessed 12 Nov 2013.
Kirchmann, H., and G. Thorvaldsson. 2000. Challenging targets for future agriculture. *European Journal of Agronomy* 12(3–4): 145–161.
Knapp, C.E. 2005. The "I-Thou" relationship, place-based education, and Aldo Leopold. *Journal of Experiential Education* 27(3): 277–285.

Krabbe, W.R. 1974. *Gesellschaftsveränderung durch Lebensreform: Strukturmerkmale einer sozialreformerischen Bewegung im Deutschland der Industrialisierungsperiode*. Göttingen: Vandenhoeck & Ruprecht.

Krebs, A. (ed.). 1977. *Naturethik. Grundtexte der gegenwärtigen tier- und ökoethischen Diskussion*. Frankfurt/M: Suhrkamp (stw).

Kristiansen, P., and C. Merfield. 2006. Overview of organic agriculture. In *Organic agriculture: A global perspective*, 1–23. Collingwood: CSIRO Publishing.

Lammerts Van Bueren, E., and P.C. Struik. 2005. Integrity and rights of plants: Ethical notions in organic plant breeding and propagation. *Journal of Agricultural and Environmental Ethics* 18(5): 479–493.

Lammerts Van Bueren, E., P.C. Struik, M. Tiemens-Hulscher, and E. Jacobsen. 2003. Concepts of intrinsic value and integrity of plants in organic plant breeding and propagation. *Crop Science* 43(6): 1922–1929.

Leopold, A. 1949. *A sand county almanac*. New York: Oxford University Press.

Libery, B., and M. Kneafsey. 1998. Product and place promoting quality products and services in the lagging rural regions of the European Union. *European Urban and Regional Studies* 5(4): 329–341.

Linse, U. 1983. *Zurück, o Mensch, zur Mutter Erde: Landkommunen in Deutschland 1890–1933*. München: Deutscher Taschenbuch-Verlag.

Linse, U. 1986. *Ökopax und Anarchie: eine Geschichte der ökologischen Bewegungen in Deutschland*. München: Deutscher Taschenbuch Verlag.

Littig, B., and E. Grießler. 2004. *Soziale Nachhaltigkeit*. Wien: Bundeskammer für Arbeiter und Angestellte.

Lochbühler, W. 1996. *Christliche Umweltethik*. Frankfurt am Main: Lang.

Lockeretz, W. (ed.). 2007. *Organic farming: An international history*. Wallingford: CAB International.

Lund, V. 2006. Natural living—A precondition for animal welfare in organic farming. *Livestock Science* 100(2): 71–83.

Lund, V., and H. Röcklinsberg. 2001. Outlining a conception of animal welfare for organic farming systems. *Journal of Agricultural and Environmental Ethics* 14(4): 391–424.

Luttikholt, L.W.M. 2007. Principles of organic agriculture as formulated by the International Federation of Organic Agriculture Movements. *NJAS-Wageningen Journal of Life Sciences* 54(4): 347–360.

Luttikholt, L.W.M., and K. Vijayalaksmi. 2004. *Principles of organic agriculture-a review validity of the most basic values of the organic movement*. Paper presented at the 6th IFOAM-Asia Scientific Conference, Yangpyung, Korea, 7–11 Sept 2004: "Benign environment and safe food".

Magnusson, M.K., A. Arvola, U.K.K. Hursti, L. Åberg, and P.O. Sjödén. 2003. Choice of organic foods is related to perceived consequences for human health and to environmentally friendly behaviour. *Appetite* 40(2): 109–117.

McEachern, M.G., and P. Mcclean. 2002. Organic purchasing motivations and attitudes: Are they ethical? *International Journal of Consumer Studies* 26(2): 85–92.

McNaughton, D.A., and J.P. Rawling. 2007. Deontology. In *Principles of health care ethics*, 2nd ed, ed. R.E. Ashcroft, 65–71. Chichester: Wiley.

Mepham, B. 2001. Foot and mouth disease and British agriculture: Ethics in a crisis. *Journal of Agricultural and Environmental Ethics* 14(3): 339–347.

Meyer-Abich, K. 2006. Physiozentrisch lebt sich's gesünder. Begründung der Umweltethik aus der Praktischen Naturphilosophie. In *Umwelt-Handeln – zum Zusammenhang von Naturphilosophie und Umweltethik*, ed. K. Köchy and M. Norwig, 29–48. Freiburg: Verlag Karl Alber.

Michelsen, J., K. Lynggaard, S. Padel, and C. Foster. 2001. Organic farming development and agricultural institutions in Europe: A study of six countries. In *Organic farming in Europe: Economics and policy*, vol. 9, ed. S. Dabbert. Stuttgart-Hohenheim: University of Hohenheim.

Minteer, B.A. 2006. *The landscape of reform: Civic pragmatism and environmental thought in America*. Cambridge, MA: MIT Press.

Minteer, B.A., and R.E. Manning. 2005. An appraisal of the critique of anthropocentrism and three lesser known themes in Lynn White's "The Historical Roots of our Ecologic Crisis". *Organization & Environment* 18(2): 163–176.

Montmarquet, J.A. 1989. *The idea of agrarianism: From hunter-gatherer to agrarian radical in western culture*. Moscow: University of Idaho Press.

Moser, P. 1994. *Der Stand der Bauern. Bäuerliche Politik, Wirtschaft und Kultur gestern und heute* [The peasantry. Peasant policy, economy and culture in the past and today]. Frauenfeld: Huber.

Naess, A. 1973. The shallow and the deep, long range ecology movement. A summary. *Inquiry* 16: 151–155.

Naess, A. 1989. *Ecology, community and lifestyle: Outline of an ecosophy.* Trans. D. Rothenberg. Cambridge: Cambridge University Press.

Nash, R.F. 1989. *The rights of nature: A history of environmental ethics*. Madison: University of Wisconsin Press.

Norton, B.G. 2008. Environmental ethics and weak anthropocentrism. *Environmental Ethics* 6(2): 131–148.

Oren, N. 1973. *Revolution administered: Agrarianism and communism in Bulgaria*. Baltimore/London: Johns Hopkins University Press.

Padel, S., and K. Gössinger. 2008. *Farmer consumer partnerships communicating ethical values: A conceptual framework*, CORE organic project series report. Tjele: CORE.

Padel, S., N. Lampkin, and W. Lockeretz. 2007. The development of governmental support for organic farming in Europe. In *Organic farming: An international history*, ed. W. Lockeretz, 93–122. Wallingford: CAB International.

Padel, S., H. Röcklinsberg, and O. Schmid. 2009. The implementation of organic principles and values in the European Regulation for organic food. *Food Policy* 34(3): 245–251.

Parrott, N., N. Wilson, and J. Murdoch. 2002. Spatializing quality: Regional protection and the alternative geography of food. *European Urban and Regional Studies* 9(3): 241–261.

Patzel, N. 2009. The soil scientist's hidden beloved: Archetypal images and emotions in the scientist's relationship with soil. In *Soil and culture*, ed. E.R. Landa and C. Feller, 205–226. Dordrecht/Heidelberg/London/New York: Springer.

Patzel, N. 2010. Der "innere Boden". In *Boden*, ed. M. Schneider, A. Fink-Keßler, and F. Stodieck, 266–274. München: Der kritische Agrarbericht.

Patzel, N., and T. Lindenthal. 2009. Der Umgang mit Böden im ökologischen Landbau. In *Handbuch der Bodenkunde*, vol. 31, ed. S. Karl, H. Blume, P. Felix-Henningsen, H. Frede, R. Horn, and G. Guggenberger. Weinheim: Wiley-VCH.

Peet, M.M. 1996. *Sustainable practices for vegetable production in the south*. Newburyport: Focus Publishing.

Pellegrini, G., and F. Farinello. 2009. Organic consumers and new lifestyles: An Italian country survey on consumption patterns. *British Food Journal* 111(9): 948–974.

Pettit, P. 1993. *Consequentialism*, vol. 6. Aldershot: Dartmouth Publishing Group.

Piaget, J. 1965. *The moral judgment of the child*. New York: The Free Press.

Pollan, M. 2007. *Second nature: A gardener's education*. New York: Grove.

Raynolds, L.T. 2004. The globalization of organic agro-food networks. *World Development* 32(5): 725–743.

Reed, M. 2002. Fight the future! How the contemporary campaigns of the UK organic movement have arisen from their composting of the past. *Sociologia Ruralis* 41(1): 131–145.

Rennings, K., and H. Wiggering. 1997. Steps towards indicators of sustainable development: Linking economic and ecological concepts. *Ecological Economics* 20(1): 25–36.

Rodale, J.I. 1948. *The organic front*. Emmaus: Rodale Press.

Rodale, J.I. 1971. *The complete book of composting*. Emmaus: Rodale Books.

Rollins, W.H. 1997. *A greener vision of home: Cultural politics and environmental reform in the German Heimatschutz movement, 1904–1918*. East Lansing: University of Michigan Press.

Ropohl, G. 1978. Einführung in die allgemeine Systemtheorie. In *Systemtheorie als Wissenschaftsprogramm*, ed. H. Lenk and G. Ropohl, 9–49. Königstein: Athenäum Verlag.

Rothschuh, K.E. 1983. *Naturheilbewegung, Reformbewegung, Alternativbewegung*. Stuttgart: Hippokrates Verlag.
Rusch, H.P. 1968. *Bodenfruchtbarkeit: Eine Studie Biologischen Denkens*. Heidelberg: Verlag K. F. Haug.
Salatin, J. 2013. *Fields of farmers: Interning, mentoring, partnering, germinating*. Swoop: Polyface.
Sandler, R.L. 2007. *Character and environment: A virtue-oriented approach to environmental ethics*. New York: Columbia University Press.
Schaefer, J. 2009. *Theological foundations for environmental ethics: Reconstructing patristic and medieval concepts*. Georgetown: Georgetown University Press.
Schlüns, J., and L. Voget. 2008. Das Inklusionsproblem und seine verschiedenen Lösungsansätze. *Forum Geoökology* 19(1): 12–16.
Schmid, O., and W. Lockeretz. 2007. Development of standards for organic farming. In *Organic farming: An international history*, ed. W. Lockeretz, 152–174. Wallingford: CAB International.
Schroth, J. 2009. Deontologie und die moralische Relevanz der Handlungskonsequenzen. *Zeitschrift für philosophische Forschung* 63(1): 55–75.
Schweitzer, A. 1976. *Die Lehre von der Ehrfurcht vor dem Leben und ihre Bedeutung für unsere Kultur*. München: Beck Verlag.
Schweitzer, A., C. Günzler, and J. Zürcher. 1999. *Die Weltanschauung der Ehrfurcht vor dem Leben: Kulturphilosophie III*. München: Beck Verlag.
Seyfang, G. 2006. Ecological citizenship and sustainable consumption: Examining local organic food networks. *Journal of Rural Studies* 22(4): 383–395.
Shaw, B. 2008. A virtue ethics approach to Aldo Leopold's land ethic. *Environmental Ethics* 19(1): 53–67.
Shearman, R. 1990. The meaning and ethics of sustainability. *Environmental Management* 14(1): 1–8.
Sieferle, R.R. 1990. *Bevölkerungswachstum und Naturhaushalt. Studien zur Naturtheorie der klassischen Ökonomie* [Population growth and ecosystem. Studies towards theory on nature and classical economy] Frankfurt: Suhrkamp.
Singer, P. 1993. *Praktische Ethik 2.revidierte und erweiterte Auflage*. Stuttgart: Reclam.
Sligh, M., and T. Cierpka. 2007. Organic values. In *Organic farming: An international history*, ed. W. Lockeretz, 30–40. Wallingford: CAB International.
Smil, V. 2004. *Enriching the earth: Fritz Haber, Carl Bosch, and the transformation of world food production*. Cambridge, MA: MIT Press.
Steiner, R. 1892. Wahrheit und Wissenschaft. *Vorspiel einer "Philosophie der Freiheit"*. Retrieved from http://anthroposophie.byu.edu. Accessed 1 Jan 2013.
Steiner, R. 1973. *Grundzüge einer modernen Weltanschauung. Seelische Beobachtungsresultate nach naturwissenschaftlicher Methode Anthroposophy*. Dornach: Rudolf Steiner Verlag.
Steiner, R. 1984. *Vom Menschenrätsel*. Dornach: Rudolf Steiner Verlag.
Stenmark, M. 2004. Überblick über einige normative ethik-prinzipien von biozentrismus und ökozentrismus. *Natur und Kultur* 5(2): 88–113.
Sterba, J.P. 2003. *Morality in practice*. Independence: Wadsworth Publishing.
Tanner, C., and S. Wölfing Kast. 2003. Promoting sustainable consumption: Determinants of green purchases by Swiss consumers. *Psychology & Marketing* 20(10): 883–902.
Taylor, P.W. 1989. *Respect for nature: A theory of environmental ethics*. Princeton: Princeton University Press.
Taylor, P.W. 2003. The ethics of respect for nature. *Environmentalism: Critical Concepts* 1(3): 61.
Thompson, P. 1988. Ethical dilemmas in agriculture: The need for recognition and resolution. *Agriculture and Human Values* 5(4): 4–15.
Thompson, G. 1998. Consumer demand for organic foods: What we know and what we need to know. *American Journal of Agricultural Economics* 80(5): 1113–1118.
Thompson, P. 2001. Land and water. In *A companion to environmental philosophy*, ed. D. Jamieson, 460–472. Malden/Oxford/Berlin/Melbourne: Blackwell Publishing.

Thompson, P. 2007. Agriculture and working-class political culture: A lesson from The Grapes of Wrath. *Agriculture and Human Values* 24(2): 165–177.
Thompson, P. 2008. Agrarian philosophy and ecological ethics. *Science and Engineering Ethics* 14(4): 527–544.
Thompson, P., and T.C. Hilde. 2000. *The agrarian roots of pragmatism*. Nashville: Vanderbilt University Press.
Thompson, P., and A. Nardone. 1999. Sustainable livestock production: Methodological and ethical challenges. *Livestock Production Science* 61(2): 111–119.
Tokar, B. 1990. Eco-Apocalypse. *New Internationalist* 210: 14–15.
Tovey, H. 1997. Food, environmentalism and rural sociology: On the organic farming movement in Ireland. *Sociologia Ruralis* 37(1): 21–37.
Tovey, H. 2002. Alternative agriculture movements and rural development cosmologies. *International Journal of Sociology of Agriculture and Food* 10(1): 1–11.
Vaarst, M., and H.F. Alrøe. 2012. Concepts of animal health and welfare in organic livestock systems. *Journal of Agricultural and Environmental Ethics* 25(3): 333–347.
Vardi, P., and P. Grosch. 1999. *The puzzle of ethics*. Revised Edition ed. Fount.
Verhoog, H., M. Matze, E. Lammerts Van Bueren, and T. Baars. 2003. The role of the concept of the natural (naturalness) in organic farming. *Journal of Agricultural and Environmental Ethics* 16(1): 29–49.
Verhoog, H., E. Lammerts Van Bueren, M. Matze, and T. Baars. 2007. The value of 'naturalness' in organic agriculture. *NJAS-Wageningen Journal of Life Sciences* 54(4): 333–345.
Vieth, A. 2008. Philosophische Ethik als symbolische Kommunikation. In *Ethik und die Möglichkeit einer guten Welt. Eine Kontroverse um die "Konkrete Ethik"*, ed. A. Vieth and C. Halbig, 227–240. Berlin: Walter de Gruyter.
Vogt, G. 2000. *Entstehung und Entwicklung des ökologischen Landbaus im deutschsprachigen Raum*. Bad Dürkheim: SÖL.
Vogt, G. 2001. Geschichte des ökologischen Landbaus im deutschsprachigen Raum [History of organic agriculture in the German-speaking region] (Teil 1). *Ökologie & Landbau* 118(2): 47–49.
Vogt, G., and W. Lockeretz. 2007. The origins of organic farming. In *Organic farming: An international history*, ed. W. Lockeretz, 14–16. Wallingford: CAB International.
von Loesch, A. 1986. Die Unternehmen der Lebensreformbewegungen. *Zeitschrift für öffentliche und gemeinwirtschaftliche Unternehmen: ZögU/Journal for Public and Nonprofit Services* 9: 449–451.
Vorster, N. 2005. Human dignity and sexual behavious–A theological perspective. *Verbum et Ecclesia* 26(3): 891–911.
Weinschenck, G. 1994. Reflections on the economic and ecological sustainability of modern plant production. In *Plant production on the threshold of a new century*, 9–22. New York: Springer.
Wenz, P.S. 2003. Leopold's novel: The land ethic in Barbara Kingsolver's prodigal summer. *Ethics & The Environment* 8(2): 106–125.
White, L. 1967. The historical roots of our ecological crisis. *Science* 10: 1203–1207.
Willemsen, A. 2008. *Die Würde der Kreatur bei Pflanzen*. Bern: Eidgenössische Ethikkommission für die Biotechnologie im Ausserhumanbereich.
Williams, O.F., and P.E. Murphy. 1990. The ethics of virtue: A moral theory for marketing. *Journal of Macromarketing* 10(1): 19–29.
Wirzba, N. 2002. *The art of the commonplace: The agrarian essays of Wendell Berry*. Washington, DC: Counterpoint.
Wirzba, N. (ed.). 2003. *The essential agrarian reader: The future of culture, community, and the land*. Lexington: University Press of Kentucky.
Zamir, E., and B. Medina. 2010. *Law, economics, and morality*. Oxford: Oxford University Press.

Part II
Standards and Certification

Overview and Observations

The three chapters in this part examine organic ethical perspectives through standards and certification processes: how the relationships between farmers and consumers can change through the certification process in Indonesia; the understanding of the organic differentiation process in the production of selected products in the UK and Tanzania; and the contribution of the IFOAM Principles to "feed the world".

In Hatanaka's 'Organic Certification and the Rationalization of Alternative Food and Agriculture: Sustainable Shrimp Farming in Indonesia', certification processes are examined as a means to change the relationship between farmers and consumers of alternative food and agricultural systems. Using the case of a shrimp-farming cooperative, the author described how both consumers and farmers lost control over their product.

The personal relationships, shared governance, and trust that characterized the early life of the food co-op were replaced by objectivity, calculability, and expert knowledge ("science-based certification"). This created a feeling that what was once a product "shared" between shrimp farmers and consumers, after certification became a product of science. When responsibility was delegated to the certifiers, it reduced the commitments from both farmers and co-op buyers.

Although this is only one example, the likelihood of declining commitments to the purpose of alternative food and agriculture and the erosion of trust underlines how certification *per se* embodies shallow forms of social justice and environmental sustainability adopted to meet commercial needs.

The issues of trust and certification create an ethical tension concerning relationships and power distribution among farmers, certifiers, scientists, and consumers. The question arises: If an ethically based framework was attached to certification and was more predominantly discussed or widely distributed among consumers, farmers and certifiers alike, would certification create more trust and relationship building?

Allison Loconto and Maarten Van der Kamp's, 'Differentiating Organics: Performing Multiple Objects to Organize Singular Markets for Organic Tea and Biscuits in the UK', use tea and cereals to show that individual enactments of organic result in a multiplicity of production, despite standardizing practices. At the same time, they show that standardizing practices, which result in a product being considered in a class by itself, or singularization, are required for trading and marketing of organic products.

They note that certification may appear uniform and that it may be necessary for consumers. However, there can be significant differentiation of production methods, market systems, interpretations of organic standards and assessment processes through certification bodies, within a particular product. Therefore, the organic standards offer the ability to see both singularity and multiplicity in the enactment of organic practices, not by defining ethical standards that attempt to govern the intentions and beliefs of producers, but rather, by defining the practices that can be used to produce organic products. These practices can, of course, contradict the IFOAM Principles or enacting them in a comprehensive manner.

This shows how singularization in the multiplicity of produced organic products allows particular organic markets to function. Thus, the authors do not support the claim that organic has 'lost its way,' because the practice of certified organic varies from place to place. Instead, the authors offer a way to see organic taking root in different contexts.

From this perspective, standards have not eroded the original concepts on which IFOAM principles are based, but only perform different interpretations or different realities of the standards. On the one hand, the IFOAM Principles and the Standards allow for different interpretations and with them, practices that are sensitive to individual circumstances. On the other hand, processing and marketing regulations (technical) are often connected to, and ask for standardization and singularization. In this way standardization is seen as supporting IFOAM principles, all of which include a diversity of requirements upon becoming organic, and specifically allows respecting cultural differences. However there is risk that such standards also open space for practices that contradict the principles.

In 'Feeding the World – The contribution of IFOAM Principles', Freyer and colleagues focus on the ethical possibilities behind the IFOAM Principles and how they can be applied to help assess global food security issues. They discuss how ethical values can connect to food supply, demand and access, in order to understand concerns associated with organic practices and the question of how "to feed the world." They suggest that the ethics embodied in the IFOAM Principles offer a framework for identifying how organic agrofood systems might contribute to food security. The authors explain how these principles have quantitatively relevant impacts on food supply, food demand and food access. Finally, they identify the relevance of ethical foundations in organic instruments such as standards, control and certification.

They therefore argue that if food issues are a common societal problem, then the principles need to be more widely addressed and integrated in mainstream food security discussions along the whole organic agrofood chain. A more systemic and holistic approach to the global food system, one in which the IFOAM Principles can ground the integration of all actors in a more systemic way, could reach the roots of todays food related problems. Consequently, there is need for discussions on how to bring them closer into standards and certification procedures, specifically when it comes to social and economic justice.

Since the introduction of organic standards and certification in 1967 by the UK's Soil Association (Schmid and Lockeretz 2007) the standards and certification schemes have generated considerable debate. On one hand, standards and certification offer a system of 'trust' between consumers who are unable to reach first-hand production. On the other hand, trust has also eroded between consumers and producers in part as the original organic values have been minimized or ignored (Alrøe et al. 2006; Goodman and Goodman 2007; Schmid and Lockeretz 2007). As Hatanaka discussed, standardization may create a situation in which it is no longer necessary to connect with the farmer but one that requires trust in an ambiguous middleman.

Hatanaka as well as Loconto and Van der Kamp raise an essential question. At what point does the reliance on scientifically based standards lead to the loss of trust between farmers and consumers? Even a well-established control mechanism (standardization through the multiplicity of production methods) is unable to ensure that farmers fulfill standards. Yet, informal control that builds consumer trust independent of external certification through direct farmer-consumer relations seems difficult to implement in an expanding and largely urban organic market. As Freyer and colleagues argue, so far standards are fulfilled, is open to whether they respect the IFOAM Principles, specifically those referring to social and economic justice.

In summary, there are three positions to consider. One is that certification only builds trust through communication and a shared discussion among farmers and consumers, a process that is excluded by mainstream certification process. The participatory guarantee system (IFOAM Norms 2012) may offer one useful instrument for building trust. An external certifier could take on primary responsibility for managing this common agreement and thereby safeguard trust. Second, there is a need to build trust and guarantees through external control processes. These could assume different forms and procedures in different parts of the world, and opens space to interpret standards and certification differently. There could be a multiplicity of approaches for assuring organic that represent an opportunity, but also a risk for principle based practices. Third, such diversity might also offer ways to bring the IFOAM Principles of fairness and care to center stage of standards and certification procedures. Specifically when reflecting on the issue of feeding the world, both Principles are indispensable.

What is obvious in all three contributions however, and in different ways, is that the IFOAM Principles provide guidance on how to interpret standards and certification procedures that is not generally acknowledged in organic practices. Bringing them more into the center stage of the debate, could serve as a value-based framework for communication and relationships among different stakeholders. Whether there is a rationalized and standardized application of standards and certification procedures, or a more open one that is build on participatory guarantee systems, the application of the IFOAM Principles builds trust among the different partners.

References

Alrøe, H.F., J. Byrne, and L. Glover. 2006. Organic agriculture and ecological justice: Ethics and practice. In *Global development of organic agriculture: Challenges and prospects*, ed. N. Halberg, H.F. Alroe, and M. Trydeman Knudsen, 75–112. Wallingford: CAB International.

Goodman, D., and M. Goodman. 2007. Localism, livelihoods and the 'post-organic': Changing perspectives on alternative food networks in the United States. In *Constructing alternative food geographies: Representation and practice*, 23–38. Oxford: Elsevier.

IFOAM. 2012. *The IFOAM norms for organic production and processing*. Retrieved from http://www.ifoam.org/en/ifoam-norms. Accessed 1 Jan 2013.

Schmid, O., and W. Lockeretz. 2007. Development of standards for organic farming. In *Organic farming: An international history*, ed. W. Lockeretz, 152–174. Wallingford: CAB International.

Chapter 3
Organic Certification and the Rationalization of Alternative Food and Agriculture: Sustainable Shrimp Farming in Indonesia

Maki Hatanaka

3.1 Introduction

Alternative forms of food and agriculture have emerged in response to problems associated with industrialized food and agriculture.[1] They consist of a variety of forms, including fair trade, organic agriculture, local markets, community-supported agriculture, and urban gardens. While the forms of alternative food and agriculture diverge, they generally share a common vision of sustainable, just, and ethical agriculture and safe and healthy food. Recently, alternative agrifood initiatives have proliferated and alternative forms of food and agriculture have become a significant component of the global agrifood system.

There are multiple ways in which alternative agrifood initiatives are governed, but one of the more prominent approaches is the use of certification (Busch and Bain 2004; Mutersbaugh et al. 2005; Taylor 2005; Hatanaka 2010b). In alternative agrifood initiatives, certification tends to be used to develop and enforce standards for upstream producers and processors. In doing so, certification distinguishes alternative products from conventional ones in the marketplace (Hatanaka et al. 2005). Today, there are a wide range of certified alternative agrifood initiatives covering nearly every aspect of food from farm to table. Among the most globally prominent initiatives are the Fairtrade International (FLO), International Federation

[1] In this chapter, the term "alternative" is used to denote a wide breadth of food and agricultural practices from farm-to-table that are oriented towards increasing the social and/or environmental sustainability of agriculture; and/or the healthiness of and/or access to food (Morgan et al. 2006).

M. Hatanaka (✉)
Department of Sociology, Sam Houston State University, Huntsville, TX, USA
e-mail: maki.hatanaka@shsu.edu

B. Freyer, J. Bingen (eds.), *Re-Thinking Organic Food and Farming in a Changing World*, The International Library of Environmental, Agricultural and Food Ethics 22, DOI 10.1007/978-94-017-9190-8_3

of Organic Agriculture Movements (IFOAM), Marine Stewardship Council (MSC), Rainforest Alliance, UTZ Certified, and 4C Association.

Given the increasing use of certification within alternative agrifood initiatives, this chapter examines how certification affects the principles, practices, and goals of alternative food and agriculture. While there is a significant body of literature on the contexts, aims, and forms of alternative food and agriculture, understudied is the ways that governance potentially impacts the character of alternative food and agriculture, such as its values and objectives. Addressing this gap in the literature, this chapter examines changes in a sustainable shrimp project in Indonesia that occurred with organic certification.

The sustainable shrimp project began in 1992 and connected Japanese consumer cooperatives (co-op) members with Indonesian shrimp farmers interested in sustainable aquaculture. Initially, democratic and shared governance, as well as a high degree of trust between shrimp farmers and co-op members characterized the project. Thus, developing respect for each other's position and needs was an integral component of the project. However, in 2002, with organic certification, formal standards, measures, and audits were introduced into the project. Consequently, my findings indicate that notions of objectivity, calculability, and expert knowledge began to replace the emphasis on shared governance, mutual understanding, and trust. The result has been the increased rationalization of the project and depersonalization of relations between project participants, most notably Indonesian farmers and Japanese co-op members.

Data on the organic shrimp project was gathered using two extensive field research site visits in 2004 and 2008. A total of 125 interviews were conducted with actors involved in shrimp farming in the region, including certified and non-certified shrimp farmers, certified and non-certified warehouse owners, project managers and organizers, hatchery owners, social movement organizations, national and regional government officials, and aquaculture specialists. Additionally, members from the Japanese consumer co-ops were also interviewed during both visits. In 2004, data was collected on the origins of the shrimp project, views of certification, the potential implications of the project, and the relationship among different actors associated with the project. In follow-up research in 2008, key informants were re-interviewed to understand the ways that the implementation of the project had progressed and changes in understandings of the project. Both sets of interview data were also supplemented by participant observation whenever possible, which focused on the interactions between actors in the project. Lastly, content analysis of archival data on shrimp aquaculture, alternative food and agriculture, and certification from websites, newsletters, and reports by Japanese co-operatives, transnational organizations, such as the Food and Agriculture Organization, national and international non-governmental organizations, and certification bodies was undertaken.[2]

[2]To maintain confidentiality, the identity of the companies involved in the organic shrimp project, as well as the project location, has been changed.

The remaining sections of this chapter are organized as follows. First, research on alternative food and agriculture is reviewed. Particular attention is given to the proliferation of certification as a governance mechanism in alternative agrifood initiatives. Second, the sustainable shrimp project in Indonesia is briefly outlined. Third is an analysis of the changes that occurred in the sustainable shrimp project with organic certification. Specifically, the focus is on the introduction of formal standards, measures, and audits that have affected the principles and practices of the project. In concluding, this chapter discusses the implications of the findings for alternative food and agriculture more generally. Specifically, I argue that certification may be producing alternative agrifood initiatives that are highly rationalized, but embody shallow forms of social justice and environmental sustainability.

3.2 Alternative Food, Agriculture, and Governance

Alternative forms of food and agriculture have proliferated in response to food safety and quality, environmental, and labor problems associated with corporate-driven, industrial agrifood production. They tend to share the view that industrialized agriculture is unethical and unsustainable, as it prioritizes the maximization of corporate profits at the expense of food safety and quality, small-farmers' livelihoods, and cultural and ecological diversity (Freidberg 2004; DeLind and Howard 2008). In general, alternative agrifood initiatives are outcomes of efforts by farmers, social movement organizations (SMOs), states, and business to reorganize the production, distribution, and consumption of food with the aim of advancing high quality, socially just, and environmentally sustainable food.

While alternative food and agriculture is characterized by significant diversity and embodies a plethora of causes (e.g., the improvement of workers' and farmers' rights, environmental protection, and animal welfare), it shares the common goal of embedding food and agriculture in networks where social, economic, and ecological relations are fair, just, and democratic (Goodman 2003; Renard 2003). In the language of conventions theory, this means that alternative forms of food and agriculture are ideally based on trust (domestic conventions) and seek to balance social and environmental responsibility (civic conventions) with price (commercial conventions), and efficiency and reliability (industrial conventions) (Reardon and Berdegue 2002). Thus, a key characteristic of alternative food and agriculture is its democratic and participatory practices that enable producers and consumers to have voice in the kinds of food produced and how it is produced (Patel et al. 2007).

One mechanism for increasing consumer participation in food and agriculture used by many alternative agrifood initiatives is to de-fetishize food by making transparent where, how, and by whom it is produced (Barham 2002; Hudson and Hudson 2003). In doing so, the idea is to empower consumers to make informed choices about the food they eat and how it is produced. Thus, in alternative food and agriculture, consumption is a key site where individuals can exercise freedom and responsibility (Barnett et al. 2005). Hence, advertising, product labeling, and standards are viewed as techniques to mobilize people as "citizen-consumers"

(Miller and Rose 1997; Lockie 2009). In this way, alternative food and agriculture often relies on what Micheletti (2003) terms "political consumerism."

As alternative food and agriculture initiatives have proliferated, a need to regulate them has become increasingly necessary. Regulatory needs include defining standards (e.g., organic, fair trade, and sustainable), standardizing standards, ensuring product integrity, and designing effective conformity assessment processes. While several approaches are used to govern alternative agrifood, certification has become the most prominent approach (Cashore et al. 2004; Bartley 2007; Hatanaka 2010b). To date, SMOs, governments, and companies are increasingly using certification to govern alternative agrifood initiatives (Mutersbaugh 2005; Mutersbaugh et al. 2005). A key factor that has driven the widespread adoption of certification is its perceived objectivity, as a result of its scientific and technical practices (Power 1997; Tanner 2000; Dunn 2005; O'Rourke 2006; Konefal and Hatanaka 2011). In particular, because of the disinterestedness of certifying bodies the product conformity assessment practice of certification is considered to produce results that are replicable and valid. Generally, the product conformity assessment practice is conducted through the use of audits by independent actors based on tangible evidence, and produces results that are independently verifiable. Thus, certification is perceived to indicate compliance with impersonal rules and calculations that exclude bias and personal preferences (Pentland 2000; Courville et al. 2003). As such, it legitimates alternative agrifood initiatives in that it provides both the actors in them and goods they produce with credibility (Hatanaka 2010a).

However, as I have argued elsewhere, in practice, certification is often not objective (Hatanaka et al. 2005; Hatanaka and Busch 2008; Hatanaka 2010c). While it consists of technical rules and procedures, which are based on scientific norms and practices (e.g., disinterestedness, replicability, and validity), such rules and procedures do not fully remove politics and particular interests from the practices of certification (Hatanaka 2010a). Rather, understandings of certification as a science-based governance mechanism obscure the ways that the practices and procedures of certification privilege some actors and forms of knowledge while marginalizing others (Konefal and Hatanaka 2011).

While research on certification has proliferated, important questions remain regarding the impacts of certification on the principles, practices, and goals of alternative food and agriculture. In the sections below, using a case study of a sustainable shrimp project in Indonesia, the ways that certification has affected an alternative agrifood initiative are examined. Specifically, the focus is on the ways that the practices and relationships change with the introduction of formal standards, measures, and audits.

3.3 The Sustainable Shrimp Project in Bojokulu, Indonesia

The sustainable shrimp project is located on the eastern coast of the island of Java, Bojokulu. Bojokulu is an area that has long been known as a milkfish and shrimp-farming site. Farmers in Bojokulu have used extensive aquaculture practices for over 300 years. Such traditional forms of shrimp aquaculture entailed a polycultural system, relied on the surrounding ecosystem to supply shrimp feed, and replaced water in shrimp ponds (*tambak)* using tidal flows. However, beginning in the 1990s, the Bojokulu Department of Marine Affairs and Fisheries began to push intensive shrimp farming through extension services. Consequently, those local shrimp pond owners who could afford to implement intensive farming practices have largely switched to such practices. While the quantity of shrimp produced increased, a myriad of problems also accompanied the intensification of shrimp farming in the region. These included the use of excessive chemical inputs, destruction of mangrove forests, loss of genetic diversity in shrimp populations, and uneven income distribution. Thus, similar to many other production sites in the global South, with the implementation of Blue Revolution technologies, there was both increases in the productivity of shrimp aquaculture and social and environmental problems (Goss et al. 2000; Stonich and Bailey 2000; Lebel et al. 2002; Barbier 2003; Environmental Justice Foundation 2003).

It is against this backdrop that a Japanese SMO, Sustainable Network (SN), developed a sustainable shrimp project in Bojokulu in 1992. Seeking to promote environmental sustainability and improve the economic welfare of farmers, the project's aim was to preserve traditional shrimp farming practices in Bojokulu by linking shrimp farmers who used traditional practices with ethically motivated co-op members in Japan.[3] Congruent with many other alternative agrifood initiatives, SN and the co-op members believed that building close relationships with farmers and developing mutual understanding were crucial if the sustainable shrimp project was going to be successful. However, SN also recognized that farmers and consumers might have different interests and understandings of sustainability, which would need to be bridged. Thus, SN viewed as one of its primary tasks bringing farmers and co-op members together in ways that would encourage dialogue and the sharing of each other's perspectives.

One of the initial areas where SN sought to bring together co-op members and farmers was in the development of the project's sustainable shrimp standards. To facilitate a collaborative standards-development process, and begin to overcome cultural and language barriers, SN officials and a group of co-op members traveled to Bojokulu multiple times. Specifically, they visited shrimp ponds and warehouses, and discussed with farmers what should be the appropriate criteria for 'sustainable' shrimp farming. Furthermore, SN hired an Indonesian aquaculture specialist who was originally from the region, and who had received his Ph.D. in aquaculture at a Japanese university, to help develop and manage the project. As a result of these efforts SN and co-op members learned that differences existed between them and farmers regarding what traditional shrimp farming and sustainability entailed.

[3]There are multiple and diverse consumer co-op organizations in Japan. The co-op organizations that have been participating in the sustainable shrimp project distinguish themselves from other co-op organizations in that they consider themselves more progressive and active in the food sovereignty movement.

While there were several differences, the most notable difference involved the use of chemicals. At the time, farmers got their fry from hatcheries where antibiotics were commonly applied. Furthermore, farmers often used chemical pesticides in their ponds to kill predatory animals such as snakes. In contrast, SN and co-op members viewed sustainable farming as entailing no chemical use. However, rather than imposing their position on farmers, SN and co-op members engaged in discussions with farmers to develop standards on chemical use that might be acceptable to both parties. The result was a compromise where the Japanese co-op members accepted the use of antibiotics at the hatcheries, given that local levels of technology greatly limited the ability to produce fry without antibiotics. In return, the farmers agreed not to use chemicals once the shrimp fry were released into the ponds. Thus, given the differences that had to be bridged, standards development was viewed as an ongoing, evolutionary process.

Implementing the standards was viewed as responsibility of farmers. Consequently, farmer compliance with the standards was not formally monitored. Rather, efforts to enforce sustainable farming practices primarily took the form of trying to build relationships between farmers and co-op members and establish trust. Such efforts included annual visits to the pond community in Bojokulu by co-op members and visits by some farmers to Japan. On the one hand, a group of representatives from the co-op would visit the pond community in Bojokulu where they met with farmers, took part in harvesting activities, and discussed the project and its goals. As the following two quotes from farmers indicate, farmers respected the co-op members for coming to Bojokulu. One farmer commented,

> At least once a year, they came here. They tried to listen to our story. They had tolerance. They at least tried to understand us, and our conditions. Therefore, we also tried to understand them.

A second farmer similarly remarked,

> We were very happy if consumers actually came to our pond and tried to know our difficulty and environment of production. It encouraged us to produce better shrimp. It's human-to-human relationship, right? We all liked that.

On the other hand, farmers from the project were invited to Japan once or twice a year. Such visits included tours to see how their shrimp were handled once they arrived in the co-op stores in Japan. Farmers would also meet with co-op members, where they would discuss co-op's principles and the shrimp project. Additionally, co-op members prepared meals using shrimp from the sustainable shrimp project, which they would then eat together. When each group returned home they were expected to share their experiences and perspectives. Thus, in its early stages the project emphasized communication and dialogue between farmers and co-op members and made significant efforts to build relationships. Given such efforts, there was a high level of trust and respect between farmers and co-op members.

In early 2000, the sustainable shrimp project began to undergo significant changes. At the time, a European buyer approached SN about purchasing sustainably produced shrimp. While interested in shrimp from the project, the European buyer preferred the project have organic certification from a third party certifier. The

European buyer viewed certification as a more reliable regulatory mechanism than suppliers' self-assessment. Thus, from the European buyer's perspective, having the project certified would better guarantee that the shrimp were produced sustainably. SN, as well as some shrimp farmers, viewed the opportunity to sell shrimp in Europe as appealing as it would allow the project to expand, which would then allow more farmers to take part in the project.

Soon after being approached by the European buyer, SN convened an open forum in Bojokulu to discuss the possibility of applying for organic certification. Local shrimp warehouse owners, pond owners, SMOs, and government officials were all invited. More than 100 people attended the meeting, and the overwhelming majority voiced enthusiasm for organic certification. While the Japanese co-op members were not interested in certification, they did not object to it. From their perspective, adding an additional layer of governance (i.e., certification) was not necessary, as they viewed their current set of relationships with farmers as sufficient for ensuring sustainable shrimp. With the support of the shrimp farming community in Bojokulu, SN and a group of shrimp farmers applied for organic shrimp certification from a well-established third-party certifier in Europe, Green Soil. In July 2002, the project became certified organic and an organic shrimp division within the sustainable shrimp project was established.

While the arrangement with the European buyer was terminated within 2 years of certification, SN and farmers maintained organic certification until 2008 when they decided not to renew it.[4] While the Japanese co-op members did not require, SN maintained certification because they considered it to be beneficial for the project. First, SN viewed the conformity assessment mechanism of certification as effective and efficient for ensuring compliance by farmers with the standards. Additionally, they had invested considerable money, time, and energy in setting up the required conformity assessment mechanism for certification, and maintaining it was not prohibitively expensive. Second, certification provided the project with a certain degree of prestige. The project was perceived as a model for sustainable shrimp production in Indonesia, and farmers from other regions in Indonesia visited Bojokulu. Once the European buyer ended its partnership with the project, shrimp produced in the organic division of the project were sold to Japanese co-ops as sustainable shrimp.

During the 6 years in which the shrimp project had organic certification, it underwent several significant changes. The most significant entailed project governance. Whereas initially there was a distinction in the governance between shrimp produced in the sustainable project and those produced in its organic division, gradually this distinction disappeared. Project managers viewed the auditing system required by certification as effective for ensuring compliance with the standards, and they planned to extend the audits and documentation requirements to the entire sustainable shrimp project. Additionally, while co-op members initially claimed that

[4]The reason was largely due to a campaign by an environmental organization that targeted Green Soil's organic standards as insufficient. For a more detailed discussion, see Hatanaka (2010b).

they preferred a system of trust-based governance, certification was convenient since it limited their need to be actively involved in the project. Thus, while a formal distinction remained between the organic division and the larger sustainable shrimp project, in practice, the principles, measures, and the audits of the organic division have begun to influence the entire sustainable shrimp project.

3.4 Organic Certification: Standards, Measures, and Audits

As described above, prior to organic certification, the sustainable shrimp project was based on relations of trust between farmers and co-op members, which were maintained largely through face-to-face relations. This arrangement fits with the objective of alternative food and agriculture of building democratic and participatory relationships in which both producers and consumers have a voice. However, with the development of formal standards, measures, and audits with certification, notions of objectivity, calculability, and expert knowledge began to displace such shared governance, mutual understanding, and trust. The three subsections below examine the ways that the development of the standards, measurement of compliance with the standards, and audits affect the original principles of the project.

3.4.1 Standards of Sustainable Farming

As noted earlier, prior to receiving organic certification, standards in the sustainable shrimp project were the outcome of continuous and ongoing dialogue and negotiation between Indonesian shrimp farmers and Japanese co-op members. Thus, both farmers and co-op members were able to express their positions and had direct control over the definition of sustainable shrimp farming. However, with certification, the standards-development process underwent considerable changes. The most notable change was the requirement that the organic standards had to be supported by scientific evidence. Put differently, what sustainable shrimp farming entailed became a product of science. This impacted the project in two significant ways. First, it constrained the degree to which farmers (and consumers) were able to participate in the development of the standards. Second, it made the standards quite rigid and thus, limited the ability of farmers to adapt their farming practices to local conditions.

Green Soil viewed input by project participants into the standards as necessary if they were to be efficacious. Thus, Green Soil used democratic practices for developing standards. For the organic shrimp project, the standard-development process entailed: first, Green Soil sharing the core principles of its organic standards with SN and shrimp farmers. Second, based on Green Soil's core principles, SN and shrimp farmers would propose organic standards for the shrimp project. Third, Green Soil would review, amend, and finalize the standards. From the perspective of

Green Soil, developing standards in such a way would enable the organic standards to be both based on scientific knowledge and be tailored to local conditions.

Given the large number of farmers who were part of the sustainable shrimp project, a committee was formed to take the lead in developing the organic standards for the project. While the standard-development process was intended to be democratic, committee members tended to feel that their positions and views were subordinated to those of Green Soil. This was especially the case in instances where there was a disagreement between the committee and Green Soil. Such marginalization was largely the result of requirements by Green Soil that the standards be based on science. Specifically, Green Soil maintained that modifications to its core principles had to be supported by scientific evidence. However, as the committee largely drew on experiential knowledge, such a requirement functioned to limit farmers' input into the standards.

A particularly contested issue between the committee and Green Soil was the degree to which mangroves should be reforested. Green Soil is a strong proponent of mangrove reforestation, as it is one of the major concerns that the international community has regarding shrimp aquaculture. Specifically, Green Soil's core principles stated that a shrimp farm should only be certified as organic if the total farm area did not exceed 50 % of the former mangrove area. Thus, former mangrove areas on farm properties must be reforested to at least 50 % within 5 years to receive organic certification from Green Soil. As the shrimp ponds on the Bojokulu delta were once mangrove forests, Green Soil stipulated that reforestation of half of the land on which Bojokulu shrimp ponds are located was necessary.

The committee disagreed with Green Soil's interpretation of mangroves in the Bojokulu region. From the committee's perspective, most shrimp ponds in Bojokulu have been in existence for 200–300 years or more, previously being used for milkfish production. Thus, committee members argued that shrimp ponds in Bojokulu have coexisted with nature for a long time and there has not been significant recent mangrove destruction as a result of shrimp farming. Given this history, the committee maintained that the standardized principles of Green Soil, which might be suitable for regions where mangrove forests had recently been cleared for shrimp production, were inappropriate for Bojokulu. The committee tried to convince Green Soil that its principle on mangrove reforestation needed to be interpreted more flexibly given the history of the ponds in Bojokulu. However, Green Soil rejected the committee's position on the grounds that their position was not supported by scientific evidence.

In an attempt to enhance the creditability of their position, the committee then enrolled an official with a European organic consultant company. Soon thereafter Green Soil agreed to enter into a renegotiation stage and the two groups were able to come to a compromise.[5] Nevertheless, despite getting Green Soil to compromise

[5]As a compromise, organic standards required that shrimp farmers plant mangrove trees at a maximum distance of 7 m around the ponds and dikes.

on mangrove reforestation, members of the committee noted that they generally felt disempowered by the process. For example, one committee member commented,

> It's bizarre. We are not allowed to decide what sustainable shrimp farming is by ourselves. It is our ponds. It is our sustainability... Only with the support and persuasion of a European consultant did they listen to our argument.

In other words, because Green Soil's position was based scientific knowledge, and the committee's position on experiential knowledge, the committee tended to feel that they were marginalized in the standards-development process. Reflecting this position, another committee member commented,

> [Certifiers] push their ideas because they have power. They develop universal standards based on their own perspectives and ideas. Then, they impose such standards on suppliers throughout the world. They just determine that their standards are the standards... They won't listen to different perspectives or opinions from us. They are not flexible.

Thus, while the standards-development process was formally democratic, in practice, farmers had limited voice because their experiential knowledge lacked credibility from Green Soil's perspective.

Additionally, the requirement that the standards be grounded in science tended to make the standards fixed and standardized, which limited their adaptability to local conditions. Commenting on what he felt was problems with the organic standards, one farmer observed,

> It is no doubt that Green Soil's standards are great and ideal. However, they also need to see and understand the local conditions. They need to understand that local farmers have long been producing shrimp in our own way and to change such method will take a long time. However, their stance is rather, "This is our standard. You have to meet our standards if you want to sell your products as organic." Their tolerance level is very low.

Thus, whereas fixed science-based standards allows for uniform products across production sites, they may also lead to standards that are inappropriate for local conditions and limit farmer ingenuity.

In sum, organic certification transformed both the standards-development process and the standards. First, while developing standards was a participatory open-ended process based on mutual understanding before certification, following certification it became a formal process based on scientific evidence. Second, the standards themselves become more standardized, fixed, and rigid. The result is that both the standards-development process and the standards themselves were based more on notions of objectivity and calculability, as opposed to democracy and trust.

3.4.2 Measures of Sustainable Farming

Prior to organic certification, there were no measures to gauge the extent to which farmers were adhering to the standards in the sustainable shrimp project. As farmers were solely accountable for adhering to the standards, how to interpret and implement the standards was their decision. Thus, compliance and accountability was completely based on trust, with Japanese co-op members trusting that farmers were implementing the standards and accurately reporting their farming practices. In large part, this fits with the aims of shared governance, empowerment, and trust that much of alternative food and agriculture seeks to promote. However, with certification, similar to the changes that occurred with the standards, how compliance with the standards is measured also became more objective and calculable. Specifically, sets of documentation templates were developed to evaluate production practices, which farmers and inspectors were required to complete.

As part of organic certification, Green Soil required the development of an internal control system (ICS)[6] and a set of documentation templates that would *effectively* and *efficiently* measure member farmer compliance with the standards. To develop and manage the ICS, a local organization, *Perlindungan Alam* (PA), was established. One of the initial tasks of PA was to develop a set of documentation templates to measure compliance by farmers with the standards. Similar to the process of developing the standards, Green Soil provided model documentation templates and asked PA to revise them as necessary. PA officials then modified and created a wide variety of documentation forms to measure farmer compliance with the standards. Green Soil then reviewed, amended, and approved the documentation templates.

The membership document template compiled data on each farmer and their pond(s) including the personal identity of farmers, registered ponds (e.g., their size, shape, and location), the official date farmers were entitled to have the pond(s), and whether farmers own the official land certificate. The production practice document template collected detailed information for each production cycle (i.e., every 3 months), including when and which hatchery the shrimp fry came from, when and how much of the shrimp and milkfish fry were released into the pond(s),[7] how much of the organic pesticide was used and when, and if any additional inputs were used and when. Similarly, the harvest document template required detailed information on the start and end time of harvesting and the quantity of shrimp and milkfish that were harvested from each pond.

As the short descriptions of the document templates indicate, first, checking farmer compliance was based on the notion of calculability. That is, the documentation templates measured tangible evidence, which could be verified by auditors. In this way, potential preferences and bias of individual inspector was minimized. Second, the measurements were 'standardized' measures so that they could be applied to all member farmers in a consistent manner. In short, constructing the ICS entailed developing a set of 'standardized' measures to monitor compliance by all farmers in the project. Thus, with certification, face-to-face relationships and trust were replaced with a contractual agreement that was enforced through a formal

---

[6]An ICS is an audit-based monitoring system designed to ensure farmer compliance with the standards.

[7]According to Green Soil's organic standards, poly-culture was a required condition for organic production.

and standardized conformity assessment process. Comparing the project pre- and post-certification, one interviewee, who was involved with the sustainable shrimp project from the beginning, and later became one of the founding members of PA, commented,

> Organic certification is very Western idea. I think organic certification is definitely a product of Western ideology. It's based on 'contractual agreements,' quite stiff, and lack of flexibility. It's based on rationality, documentations, contracts, and signature. There is no human-to-human relationship entailed.

In other words, personal relationships and relations of trust were largely replaced with notions of calculability and objectivity, which were enforced through formal contracts.[8]

3.4.3 Audits of Sustainable Farming

Prior to organic certification, there was no active enforcement of the standards. As noted above, efforts by SN and co-op members to ensure compliance with the standards by farmers were largely based on relations of trust. The belief was that in building personal relationships, each party would come to understand and respect each other's priorities. Thus, in its original form, the project permitted considerable autonomy to farmers, allowing them to be solely responsible for compliance with the standards. Such an approach, which embodies trust, respect, and personal relationships, is congruent with the aims of alternative food and agriculture.

Similar to how the standards and measures of compliance with them changed, how the standards are enforced also changed significantly with organic certification. Specifically, farmers have become integrated into a multi-layered *audit system*. In the organic project, there were two types of audits: (1) audit of the production site and (2) audit of the certification process. In both sets of audits, the emphasis was on objectivity and calculability. That is, the audits should be free of bias, based on tangible evidence, and replicable.

Auditing of the production site entailed checking compliance by farmers with the standards and included both internal and external audits. PA, as part of the ICS system, conducted internal audits. The ICS stipulated that each pond must have at least one unannounced inspection per production cycle (i.e., 90 days). Inspectors were also required to be at ponds during harvest time (twice a month, 1 week per time) to oversee the shrimp harvest. During audits, inspectors examined the documentation completed by member farmers, questioned farmers as to their practices, completed inspection documents, and searched the huts and ponds for chemical input packages or containers. Following inspections, farmers were evaluated and were assigned a pass or fail grade. If they passed, they could sell

[8]When approved to become a member farmer in the organic project, farmers had to sign a membership contract in which all the standards were listed. In signing the contract farmers acknowledged their obligation to conform to the standards.

their shrimp as organic and receive a price premium. Additionally, audit results were documented and if a farmer failed three consecutive audits they faced termination of their membership. Thus, with the introduction of audits, formal sanctions also became part of the project.

The production site also underwent an annual external audit by Green Soil. Specifically, a team of the auditors visited Bojokulu annually and audited the efficacy of the ICS. External auditors stayed in Bojokulu for approximately a week and audited a random sample of the documentation filled in by both member farmers and PA's inspectors. The team of external auditors also visited a random sample of member ponds and questioned member farmers. Similar to the internal audit, the external audit also used formal audits. That is, if the project passed the audit, then certification was renewed and shrimp produced in Bojokulu could continue to be labeled as organic.

In addition to audits of the production site, the certification process itself was also audited. Specifically, Green Soil was audited to ensure its capability as an independent governance body. Commonly referred to as accreditation today, independent accreditor organizations audit certifiers to ensure their efficacy and technical competency in undertaking certification.[9] Thus, Green Soil was annually audited by the International Federation of Organic Agriculture Movements (IFOAM) to ensure that its certification process was objective, transparent, and efficient.[10]

In sum, with organic certification, a formal, multi-layered audit system replaced relations of trust in the sustainable shrimp project. In contrast to relations of trust, which stress personal relationships and respect, the use of audits embodies notions of objectivity and calculability. Furthermore, certification also introduced the processes of discipline and control into the sustainable shrimp project.

3.5 Conclusion

This chapter examines an alternative agrifood initiative that sought to advance environmental sustainability and the economic welfare of shrimp farmers in Bojukulu, Indonesia. The focus is on how organic certification affected the principles and practices of the project. Specifically, I argue that with organic certification and the

[9]Accreditation emerged largely in response to the proliferation of third-party certifiers and kinds of certification. For example, multiple third-party certifiers certifying to the same set of standards, but using different audit practices may lead to inconsistencies in the certification process. Moreover, suppliers of a good or service may be financially burdened by multiple audits, if different purchasers or users require different certifiers/certifications. Thus, accreditor organizations were developed to "regulate" third-party certifiers and address these potential problems.

[10]Furthermore, it should be noted that accreditor organizations are often also subjected to audits. Accreditor organizations may have their own operations audited through membership in a global organization of accreditors—i.e., an accreditor association. Thus, similar to how accreditor organizations set standards for certifiers, accreditor associations tend to set standards for accreditor organizations and audit them accordingly (see Hatanaka and Busch 2008).

introduction of formal standards, measurements, and audits the original principles and practices of the project began to erode. Most notably, the personal relationships, shared governance, and trust that characterized the early stages of the project were replaced by objectivity, calculability, and expert knowledge.

One consequence of this shift was a significant decline in contact between Indonesian farmers and Japanese co-op members. Describing how the role of Japanese co-op members have changed, one PA official who has been involved with the project from the beginning commented:

> [Japanese] co-op members have definitely changed over time with the shift to certification. Their awareness has become low . . . no more visits, little understanding of what is going on at the production site . . . There is no dialogue between consumers and producers now. What we have now is a very impersonal network.

In short, as Japanese co-op members came to rely on the formal standards and audits, they became disengaged from the project. At the same time, having been marginalized in the standards-development process, and seeing declined engagement by co-op members many farmers' commitment to the project also lessened. Speaking of the changes among farmers, an SN official remarked,

> Under certification, member farmers have become puppets. They just follow the rules that were set up by the certifier. They neither think, nor try to improve their farming practices. They just passively follow all the rules and procedures that were prescribed. They stopped thinking, stopped participating in the project. Thus, their morale is low.

In short, as the project became more formalized with certification, the project became increasingly depersonalized and both farmers and co-op members became less committed to both the project principles and each other. Furthermore, not only did the co-op members' relationship with farmers change, but how farmers were understood also changed. Prior to certification, co-op members and SN viewed farmers as a partner in a joint project to achieve mutual goals. However, with certification, how farmers were understood largely changed to that of a subject who had to be disciplined and controlled.

With the depersonalization of co-op member and farmer relations and the consequent re-conceptualization of farmers as subjects, the project began to encounter more disagreements, conflicts, and antagonisms. Most notable is that some farmers began not to fully comply with standards and falsify documents in order to remain in the project.[11] Thus, with certification, as the project became increasingly rationalized there was an erosion of both commitment and trust. For these reasons, SN and PA decided not to renew organic certification in 2008. Feeling that the original principles and practices of the project had largely become lost, they decided to return to the original project design and focus on trying to rebuild relationships and trust between farmers and co-op members.

While the findings in this chapter are based on a single case, they raise important questions regarding the effects of certification on alternative agrifood initiatives.

[11] See Hatanaka (2010b, c) for elaboration of this point.

As alternative food and agriculture expands, and becomes an increasingly well-established component of the global agrifood system, there is a trend towards increased rationalization and standardization. This chapter indicates that such developments may constrain the original intent and aims of alternative food and agriculture. Specifically, the emphasis on objectivity, calculability, and expert knowledge that characterize certification may limit both farmer and consumer participation and voice in alternative food and agriculture. As the sustainable shrimp project indicates, with certification there is the danger that farmers will become subjects to be controlled and consumer activity will be limited to just that of purchasing practices. Under such circumstances, the likelihood of a declining commitment to the aims of alternative food and agriculture and erosion of trust would be significant. Thus, with certification I argue that the trend may be towards alternative agrifood initiatives that are increasingly standardized, objective, calculable, but embody shallow forms of social justice and environmental sustainability. Put differently, building on Georg Ritzers (2010) notion of McDonaldization, what may be taking place in alternative food and agriculture is what I refer to as "McSustainability" or "McJustice."

References

Barbier, E. 2003. Habitat-fishery linkages and mangrove loss in Thailand. *Contemporary Economic Policy* 21(1): 59–77.

Barham, E. 2002. Towards a theory of values-based labeling. *Agriculture and Human Values* 19: 349–360.

Barnett, C., P. Cloke, N. Clarke, and A. Malpass. 2005. Consuming ethics: Articulating the subjects and spaces of ethical consumption. *Antipode* 37(1): 23–45.

Bartley, T. 2007. Institutional emergence in an era of globalization: The rise of transnational private regulation of labor and environmental conditions. *American Journal of Sociology* 113(2): 297–351.

Busch, L., and C. Bain. 2004. New! Improved? The transformation of the global agrifood system. *Rural Sociology* 69(3): 321–346.

Cashore, B., G. Auld, and D. Newsom. 2004. *Governing through markets: Forest certification and the emergence of non-state authority*. New Haven: Yale University Press.

Courville, S., C. Parker, and H. Watchirs. 2003. Introduction: Auditing in regulatory perspective. *Law & Policy* 25(3): 179–184.

DeLind, L., and P. Howard. 2008. Safe at any scale? Food scares, food regulation and scaled alternatives. *Journal of Agriculture and Human Values* 25: 301–317.

Dunn, E. 2005. Standards and person-making in East Central Europe. In *Global assemblages: Technology, politics and ethics as anthropological problems*, ed. A. Ong and S. Collier, 173–193. Malden/Oxford: Blackwell Publishing.

Environmental Justice Foundation. 2003. *Smash and grab: Conflict, corruption and human rights abuses in the shrimp farming industry*. London: Environmental Justice Foundation Charitable Trust.

Freidberg, S. 2004. *French beans and food scares: Culture and commerce in an anxious age*. Oxford/New York: Oxford University Press.

Goodman, D. 2003. The quality 'turn' and alternative food practices: Reflections and agendas. *Journal of Rural Studies* 19: 1–7.

Goss, J., D. Burch, and R.E. Rickson. 2000. Agri-food restructuring and third world transnationals: Thailand, the CP Group and the Global Shrimp Industry. *World Development* 28(3): 513–530.
Hatanaka, M. 2010a. Assessing rule-based governance mechanisms in an era of scientism. *Journal of Rural Social Sciences* 25(3): 141–159.
Hatanaka, M. 2010b. Trust, certification, and partnership in an organic shrimp network: Rethinking transnational alternative agrifood networks. *World Development* 38(5): 706–716.
Hatanaka, M. 2010c. Governing sustainability: Examining audits and compliance in a third-party-certified organic shrimp farming project in rural Indonesia. *Local Environment* 15(3): 233–244.
Hatanaka, M., and L. Busch. 2008. Third-party certification in the global agrifood system: An objective or socially mediated governance mechanism? *Sociologia Ruralis* 48(1): 73–91.
Hatanaka, M., C. Bain, and L. Busch. 2005. Third-party certification in the global agrifood system. *Food Policy* 30(3): 354–369.
Hudson, I., and M. Hudson. 2003. Removing the veil? Commodity fetishism, fair trade, and the environment. *Organization and Environment* 16: 413–430.
Konefal, J., and M. Hatanaka. 2011. Enacting third-party certification: A case study of science and politics in organic shrimp certification. *Journal of Rural Studies* 27(2): 125–133.
Lebel, L., N.H. Tri, A. Saengnoree, S. Pasong, U. Buatama, and L.K. Thoa. 2002. Industrial transformation and shrimp aquaculture in Thailand and Vietnam: Pathways to ecological, social, and economic sustainability? *Ambio* 31(4): 311–323.
Lockie, S. 2009. Responsibility and agency within alternative food networks: Assembling the "citizen consumer". *Agriculture and Human Values* 26(3): 193–201.
Micheletti, M. 2003. *Political virtue and shopping: Individuals, consumerism, and collective action*. Houndsmills/Basingstoke/Hampshire/New York: Palgrave Macmillan.
Miller, P., and N. Rose. 1997. Mobilizing the consumer: Assembling the subject of consumption. *Theory, Culture, and Society* 14(1): 1–36.
Morgan, K., T. Marsden, and J. Murdoch. 2006. *Worlds of food: Place, power, and provenance in the food chain: Place, power, and provenance in the food chain*. Oxford/New York: Oxford University Press.
Mutersbaugh, T. 2005. Just-in-space: Certified rural products, labor of quality, and regulatory spaces. *Journal of Rural Studies* 21(4): 389–402.
Mutersbaugh, T., D. Klooster, M.-C. Renard, and P. Taylor. 2005. Certifying rural spaces: Quality-certified products and rural governance. *Journal of Rural Studies* 21: 381–388.
O'Rourke, D. 2006. Multi-stakeholder regulation: Privatizing or socializing global labor standards? *World Development* 34(5): 899–918.
Patel, R., R. Balakrishnan, and U. Narayan. 2007. Transgressing rights: La Via Campesina's call for food sovereignty/Exploring collaborations: Heterodox economics and an economic social rights framework/Workers in the informal sector: Special challenges for economic human rights. *Feminist Economics* 13(1): 87–116.
Pentland, B.T. 2000. Will auditors take over the world? Program, technique and the verification of everything. *Accounting, Organizations and Society* 25: 307–312.
Power, M. 1997. *The audit society: Rituals of verification*. Oxford/New York: Oxford University Press.
Reardon, T., and J. Berdegue. 2002. The rapid rise of supermarkets in Latin America: Challenges and opportunities for development. *Development Policy Review* 20(4): 371–388.
Renard, M.-C. 2003. Fair trade: Quality, market and conventions. *Journal of Rural Studies* 19: 87–96.
Ritzer, G. 2010. *The McDonaldization of society 6*. Thousand Oaks: Pine Forge Press.
Stonich, S.C., and C. Bailey. 2000. Resisting the blue revolution: Contending coalitions surrounding industrial shrimp farming. *Human Organization* 59(1): 23–36.
Tanner, B. 2000. Independent assessment by TPC bodies. *Food Control* 11: 415–417.
Taylor, P.L. 2005. In the market but not of it: Fair trade coffee and forest stewardship council certification as market-based social change. *World Development* 33(1): 129–147.

Chapter 4
Differentiating Organics: Performing Multiple Objects to Organize Singular Markets for Organic Tea and Biscuits in the UK

Allison Loconto and Maarten Van der Kamp

4.1 Introduction

When shopping for 'sustainable' products for their national habit of tea and biscuits, consumers in the United Kingdom (UK) can choose from a range of differently certified items and brands. This choice illustrates a conceptual and material separation between 'conventional' products and 'sustainable' products and has been qualified as a political choice (DuPuis 2000). Organic is one such certification of products that is built on the principles of health, ecology, fairness, and care (IFOAM 2009a). Yet as an organizing concept, organic encompasses a multitude of different sanctioned organic practices. For example, the International Federation of Organic Agriculture Movements (IFOAM) norm serves to coordinate cross-recognition of the 71 national regulations currently existing around the world (IFOAM 2009b; Willer and Kilcher 2011). This large number of regulations reflects the multiplicity of organic practices since these practices were first brought together formally within

This chapter is based on work supported by the National Science Foundation (Grant # SES-0924202) and by an Institute of International Education Fulbright Fellowship, funded by the US Department of State. Any opinions, findings, and conclusions or recommendations expressed in this material are the authors' and do not necessarily reflect the views of the funding agencies.

A. Loconto (✉)
Institut National de la Recherche Agronomique, Sciences en Société (INRA-SenS), Institut Francilien Recherche Innovation et Société, Université Paris-Est Marne-La-Vallée, 5 boulevard Descartes, Champs-sur-Marne 77420, France
e-mail: amloconto@versailles.inra.fr

M. Van der Kamp
Value in Enterprise Limited, 60 Queen Edith's Way, Cambridge CB1 8PW, UK
e-mail: maarten@valueinenterprise.co.uk

B. Freyer, J. Bingen (eds.), *Re-Thinking Organic Food and Farming in a Changing World*, The International Library of Environmental, Agricultural and Food Ethics 22, DOI 10.1007/978-94-017-9190-8_4

the construct of a social movement. These multiple practices should be expected considering how the movement was built upon a number of experiments in diverse geographic locales (e.g., Balfour 1978; Steiner and Gardner 1993; Howard 2006; Northbourne [1940] 2003) and slowly consolidated into standards of practice and then formal regulations.

The first organic standards (and accompanying certification systems) were introduced in the 1960s and 1970s as good agricultural practices to teach (mostly small) farmers how to farm according to the organic principles (Guthman 2004). Initially, these were only remotely connected to consumers, but over time, and with the involvement of supermarkets (offering more choice) and governments (protecting consumers from fraud) the standards came to be at the heart of organizing organic markets and consumption. For example, the European Union (EU) regulation is used by producers around the world as the EU and United States (US) markets combined comprise of 97 % of the global demand for organic products (Willer and Kilcher 2011). While there are a large number of certification bodies that are certifying against the EU organic regulation, the majority of products on UK supermarket shelves carry the Soil Association certification mark. The Soil Association accounts for around 80 % of the certified organic food sold in the UK (Soil Association 2011) and is noted for its own standard that predates and goes beyond the EU requirements. The standardized practices called for by the Soil Association suggest that all of those organic products that carry the Soil Association seal are produced under the detailed conditions embodied by the rules in the standard. However, the literature suggests that what is prescribed in a standard doesn't always translate smoothly into practice (e.g., Power 1997). In fact, recent research suggests that the practices of complying with sustainability standards are influenced by factors that are not necessarily included in the written standards themselves (Gibbon et al. 2010; Loconto 2010). In other words, there is an unresolved tension between the singularity of practices proposed by a standard and multiplicity enacted in practice.

In this paper we examine the notions of singularity and multiplicity in organic practice, arguing that the concept of organic is indeed enacted in multiple ways while the process of singularization is fundamental to the organization of a market for organic. As Guthman (2004) argues, the diversification of the organic consumer has been accompanied by the diversification of the producer, thus the standards that we see in practice today are used by both large agribusinesses and small farmers. While this could be seen as an ontological shift away from the foundational principles of organic agriculture (Guthman 2002; Jaffee and Howard 2009), we argue that the performances of individual producers are still multiple. Indeed, they are necessarily so, due to the local and historical situated-ness of organic practices and are only singularized at the point where 'organic products' circulate. We use the notion of 'performativity' to analyze how the practice of organic farming is at once similar and dissimilar based on the contexts in which a product is grown and traded. Here we utilize the notion of the standard as a calculative device (Callon 1998) to explore how organic tea grown in Tanzania and organic cereals grown in the UK are rendered 'singular' in the UK market, yet 'multiple' in the practices of production. We conclude that despite the use of standards as market devices to create a singular organic market, the practice of organic farming remains diverse. Thus, these cases illustrate Mol's (2002) and Law's (2008) arguments that reality, in this case organic farming, remains 'multiple', despite attempts to singularize practices through standardization.

4.2 Conceptualizing the Performances of Organic

In recent years performativity analysis has gained traction in economic sociology, critical management studies, science and technology studies (STS), and in rural sociology (Mol 2002; Busch 2007; Callon 2007; MacKenzie et al. 2007; Law 2008; Loconto 2010). Historically, however, it has inspired two ontologically disparate interpretations in the literature. The first notion of *performativity* suggests a separation between observed action and the knowledge about that action (Goffman 1974; Hilgartner 2000; Mead [1934] 1962). Goffman (1974) claims that 'standard-making routines' can be "performed with ease or clumsiness, awareness or not, guile or good faith" (p. 75) but nonetheless they must be enacted to be realized. This vision of performance "focuses attention not only on the rhetoric and narrative of the performance itself but also on the way performance expresses – and is embedded in – modes of information control" (Hilgartner 2000, p. 11). In other words, the observed enactments are representations of reality, rather than reality itself that is perceived.

Other theorists argue that action cannot be separated from the knowledge about it. In this second sense, the word performativity has been traced back to Austin's (2004 [1962]) notion of *illocution,* or 'performative utterance', where "the issuing of the utterance is the performing of an action" (p. 163). The main point that separates Austin's version of performance from Goffman's is that it insists that an 'inward' performance, or backstage, is not necessary for an 'outward' performance to occur. Put differently, "the inside is merely a fold of the outside" (Deleuze 1988 cited in Bell 2007, p. 14). It is this "interweaving of 'words' and 'actions' – of representations and interventions – that the concept of 'performativity' is designed to capture" (Muniesa 2007, p. 5). In other words, real objects and subjects and the representations of both are enacted simultaneously (Law 2008). Moreover, these performances are constituted by both the performers and the audiences who evaluate the performances (Boltanski and Thévenot 2006 [1991]). This interpretation of performances, or practices, enables an analysis of how values are enacted through both actions and discourse (cf. Callon 1998; Law 2004; MacKenzie et al. 2007; Muniesa 2007).

Human – nature interactions have been illustrative of the variation and difference that emerge when paying attention to performances (Szerszynski et al. 2003). Mol (2002) shows that "attending to enactment rather than knowledge has an important effect: what we think of as a single object may appear to be more than one (...) objects come into being – and disappear – with the practices in which they are manipulated. Since the object of manipulation tends to differ from one practice to another, reality multiplies" (p. vii, 6). Law (2008) continues that these multiple performances can be, and frequently are, inconsistent. This may be contributed to the multiple performances that actors are constantly engaging. Boltanski and Thévenot (2006 [1991]) explore this in their observation that while philosophers construct 'pure' polities for analysis, life demands that we inhabit multiple polities on a daily basis.[1] The ability to avoid situations or 'play along' with them is "at work whenever persons have to complete the passage between situations arising from different worlds; in a complex society that includes multiple arrangements, this capacity is thus indispensable to the normal conduct of daily life" (Boltanski and Thévenot 2006 [1991], p. 234). Therefore, the ability to continuously manage these multiplicities in order to produce coherent performances results in different interpretations of these interactions. This version of performativity reveals what is at stake with this type of analysis: "since the real is relationally enacted in practices, if those practices were to change the real would also be done differently" (Law 2008, p. 635).

The reality of the notion of 'organic' is that the objects and subjects (e.g., standards, farming practices, producers, and consumers) are simultaneously performed in the creation of an organic market. We turn to research in economic sociology to explain these processes better. Callon (1998) claims that economics is performative: "economics, in the broad sense of the term, performs, shapes and formats the economy, rather than observing how it functions" (p. 2). Callon argues that the existence of a market implies the circulation of merchandise, that is, the existence of goods that are transformed into things that can be exchanged. This circulation is simultaneously a process of production and qualification. This process attaches products to users in ways that make the products essential to the identification of users – i.e., entanglement and disentanglement (Callon et al. 2002). In other words, it is not external processes of rules selection that make markets, rather, the construction of markets is "through the emergence of the economic actors, which are their participants" (Allaire 2009, p. 25). Here the focus is not only on the people, but the socio-technical devices as well. In other words, the organic standards work as *agencements*, which are the collectives of people, technical devices, written standards, etc. These collectives are mobilized through interactions with institutions, conventions, groups, etc. and thus have 'the capacity to act and give meaning to action' (Callon et al. 2002).

One outcome of the use of calculative devices, in our case organic standards, in the construction of an economy of 'organic' quality is the singularization of organic products in opposition to 'conventional' products. The process of singularization is said to create attachment between consumers and specific products as the market for these singularized (organic) products is created. However, in order for one product to be singularized over others, there must be an underlying similarity as

[1] Boltanski and Thévenot originally described six polities or worlds of worth to which they claim that people appeal when they justify their actions. These polities are: market, civic, inspirational, domestic, industrial, and opinion. Additional research has added the following: environmental, information, and project-based.

a point of comparison (Callon et al. 2002). For example, organic tea must be similar enough to the quality profile of conventional tea in order to convince consumers that the organic product is just as good as the conventional product, while providing additional environmental benefits (some also claim health benefits) and thus singular and desirable. However, with the case of the use of standards in this process we see another layer of singularization occurring through the process of standardized differentiation (Hatanaka et al. 2006). Here, singular products have been systematically differentiated through the process of standardization of practices and qualities. In other words, the singularity is based on an underlying sameness of standardized organic practice in order to differentiate itself from the similarities shared with conventional practice. This notion of singularization is particularly important when we consider the ontological connotations of performativity analysis. How does a single notion (i.e., organic) singularize products if its performance in practice is multiple?

Understanding the role that standards play in this process has emerged recently as an object of study. For example, standards have been explained as market devices that perform particular market arrangements (Busch 2007; Kamp 2010; Loconto and Busch 2010; Konefal and Hatanaka 2011). However, it is not the standard itself that creates the market. It is the mobilization of the standard as a socio-technical device to 'assemble' different actors in a market network. Organic is a case in point. The EU regulation did not create a market for organic. Rather, it was the network of actors – including farmers, researchers, activists, journalists, consumers and policy-makers – that mobilized a concept of organic (including its techno-scientific artifacts) and a market for organic products. The EU regulation is the outcome of this mobilization, yet now stands as a stabilized, singular, enactment of organic for the EU market. However, if we take the concept of performativity seriously, we see that there are multiple enactments of organic farming that are constantly circulating in practice. This paper attempts to capture some of these.

4.3 Methods

To examine how enactments of organic farming can differ in practice, we compare two sets of farming practices that are in various ways opposites. On the one hand we explore how tea is grown as a crop from perennial plants by Sub-Saharan African producers for the specific purpose of export to European markets; on the other we describe the cropping practices of cereal produced from annual plants by UK farmers for the home market. As we will show, these different settings allow us to explore differentiations in certification regimes as well as cropping practices. Using a case study methodology of organic practices, we have relied upon four qualitative data collection methods: (1) semi-structured interviews, (2) semi-structured focus groups, (3) field observations and (4) document content analysis (e.g., standards, websites, published material, and databases). Data collection by the authors occurred in Tanzania, Kenya, Germany and the UK between June 2008 and

July 2010. In total, 60 semi-structured interviews were conducted with organic operators, processors, traders, buyers, certifiers, standards developers, non-governmental organization (NGO) officers, and government officials. Four semi-structured focus groups were conducted in Tanzania with farmers, farm workers and farm/factory managers in Swahili by the first author. Interviews and focus groups were audio recorded and transcribed.[2] Observations were recorded in field notes and memos. Individuals participated in the research by informed consent and the names of some organizations have been changed to protect anonymity. Together, these data were analyzed to infer common themes around the performances of organic.

4.4 The Singular UK Market for Organic

Through the practices of producing a concept of organic in consumer and policy contexts we see a case for singularity emerging whereby organic is a cohesive concept that can frame consumer and political action. Didier (2007) has illustrated how statistics characterize and transform the objects that they describe. This is helpful to our case, particularly in the way that some *expressions*[3] of statistics appear singular. This is the case with the use of statistics at the aggregate level to assemble a singular market for organic food. Let us explain with examples of aggregated statistics. The World Watch Institute's publication "Vital Signs" ran the headline "Organic Agriculture Sustained Through Economic Crisis", yet the byline qualifies:

> The global organics market is recovering gradually from the economic recession, although trends vary widely by country. Growth in the European Union's organics market slowed overall in 2009, declining nearly 13 percent in the United Kingdom and remaining stagnant in Germany. Yet sales of organics continued to grow by double digits in France, Switzerland, and Sweden (Beck 2011).

This shows a global trend that is demarcated by country and indeed the UK market for organic declined. It is this aggregation of statistics at specific scales that carve out a singular market for organic. In the aggregate, organic becomes singular. It is no longer organic tea, or cereals or milk but organic sales and an organic market.

Much of the singularization of an organic market in the UK has been accomplished through efforts made by the Soil Association through marketing and public awareness campaigns. This holistic notion of organic is expressed in the Soil Association's framing of organic farming as the solution to a multitude of social and environmental problems as part of their policy work. For example, in their 2009 Organic Market Report, the Soil Association's policy director asks: "The question

---

[2]The Swahili was transcribed and translated into English by the first author together with a Tanzanian research assistant.

[3]Didier suggests that the term expression is more suited to describe the theory of performativity as he claims that performativity remains linguistic in its ability to explain the assembly of subjects and objects in reality.

we should really be asking is not 'can the public afford organic food?' But 'can our policy makers afford to carry on ignoring the potential of organic farming?'" (Soil Association 2010). This is reinforced by the rather fragmentary associations of organic with naturalness, health, various environmental benefits, animal welfare, absence of pesticides or genetically modified organisms (GMOs), and taste by consumers (Soil Association 2010). Until the global economic downturn, this was a successful strategy, with the UK organic market growing from £100 m in 1994 to £2.1b in 2008. However, with slumping sales in 2009 and 2010, various actors active in the organic sector coordinated marketing campaigns in which the benefits of organic production were simplified and harmonized.

A common technique is to pitch organic against 'conventional', often associating it with the use of GMOs. For example, in 2009, organic farming was practiced on 37.2 million hectares worldwide, a 5.7 % increase from 2008 and a 150 % increase since 2000. This includes land that is transitioning to organic production. The organic area amounted to 0.85 % of global agricultural land in 2009 (by comparison, producers seeded 2 % of agricultural land worldwide with genetically modified crops) (Beck 2011). While this is in-line with criteria specific to the EU organic standard, what is being capitalized on in these types of campaigns is the holistic ontological assumption that organic is a way of life and moreover, a singular way of farming (Guthman 2004).

We have taken the cases of tea and cereals as illustrative of the way through which these processes of singularization and multiplicity are enacted as each of these products are processed and blended to reach the state in which they appear in the consumer market. They also represent core sectors of the organic market that reflect the changes felt in sales in 2009. For example, organic tea consumption increased by 1.2 % while those foods based on cereals (breakfast cereals, biscuits, bread and bakery) declined by 8.9 %, 19 %, and 39.8 % respectively (Soil Association 2010). In the next sections we explore how this singular notion of organic is enacted in the production of tea and cereals.

4.5 Multiple Organics – The Case of Tanzanian Tea

Tanzanian tea is not the most easily recognized, nor the most sought after tea available on the conventional market – much less on the organic market. However, this relative obscurity has resulted in a rather self-contained industry and provided a unique opportunity to gain access to the entire organic tea industry in the country. This access has allowed exploration of the concept of organic amongst a very committed network of actors – which is microcosmic of the larger tea industry. Tea (black, green, white, and oolong) is made with the leaves of the *Camelia Sinensis* tree, which is grown at high altitudes (800–2,200 m) in the tropical regions of the world. The tea bush is a perennial crop with a productive life of at least 100 years if 'properly' maintained. Tea is plucked every 7–20 days year round, providing constant income and requiring constant labor. Tea is processed in geographically defined catchment areas of a processing factory, as 'quality' tea must reach the factory in less than 12 h from the time it is plucked. In Tanzania tea is grown on estates (>200 ha) and on outgrower farms (smallholder, average 0.37 ha; medium-scale, average 16 ha) (Simbua and Loconto 2010).

The distinction between estate and outgrower is based on the ownership of the farms. If owned by the tea processing company, they are estates, while outgrowers are individual farmers who sell their tea to the factory. The tea industry maintains the traditional practice of tracing tea from individual fields to single estate lines that are traded in a non-blind auction, which lends itself to the transparency requirements that are part of the standardized notion of organic. In Tanzania, there are only two estates and factories that are certified organic.

The move to organic was made in the late 1980s as the organic certified companies claimed that they first joined the system because they had acquired overgrown fields and the costs for conventional rehabilitation were much higher compared to the Organic certification. In other words, an economic calculation was the basis for joining the organic system, and similar justifications have kept these two estates in this system for the past 20 years. However, the economic calculation is only a part of the multiple enactments of the organic standard in Tanzania, which are best illustrated by three examples: *organic by default*, *parallel organics*, and *cross-certified*.

4.5.1 Organic by Default

In describing the organic sector in Tanzania, the International Trade Center notes: "a number of other crops continue to be grown organically "by default" without being certified" (International Trade Center (ITC) 2011). Every research participant interviewed in the Tanzanian organic system discussed the current farming practices in Tanzania as *organic by default*. While this notion is a highly contested claim within the organic movement (Scialabba 2000; Mansfield 2004; Vogl et al. 2005), it has been categorized as a type of farming system in Africa (cf. Hillocks 2002; Bolwig et al. 2009). We argue that the performance of *organic by default* has more to do with the relative poverty of Tanzanian farmers and the discourses surrounding 'green revolution' technologies than as a reflection of the organic movement.

One inspector claimed that "we cannot discuss *organic by default* because tea consumes so much nutrients, mostly Nitrogen (N), you must use synthetic fertilizers – everyone does." Indeed, in conventional tea farming, the estates are applying about 300 kg of N per hectare of tea and smallholders have access to between 30 and 100 kg of N per hectare of tea through their contracts with the estate factories. Respondents claimed that this synthetic fertilizer was often shared among all of the smallholders' crops; therefore the actual amount used on tea by a smallholder was difficult to determine. The low tea yield of smallholders compared to the estates was indicated as an example of the lack of synthetic fertilizer use. This was an often-cited example of how the smallholders were *organic by default*, since synthetic insecticides are not used in Tanzania and synthetic herbicides (glyphosate)[4] also fall into the same category of low access to inputs that smallholders face. While the smallholders were *organic by default* in practice, they were not certified as such again due to their financial situation.

There were clear distinctions in the organic practices between the two certified estates that also reflect the resource constraints faced even by the estate sector. One estate was specifically categorized as *organic by default* based on the perceived neglect in its farming practices. A manager explained that the main constraints for maintaining organic practices were the procurement of proper composting materials and conducting the manual weeding. While using tea waste from the factory as a fertilizer is common practice in conventional farming, it also constitutes an organic compost material. While the use of sunflower cake, black wattle waste, cow manure and spent brewer grains are also available in Tanzania as organic composting materials, the costs of procuring enough to cover the entire estate was explained as being prohibitive and thus it was common practice to only use tea waste by the *organic by default* estate. This was perceived as a 'default' practice, often because of the low levels of N in the tea waste. The presence of weeds is a second indicator for *organic by default*. Weeds cause problems for young tea during the first 4 years after the initial planting. After this, the tree bush is strong enough to withstand the stress and the canopy cover created by the tea bushes reduces the frequency of weeding later in the bush's life. Forking is the preferred method for weeding in the organic tea estates, but as the *organic by default* estate noted, this requires a lot of man-days, which are difficult to pay for on a small budget. Therefore, weeding was completed infrequently. This practice was noted as being the key indicator of their 'organicness': "all the inspector has to do is see the weeds and he knows that we are Organic" (Estate Manager).

These performances of resource-poor *organic by default* have an additional angle in their enactments in Tanzania. This is the displacement of traditional agriculture by conventional agriculture in the Tanzanian context. One inspector explained:

> Organic is quite good. In Iringa, in the seventies we used to harvest a lot of maize from one hectare, 26 bags, 90 kg per bag. Now you go around nobody is getting 26 bags per hectare anymore, they are getting 5–6 bags, it has gone down like that. Why? It is because we were doing organic before, I think, and later we added a lot of artificial fertilizers on the soils and the soil structure changed completely. So that way you add the same amount of artificial fertilizer and it seems it is leaking, it goes down and the soil cannot contain the artificial fertilizer, so maybe the roots cannot get the artificial fertilizer. So as a result, even if you add more, maybe because the structure is gone, and maybe the acidity has gone up and the acidity has killed some beneficial organisms. You see it is very bad to be conventional. So something good for the world now is that we are going organic.

---

[4]Glyphosate is the active ingredient in Monsanto's non-specific herbicide Roundup. It is generally considered to be of low toxicity to humans (lethal to most plants), but recent research has shown detrimental health effects for humans.

The tension that remains within the organic sector in Tanzania is the difficultly of introducing the western organic practices within a system that has spent the past 30 years convincing farmers that their traditional practices were inferior. A member of the organic movement elaborates:

> You even find some farmers who make a little more money and they decide even to shift back to spraying because of the mentality which has been there. That if you do this spraying, if you do the chemical fertilization, you are modern, you are really advanced. You are farming now like western people. So this other way of farming is primitive. And then the middle class people now they are translating as primitive. [. . .] They are being told either your own primitive way and if you use fertilizers and pesticides then you advance.

In other words, *organic by default* is enacted as the inability of farmers to effectively practice 'conventional' agriculture. While this is based on financial resource constraints, it is not necessarily by default uncertified organic. The certified estate managers justified their involvement in the certification systems on financial grounds. It was claimed that the costs of conventional rehabilitation of the tea estates was more expensive than organic certification. It is thus this failure to conform to the science-based notions of conventional and organic farming that enacts the notion of *organic by default* in Tanzania.

4.5.2 Parallel Organics

The way in which *organic by default* is further diversified in the Tanzanian context is what we term *parallel organics*. Mutersbaugh (2005) claims that certifications represent *parallel production*; whereby large corporations certify a small portion of their production and use this as a way to produce an image of 'sustainability' to their customers. What we propose as *parallel organic* refers to the nature of the regulation of organic practices in Tanzania and the discourses that accompany it. What we find is an externally funded, nationally supported, policy environment for organic and a parallel international commercial system for organic. These two systems are both focused on the construction of a notion of organic in Tanzania: the policy environment is constructing an institutional infrastructure for organic standards, while the commercial systems are responding to market information communicated to them through their value chains. Different European certifiers certify the two Tanzanian estates. The decisions to use a specific certification body are made by factory management based on the instructions given by their international buyers who inform them of the certifier that must be used. For example, one buyer specifically requests IMO (Institute of Marketecology) certified organic, rather than only certification against the EU regulation for organic.

While this commercial system has been in practice in Tanzania since the 1980s, a policy environment for organic was created in the 2000s by European donors (EPOPA program),[5] including an East African organic standard, a national Tanzanian organic agriculture movement (TOAM) and a certification agency, as a *parallel regulation* for organic without the involvement of the already existing actors in the organic tea sector (Mbiha and Ashimogo 2010). Moreover, these two 'organics' are working largely in isolation of each other. Illustrative for this isolation is the issue that the local organic certification body is not accredited to certify for the EU regulation, but only for the East African organic standard, which is not recognized as equivalent by the EU and thus irrelevant for export production. This parallelism is illustrated in TOAM's justification of its creation: "At the time of its [TOAM's] formation, there were a number of actors in the sector whose valuable interventions were neither coordinated nor well publicized. The actors also had little knowledge of each other and each other's activities" (Organic Africa Pavilion 2011). Indeed, TOAM has consistently included one of the tea estates in its promotional material, despite no formal participation of the estate in TOAM's activities. We are beginning to see efforts at coordinating a national Tanzanian organic sector, but this process is far from complete. Therefore, what we see in terms of 'parallel organics' is an organic movement, rather than a commercial enterprise, which enacts an image of 'sustainability' reflective of only a small portion of the story of organic in Tanzania.

4.5.3 Cross-Certified

Cross-certification between organic and Fairtrade standards is well noted in the literature (Parrish et al. 2005; Raynolds et al. 2007; Loconto 2010). This duplicity is an enactment of organic that is found often in tropical commodities, both in the consumer market and in production practices. On the one hand, the consumer labels in the UK market clearly demarcate borders between the Fairtrade and Organic attributes of a cross-certified product:

Organic is your assurance that this product is sourced from growers who avoid the routine use of pesticides to give you high product quality while enriching the environment through organic farming methods. Fairtrade is an alternative to conventional international trade, providing the security of long term contracts as well as the payment of a social premium to allow for democratically agreed investment in projects such as education, water supplies and medical facilities (Co-operative 2010).

The focus is on communicating to consumers what each of these concepts should mean to producers. On the other hand, this differentiation was not found to be so clear in the practice of organic farming in Tanzania. The two estates that are

[5]EPOPA stands for Export Promotion of Organic Products from Africa. The program was funded by Sida (Swedish International Development Co-operation Agency) and was implemented between 1997 and 2008 by Grolink and Agro Eco.

certified Organic are also certified Fairtrade, Ethical (Ethical Tea Partnership) and one estate had begun the Rainforest Alliance certification process. Indeed, when trying to distinguish between the practices that were conducted on farm in order to comply with the organic standard and those that are intended to comply with the Fairtrade standard were often difficult to distinguish. There was general agreement that organic was good for the workers and the environment, while Fairtrade provided funds; however the boundaries of these two concepts were often overlapping. The following is an excerpt from a focus group with farm workers, which illustrates this blurred boundary well.

> Q: Do you think that organic is a part of Fairtrade or are they two different things?
>
> A: They are not different, first organic with today's knowledge is indeed that which we have wanted. Because we avoid a number of different environmental damages ranging from serious damage that can occur with placement of fertilizer that visits plants, animals and other things that require the vegetation like trees, water sources and other things. Therefore, organic farming is important and [it is] not the difference whether or not organic farming is not very important nor as important as conventional farming (kilimo cha mbolea ya chumvichumvi).[6] By now we encourage it even, now we criticize conventional farming with urea, because it encourages that man move. He has to take deliberate steps to change the system where it was changed [in the past] to provide synthetic fertilizers [urea]. Change synthetic for natural fertilizers to preserve this environment and prevent pollution of these things.

There has also been institutional boundary work that facilitates the multiple enactment of organic with other standards in Tanzania. Specifically, this refers to the concessions made for organically produced tea in the other standards systems. For example in Fairtrade, "FLO[7] encourages companies to work towards organic practices where socially and economically practical" (FLO 2009, p. 30). The Rainforest Alliance has reduced the distance required to separate farms and territorial ecosystems by more than 50 % for organic farms. Also, "The farm must give priority to organic fertilization using residues generated by the farm" (RA 2009, p. 36), thus prioritizing organic agriculture techniques in their standard. The Ethical Tea Partnership does not explicitly mention organic techniques, but does so indirectly through its Memoranda of Understanding with FLO and Rainforest Alliance. In terms of organic practices in Tanzania, this encouragement was enacted through the audit practices, an inspector explains:

> The key benefits of all the certifications, is they are forcing the world to go organic. They would all prefer someone who is organic. They are saying: if you are organic, then I don't check this. The move of the world now is to have something that is organic.

This is particularly salient as the independent contracting nature of inspectors in East Africa translates into the same small network of inspectors conducting audits for different standards. For example, it is common to find the same inspector conducting audits for FLO, IMO, Starbucks' C.A.F.E. Practices, and Utz certified. As explained

---

[6] *Kilimo cha mbolea ya chumvichumvi* translates literally into Farming with 'salt salt' fertilizer, meaning farming with urea.

[7] Fairtrade Labelling Organizations International.

by an inspector, this is the nature of a free-lance inspector. The more trainings and diplomas (certifications) the inspector has, the more jobs s/he can take on per year. This means that the same inspector may return to the same farm for two or three different standards. Put differently, the enactment of cross-certified organic consists of both farms and inspectors.

4.6 Multiple Organics – The Case of UK Cereals

In contrast to Tanzanian tea, finding organic cereals grown in the UK on supermarket shelves is easier. They are mostly found in a processed form, such as in a breakfast cereal, muesli, biscuit or bread, or, less visible, in meat, dairy and other animal-based products (most cereal production is for livestock feed). This makes cereal an excellent case to consider the organization of markets in an organic context: as commodities which serve as inputs to many agricultural and processing practices, their organic status travels with them throughout a multitude of possible supply chains, and therefore illustrate the tensions between singularized organic goods and traded commodities.

As this type of agriculture emerged out of a rejection of agricultural orthodoxy, the notion of an *organic by default* as described in the Tanzanian case does not exist in the UK. Indeed, for most of the past century agricultural policy in the UK and the EU has favored so-called high-input-high-output systems at the expense of alternative modes of farming (policy support for organic farming was introduced in 1994). Thus, it is a conscious choice for farmers to move from highly industrialized and intensive farming systems based on agrochemicals to a low-input system that is often and mostly extensive. Traditionally, this meant that small, family-run farms would convert to organic farming and their products would be marketed at a price premium to consumers who shared the organic ideals (Conford and Holden 2007). However, with the rapid expansion of the organic markets after a number of food scares in the 1990s and the introduction of financial support for conversion and for the environmental benefits delivered through farming, organic farming became an interesting economic proposition to a wide variety of farm arrangements and sizes.[8] Thus, the performances of organic farming range widely from (a diminishing number of) traditional, small niche organic farms to extensive single enterprise operations and large, industrialized and relatively intensive businesses. Moreover, especially large estates converting to organic production adopt a parallel production process (cf. Mutersbaugh 2005) where only about half of the farm is converted to strategically spread risk between variability in organic production and fluctuating oil prices. Yet, what all of these performances have in common is their approach to growing cereal crops.

---

[8]Arrangements range from stockless arable and a wide variety of mixed farms, as well as combinations of arable and horticultural enterprises, and from 2 ha for small hobby farms to over 3,000 ha for large estates.

Cereals, as arable crops, constitute a group of annual plants that can be grown under a wide variety of conditions. To maximize yields, agricultural orthodoxy is based on standardizing these conditions through the use of artificial fertilizers, herbicides and pesticides. In organic production systems, the use of these agrochemicals is prohibited, and farmers therefore need to manage yields and weeds by shaping their farming system around the local, material, social and economic conditions of their farm. Barring very few exceptions, this usually requires the establishment of a rotation, which is a system in which a succession of different crops is grown on a field. This constitutes an active use of the agronomic properties of plants to manage the fertility of the soil: different plants use, reintroduce and make available different nutrients, different root systems have different impacts on soil structure, and the variety of crops minimizes disease pressure for any single species and reduces the potential for any one type of weed to become dominant (for a detailed description of these processes, please refer to Lampkin 1990). Also, managing fertility and weeds often involves an active mobilization of livestock to provide manure and to selectively graze or uproot weeds.[9] These management processes are very specific: fields, and often sections in a field, have their own geographic and agronomic characteristics. These affect how the soil can be cultivated and what is required to maintain fertility, and therefore what crops can be grown at any given time. Moreover, as one adviser explained, the social organization of a farm (e.g., workforce) and its economic conditions (required level of profitability) shape the rotation:

> Every ... the challenge is every farming business and every farmer is different because they all have different requirements. Some farmers will want a modest level of productivity and a modest level of profitability and income because maybe they're owner-occupiers and they don't have mortgages to pay; maybe they have other business interests and other sources of income; whereas other farmers will want to maximize production and productivity because they need it. So your response to that scenario is going to be different for every farm, added to which you've then got the complexities of different climatic locations and also different soil types. So, it's a 3D jigsaw, with a blindfold on, all the time.
>
> As such, performing an organic arable enterprise requires not only compliance with the standards, but also careful planning—as another adviser and a farm manager commented separately, in an organic system it is very difficult to respond to problems when they occur:
>
> [...] if you're growing crops conventionally then you actually spoon feed them, you give them what they need when they need [it]. In terms of an organic system you've got to get the seedbed right because you've got one chance. You put the crop in the ground and that's really very little you can do to manipulate that. (Adviser 2)
>
> No, what we have had out of the organic farming is it's a heck of a lot more complicated than conventional farming and if something goes wrong you can't just sort it with a spray. So, it needs to be planned, worked out, and thought about much more carefully. (Farm manager)

Yet, this does not mean that rotations, once drawn up, become rigidly adhered to schemes to manage the fields to which they apply. In fact, all cereal farmers

[9]This reflects the origins of organic farming as a low-input system: the archetypal organic farm integrates arable and livestock enterprises to minimize the use of inputs into those enterprises that have not been generated on that farm.

in the sample recounted how they regularly tinkered with rotations to respond to changing conditions and to address emergent problems. Thus, the growing practices for cereal crop production are internally highly differentiated, requiring a rigidly adhered to regime of flexible planning to ensure that the production is technically and economically feasible while remaining within the limits set by the standards and the material conditions of the farm site.

This means that each performance is different: no two farms (and even fields) are ever the same. Yet, regardless of these differences in how organic is performed on any particular farm, they are all governed by a single certification regime. In contrast to the Tanzanian case, there is a single policy and regulatory context in which certification bodies operate. Due to specific historic processes, there are seven certification bodies[10] that offer organic certification. While these organizations compete with each other over fees and additional certification services, for example, the Farm Assurance scheme or the British Retail Consortium certification, their organic certification is governed by the European Union regulation on organic farming. This regulation is administered by the relevant ministry (DEFRA),[11] which has formally devolved the day-to-day certification of licensees to authorized private certification bodies. To receive authorization, these bodies need to organize themselves and their activities according to a standard for certification systems (ISO65,[12] EN45011[13]), against which they are accredited by the independent UK Accreditation Service. Moreover, these bodies are required to coordinate common interpretations of the regulations so that each performance by a licensee is assessed according to the same criteria, irrespective of which certification body they are licensed with. Thus, this regulatory regime is aimed at harmonizing how the differentiated growing practices of licensees are assessed and certified.

4.7 Conclusions on Differentiating Organics

In the beginning of this paper, we proposed the question of how a singular notion can singularize products if its practice is multiplicitous. We argue that by conceptualizing standards as calculative devices in *agencements*, performativity analysis allows us to answer this oxymoronic question. We have shown that despite the EU regulation as an organizing device, we see a number of different practices around singularization and homogenization that render the different organic characteristics

[10]These are the Soil Association, Organic Farmers and Growers, the Organic Food Federation, the Biodynamic Agricultural Association and Ascisco as national schemes, and Quality Welsh Food Certification, the Scottish Organic Producer Association and the Irish Organic Farmers and Growers Association as regional schemes.

[11]Department for Environment, Food and Rural Affairs.

[12]See International Standardization Organization (1996).

[13]See CEN (Comité Européen de Normalisation) (1998).

(in)visible both before and after the 'organic' product appears to the consumer in the supermarket. What we have done in this paper is present two cases of organic practices that illustrate multiplicity, despite the singularization of an organic market. We characterized these multiple practices based on the themes of *organic by default*, *parallel organics* and *cross-certification* to illustrate that not only is organic differently enacted in Tanzania and in the UK contexts, but also within each of these contexts.

While the use of statistics and promotional campaigns help to create a single, measurable global market for organic in opposition to the conventional market, the pitting of organic against conventional is mobilized also in the practice of farming organically. In Tanzania this takes the form of *organic by default* constituted by a mix of traditional practices and financial constraints faced by the farmers. In the UK, this is enacted as a commitment of transforming a part (sometimes all) of conventionally farmed land into a (sometimes) scientifically calculated organic practice that responds to the annual plants. *Parallel organics* carry parallel interpretations. In the case of the UK, we see a reinforcement of Mutersbaugh's (2005) finding, while in Tanzania we use the term to refer to the parallel regulatory regimes. Finally, *cross-certification* refers to the multiplicity of standards, certifications and accreditations that are circulating to assemble the network of certified production.

This type of analysis reminds us of the necessity of context in understanding how standards are enacted. First, unexpected things happen when living beings are 'subjected to standardization'. Thévenot (2009) concludes that "standardization provides a guarantee by attributing properties to standardized objects, while all the while the very notion of engagement accounts for guarantees that rest on a dependency between agent and environment that goes against any such attribution" (p. 805). This quote encapsulates our argument regarding enactments. We suggest that the process of standardization creates increasing rigidity to the standards themselves – while we see the practice of organics continues to maintain flexibility of interpretation – which is not always written into the standards. The multiplicity of organic practices allows us to see both singularity and multiplicity in enactments. As a calculative device, the standard enables this by making certain classes of performance compatible. Not by defining ethical standards which attempt to govern the intentions and beliefs of producers, but rather more tangible standards by defining the practices which can be mobilized to produce 'organic' products. For example, the production of organic tea is characterized not so much by internal differentiations in the practices of growing, harvesting and processing the crop, but all the more in how those practices are certified under different regimes. In contrast, organic cereal production is characterized by a more homogeneous certification regime and more internally differentiated growing practices.

The analysis in this paper suggests that the processes of third party certification provide one mechanism of achieving commensurability between different enactments; they provide a device "'to abstract', that is, to transport, transform and displace an action into a formal, calculative space" (Loconto and Busch 2010, p. 527). This means that all site-specific enactments of the standards are abstracted into a formal space where they are made commensurable: each enactment undergoes an external verification of compliance with the rules in the standards based on a standardized and accredited certification process and hence becomes available in the organic space that is thereby created. The uniformity of organic 'stuff' is not located in the individual enactments through which it was produced, but in the way these enactments are made commensurable. It is this particular uniformity, then, that provides the basis on which goods can be singularized as products having certain qualities (Callon et al. 2002) and which can be integrated in pre-existing practices of consumption (e.g., Gronow and Warde 2001; Halkier 2009).

Therefore, in the light of our analysis we can contribute to the discussion of an ontological shift in organics in the following way. Performativity analysis has revealed a shift from ideas about holistic ideals to circumscription of practices, i.e. from complete ways of life to a practical organization of markets. While this does constitute a shift, we should not try to classify this in moral terms – especially since what 'organic' (in the singularized version) means has changed as well. Rather, the power of our argument is to show how the singularization of multiplicitously produced products in the here and now (as the outcome of contingent, historical processes involving many actors, not just the initiators of the organic movement) allows markets for organic to function. We cannot support the claim that organic has lost its way because the practice of certified organic varies so much from place to place. But rather as we begin to see organic taking root in different contexts we can reveal with performativity analysis that the standards have not eroded the original concepts but are only performing different interpretations, and simultaneously different realities, of them. We argued in this paper that despite organizing (standardizing) practices, individual performances of organic result in a multiplicity of organics that are necessary for dealing with the complexities of farming. Simultaneously, the standardizing practices that result in singularization are required for the trading and marketing of organic products.

References

Allaire, G. 2009. *Economics of Conventions and the New Economic Sociology and our Understanding of Food Quality and New Food Markets and Trade Institutions: What are markets that pure economics does not know?* Paper presented at the 27th annual conference of Agricultural Economists (IAAE), Beijing, China, Aug 2009.

Austin, J.L. 2004 [1962]. How to do things with words. In *How to do things with words – Austin J. L*, ed. J. Rivkin and M. Ryan, 162–177. Oxford: Blackwell Publishers.

Balfour, E.B. 1978. *The living soil and The Haughley experiment*, New rev. ed. London: Faber and Faber.

Beck, E.L. 2011. *Organic agriculture sustained through economic crisis vital signs: Global trends that shape our future*. Washington, DC: World Watch Institute. Retrieved from http://vitalsigns.worldwatch.org/vs-trend/organic-agriculture-sustained-through-economic-crisis. Accessed 27 May 2011.

Bell, V. 2007. *Culture and performance: The challenge of ethics, politics and feminist theory*, English ed. New York: Berg.

Boltanski, L., and L. Thévenot. 2006 [1991]. *On justification: Economies of worth*. Princeton: Princeton University Press.

Bolwig, S., P. Gibbon, and S. Jones. 2009. The economics of smallholder organic contract farming in tropical Africa. *World Development* 37(6): 1094–1104.
Busch, L. 2007. Performing the economy, performing science: From neoclassical to supply chain models in the agrifood sector. *Economy and Society* 36(3): 437–466.
Callon, M. (ed.). 1998. *The laws of the markets*. Oxford: Blackwell.
Callon, M. 2007. What does it mean to say that economics is performative? In *Do economists make markets? On the performativity of economics*, ed. D. MacKenzie, F. Muniesa, and L. Siu, 129–163. Princeton: Princeton University Press.
Callon, M., C. Méadel, and V. Rabeharisoa. 2002. The economy of qualities. *Economy and Society* 31(2): 194–217.
CEN (Comité Européen de Normalisation). 1998. *EN 45011:1998 General requirements for bodies operating product certification systems* (ISO/IEC Guide 65:1996). Brussels: CEN-CENELEC.
Conford, P., and P. Holden. 2007. The soil association. In *Organic farming: An international history*, ed. W. Lockeretz, 187–201. Wallingford: CAB International.
Co-operative. 2010. *The co-operative organic fairtrade tea carton*. Manchester: Co-operative Group Ltd.
Didier, E. 2007. Do statistics "perform" the economy? In *Do economists make markets? On the performativity of economics*, ed. D. MacKenzie, F. Muniesa, and L. Siu, 245–276. Princeton: Princeton University Press.
DuPuis, M. 2000. Not in my body: RBGH and the rise of organic milk. *Agriculture and Human Values* 17(3): 285–295.
FLO. 2009. *Generic fairtrade standards: Small producers' organizations – Current version: 15.08.2009*. Bonn: Fairtrade Labelling Organizations International.
Gibbon, P., E. Lazaro, and S. Ponte. 2010. *Global agro-food trade and standards: Challenges for Africa*. Houndmills/Basingstoke/Hampshire/New York: Palgrave Macmillan.
Goffman, E. 1974. *Frame analysis: An essay on the organization of experience*. New York: Harper & Row.
Gronow, J., and A. Warde. 2001. *Ordinary consumption*. London/New York: Routledge.
Guthman, J. 2002. Commodified meanings, meaningful commodities: Re-thinking production-consumption links through the organic system of provision. *Sociologia Ruralis* 42(4): 295–311.
Guthman, J. 2004. *Agrarian dreams: The paradox of organic farming in California*, vol. 11. Berkeley/Los Angeles/London: University of California Press.
Halkier, B. 2009. A practice theoretical perspective on everyday dealings with environmental challenges of food consumption. *Anthropology of Food* S5(September): 2–13.
Hatanaka, M., C. Bain, and L. Busch. 2006. Differentiated standardization, standardized differentiation: The complexity of the global agrifood system. In *Between the local and the global: Confronting complexity in the contemporary agri-food sector*, ed. T. Marsden and J. Murdoch, 39–68. Oxford: Elsevier.
Hilgartner, S. 2000. *Science on stage: Expert advice as public drama*. Stanford: Stanford University Press.
Hillocks, R.J. 2002. IPM and organic agriculture for smallholders in Africa. *Integrated Pest Management Reviews* 7(1): 17–27.
Howard, A. 2006. *The soil and health: A study of organic agriculture*. Lexington: University Press of Kentucky.
IFOAM. 2009a. *The principles of organic agriculture*. Retrieved from http://www.ifoam.org/about_ifoam/principles/index.html. Accessed 23 Jan 2009.
IFOAM. 2009b. *The IFOAM norms for organic production and processing – Version 2005*. Bonn: International Forum for Organic Agriculture Movements.
International Standardization Organization. 1996. *ISO/IEC Guide 65:1996 General requirements for bodies operating product certification systems*. Geneva: International Standardization Organization.
International Trade Center (ITC). 2011. *Organic & natural foods & products farming country profile*, Tanzania. Retrieved from http://www.intracen.org/organics/Country-Profile-Tanzania.htm. Accessed 27 Mar 2011.

Jaffee, D., and P. Howard. 2009. Corporate cooptation of organic and fair trade standards. *Agriculture and Human Values* 27(4): 387–399.
Kamp, M.v.d. 2010. Coexisting organic standards: Making boundaries, navigating multiplicity and enacting commonality. In *System innovations, knowledge regimes, and design practices towards transitions for sustainable agriculture*, ed. M. Barbier and B. Elzen, 69–85. Paris: INRA – Science for Action and Development.
Konefal, J., and M. Hatanaka. 2011. Enacting third-party certification: A case study of science and politics in organic shrimp certification. *Journal of Rural Studies* 27(2): 125–133.
Lampkin, N. 1990. *Organic farming*. Ipswich: Farming Press.
Law, J. 2004. *After method: Mess in social science research*. London/New York: Routledge.
Law, J. 2008. On sociology and STS. *Sociological Review* 56(4): 623–649.
Loconto, A. 2010. Sustainably performed: Reconciling global value chain governance and performativity. *Journal of Rural Social Science* 25(3): 193–225.
Loconto, A., and L. Busch. 2010. Standards, techno-economic networks, and playing fields: Performing the global market economy. *Review of International Political Economy* 17(3): 507–536.
MacKenzie, D.A., F. Muniesa, and L. Siu. 2007. *Do economists make markets?: On the performativity of economics*. Princeton: Princeton University Press.
Mansfield, B. 2004. Organic views of nature: The debate over organic certification for aquatic animals. *Sociologia Ruralis* 44(2): 216–232.
Mbiha, E.R., and G.C. Ashimogo. 2010. Challenges and opportunities of organic agriculture in Tanzania. In *Global agro-food trade and standards: Challenges for Africa*, ed. P. Gibbon, E. Lazaro, and S. Ponte, 101–119. Houndmills/Basingstoke/Hampshire/New York: Palgrave Macmillan.
Mead, G.H. [1934] 1962. *Mind, self and society*. Chicago: University of Chicago Press.
Mol, A. 2002. *The body multiple: Ontology in medical practice*. Durham: Duke University Press.
Muniesa, F. 2007. Market technologies and the pragmatics of prices. *Economy and Society* 36(3): 377–395.
Mutersbaugh, T. 2005. Just-in-space: Certified rural products, labor of quality, and regulatory spaces. *Journal of Rural Studies* 21(4): 389–402.
Northbourne. [1940] 2003. *Look to the land*, 2nd, rev. special ed. Hillsdale: Sophia Perennis.
Organic Africa Pavilion. 2011. *Exhibitors of the Africa Pavilion: Tanzania Organic Agriculture Movement (TOAM)*. Retrieved from http://organicafricapavilion.com/exhibitor-menu?sobi2Task=sobi2Details&catid=3&sobi2Id=8. Accessed 11 April 2011.
Parrish, B.D., V.A. Luzadis, and W.R. Bentley. 2005. What Tanzania's coffee farmers can teach the world: A performance-based look at the fair trade-free trade debate. *Sustainable Development* 13(3): 177–189.
Power, M. 1997. *The audit society: Rituals of verification*. Oxford/New York: Oxford University Press.
RA. 2009. *Sustainable agriculture standard*. San José: Sustainable Agriculture Network, Rainforest Alliance (RA).
Raynolds, L., D. Murray, and A. Heller. 2007. Regulating sustainability in the coffee sector: A comparative analysis of third-party environmental and social certification initiatives. *Agriculture and Human Values* 24(2): 147–163.
Scialabba, N. 2000. *Factors influencing organic agriculture policies with a focus on developing countries*. Paper presented at the IFOAM (International Federation of Organic Agriculture Movements) scientific conference, Basel, Switzerland, 28–31 Aug 2000.
Simbua, E.F., and A. Loconto. 2010. *Sustainability of smallholder tea production in the United Republic of Tanzania*. Rome: Food and Agriculture Organization of the United Nations.
Soil Association. 2010. *Organic market report 2010*. Bristol: Soil Association.
Soil Association. 2011. *Our history*. Retrieved from http://www.soilassociation.org/Aboutus/Ourhistory/tabid/70/Default.aspx. Accessed 20 Mar 2011.
Steiner, R., and M. Gardner. 1993. *Spiritual foundations for the renewal of agriculture: A course of lectures held at Koberwitz, Silesia, June 7 to June 16, 1924*. Kimberton: Bio-Dynamic Farming and Gardening Association.

Szerszynski, B., W. Heim, and C. Waterton. 2003. *Nature performed: Environment, culture and performance*. Oxford/Malden: Blackwell Pub./Sociological Review.
Thévenot, L. 2009. Governing life by standards: A view from engagements. *Social Studies of Sciences* 39(5): 793–813.
Vogl, C.R., L. Kilcher, and H. Schmid. 2005. Are standards and regulations of organic farming moving away from small farmers' knowledge? *Journal of Sustainable Agriculture* 26(1): 5–26.
Willer, H., and L. Kilcher (eds.). 2011. *The world of organic agriculture – Statistics and emerging trends 2011*. Bonn/Frick: IFOAM/FiBL.

Chapter 5
Feeding the World – The Contribution of IFOAM Principles

Bernhard Freyer, Jim Bingen, Milena Klimek, and Rebecca Paxton

5.1 Background and Objectives

The kind of growth in food production needed to feed the world is one of today's most widely discussed topics (Seufert et al. 2012; Ehrlich and Ehrlich 2013). In the context of feeding the world, there are significantly different ideas about how to promote sustainable agriculture (Rigby and Cáceres 2001) or how to think about agricultural growth (Pretty 2008). One of the enduring issues in this debate, and the focus of this chapter involves the question of whether organic production can feed the world, especially when confronted by the overwhelming dominance of high input (such as synthetic chemicals and bio-technology) production on which most food polices and high-profile world hunger programs depend.

In order to address such an issue, this chapter examines how the IFOAM Principles (IFOAM 2009) offer an ethical foundation for thinking differently about the quantitative potential of organic feeding the world and contributing to world food security and food sovereignty (cf. Pinstrup-Andersen 2009). Following an overview of what is called the high- and low-input models for food production, we present our position that achieving food security and sovereignty, or feeding the world, is not only a question of technology, logistics, science or knowledge, but a question of the ethical values that frame this global goal (cf. Runge et al. 2003).

B. Freyer (✉) • M. Klimek • R. Paxton
Department of Sustainable Agriculture Systems, Division of Organic Farming, University of Natural Resources and Life Sciences (BOKU), Vienna, Austria
e-mail: Bernhard.Freyer@boku.ac.at; Milena.Klimek@boku.ac.at; Rebecca.Paxton@boku.ac.at

J. Bingen
Department of Community Sustainability, Michigan State University, East Lansing, MI, USA
e-mail: Bingen@msu.edu

B. Freyer, J. Bingen (eds.), *Re-Thinking Organic Food and Farming in a Changing World*, The International Library of Environmental, Agricultural and Food Ethics 22, DOI 10.1007/978-94-017-9190-8_5

Specifically, we suggest that the IFOAM Principles offer a framework for examining previously overlooked alternative designs for organic agrofood systems that may open new approaches for addressing food. Food security here, is described through four terms: food supply, processing, distribution / trade and demand. We use food security as a shorthand term to refer to the full range of issues that arise in thinking about the topic including food access and food sovereignty. Furthermore, food sovereignty is highly linked with the contemporary food crisis (Hunt et al. 2009; 2011) and is seen here as a relevant precondition of food security (IFOAM 2009; McIntyre et al. 2009; Meat & Wool New Zealand 2010).

Our working hypothesis is that considering ethical issues offers a new perspective on debates that have centered largely on quantitative indicators. We argue that the IFOAM Principles offer explicit guidance that directly informs the more quantitative dimensions of food production and food security including access to food. Moreover, we hypothesize that operationalizing or deliberately putting ethics into practice contributes in the long run to increasing world food supply and to more resource efficient practices that help to increase food access. We recognize that current worldwide organic practices do not always addresses these values. As a result, we turn to the IFOAM Norms as a basis for reviewing the kind of instruments that administer, describe and regulate organic practices.

For heuristic purposes, we present two paradigms – high and low-input agriculture – as a point of departure for a critical review of the feeding the world debate. Following a brief discussion of the relevance of ethics to the approach toward food security in both paradigms, we examine how the IFOAM Principles, if applied quantitatively, could contribute to food security. Finally, we ask if the application of the IFOAM Norms – specifically the IFOAM Standards Requirements, the IFOAM Standards for Organic Production and Processing and the IFOAM Accreditation Requirements for Bodies Certifying Organic Production and Processing – are sensitive to the IFOAM Principles' ethics to contribute to achieving food security. We conclude with a critical review of our discussion and recommendations.

5.2 The Debates on Feeding the World

This section outlines the relevance of two dominant agricultural paradigms (high-input and low-input) (see Beus and Dunlap 1990) to "feeding the world." We focus on the issues of production, ecology and food consumption patterns. Examining these paradigms lays the groundwork for developing an ethical perspective on how to feed the world.

5.2.1 The High-Input Paradigm

The high-energy input agrofood-paradigm is characterized by a common reliance on crop biotechnology on both large- and small-scale farms, as well as the corporate concentration in processing and marketing/distribution (Grievink 2002; Humphrey and Memedovic 2006, p. 35). Equally important, both foreign public and private capital investments in food production continue to be based on this high input paradigm. This production approach is widely promoted as the most effective way to produce the food needed to feed the world. However this rapid intensification of agricultural production is accompanied by many negative consequences for environmental, social and economic justice (e.g., Stoate et al. 2001; Trigo and Cap 2004).

Several studies identify numerous environmental problems related to the use of biotechnological cropping (e.g., Tscharntke et al. 2012). Establishing high-input agriculture on mainly fragile soils will lead to soil degradation and an irreversible loss of soils, loss of water qualities and quantities (Rockström et al. 2007; Godfray et al. 2010), loss of biodiversity as well as negative impacts on the climate (Galloway et al. 2008). Of equal concern is the increasing evidence of the relationships between crop biotechnology that facilitates large-scale monocropping (more than 10 % of the agricultural land in the US and Canada) and cropping for bioenergy production (Hendrickson and James 2005). As some might suggest, this is like "putting top soil in our gas tanks" with negative consequences on soil fertility and water quality (e.g., see Goolsby and Battaglin 2000; Nellemann 2009), not to mention the competition posed to food production (see Government Office for Science 2011). This is not to generally put in question bioenergy production, but the production methods of the high-input paradigm, and their negative influence on ecology and the socio-economy, similar to that of food production (Searchinger et al. 2008). Finally, others report on human and animal health problems related to the use of food products made from genetically modified crops (Hendrickson and James 2005; Gurian-Sherman 2009; Zacune 2011) and critical power relations of GMO businesses (Walters 2006).

This paradigm specifically promises the production of the large quantities of food required to meet the projected increase in demand for food from a growing world population. Based on Neo-Malthusian thinking, this approach holds that world population growth will outstrip the ability to produce enough food. This argument has regularly cycled through numerous public debates for decades as an argument for high-input agriculture (e.g., see Ehrlich 1968). But simply to increase production does not assure either widespread or equitable access to food (Thompson 2008). Furthermore, this approach pays scant attention to the political economy of population growth, or to questions related to corporate power in food production, including recent concerns over "land grabbing" via corporate, multi-national and national investments (Cotula 2009; Kachika 2010).

The increasingly frequent "land grabs" throughout Asia, Africa and Latin America (Volpi 2007; Daniel 2011; McMichael 2012; Tscharntke et al. 2012; Oliveira 2013) are all based on the use of high input methods to produce crops consistent with Western diets, and at the cost of unchecked deforestation in developing and emerging countries (van Solinge 2010). Further deforestation in the South for gaining agricultural land will have negative consequences for the world climate, and is therefore not a sustainable way of food procurement (Malhi et al. 2008).

Since the 1970s, food consumption habits, especially in the developing countries, have been moving toward heavily meat-based diets (Pinstrup-Andersen et al. 1999). This is projected to continue due to and is particularly exasperated through the many relaxed trade barriers of many countries, concerning imported meat and poultry (Delgado 2003; Speedy 2003). The underlying assumption that the world should adopt Western meat-based food consumption patterns (Drewnowski and Popkin 1997; Gerbens-Leenes and Nonhebel 2002; Imhoff et al. 2004; Delgado 2005; McAlpine et al. 2009) calls for dramatic cereal crop yield increases relying on the use of intensified agricultural practices. However, meat-based diets (Tilman et al. 2002), in contrast to grain-based diets (Carlsson-Kanyama and González 2009; Godfray et al. 2010), do not meet basic nutrition recommendations (Mancino 2005; Wang et al. 2008; United States Department of Agriculture 2009). Instead, they require the adaptation of current food consumption patterns (Smil 2002; Dietz et al. 2007; Godfray et al. 2010, p. 816; Aiking 2011). In addition, the health risks (Duchin 2005) of meat-based diets offer sound reasons for revising agricultural and food policy that focuses on the promotion of such a diet (Willett et al. 1995; Martínez-González et al. 2002). Related meat production leads to increased consumption of the limited natural resource of phosphorus (Cordell et al. 2009). Intensive livestock production predominantly produced for a wealthy North, provokes food insecurity through desertification, sea-level rise, increased extreme weather occurrences, land shortage, etc. As a result it will dramatize societal and political conflicts accompanied by a lack of food access (McMichael et al. 2007).

Issues concerning food supply revolve not only around questions of production and controlling post-harvest losses. On a global scale, a growing number of reports express concerns about the ways in which the global "biotech agrofood system" contributes to the creation of dependency of local food systems on seed material and world market prices and the destruction of local economies (Hendrickson and James 2005). Together with the climate change induced production crisis and the increase of food prices (Trostle 2008), ventures at the stock market (Wahl 2009), food access and food sovereignty particularly in the South, becomes insecure (Rosset 2003; McMichael 2009).

In addition to high-input farming, special agricultural programs often promoted by international and national agencies promise "new" approaches to feeding the world (Setboonsarng 2006).[1] The AGree program (AGree 2012) for example, promises to transform current food and farming by promoting a broad mix of practices that are derived largely from the high-input practices. This approach, commonly called integrated agriculture or ecofriendly intensification, is widely practiced in Europe and "bridges" conventional and organic farming. The program acknowledges the growth in organic and the popularity of local and regional food systems activities (AGree 2012), but the "path ahead" that it charts neglects the potential contribution of organic farming to transforming current policies and

[1] See former and latest reports for example: http://www.saa-safe.org/; or, www.feedthefuture.gov/. The German-FAO initiative for "more responsible investments in Agriculture to eradicated hunger and poverty" http://www.fao.org/news/story/en/item/168390/icode/

practices. As such, the AGree program reflects a "more of the same" focus on increasing food production in the North to feed the South. Such approaches ignore the social and economic implications that negatively affect food security in the South through increased dependency on the North. Similar observations were made on the "Strategy for Sustainable Farming and Food" (Barr 2003; Winter and Lobley 2009).

Thus, the extent to which the high-input paradigm contributes to feeding the world continues to be speculative at best, especially when the negative ecological, economic, health and social implications (external costs) of this kind of food production are not considered in any calculation within food security (Pretty et al. 2000; Tegtmeier and Duffy 2004, p. 322).

5.2.2 Organic Paradigm

There is evidencethat the current paradigm of growth is full of conflicts and negative impacts, e.g., on climate, biodiversity, food security, water or access to land for the poor (Miller and Rose 1997; Barham 2002; Padel 2002; Lunneryd 2003; Renard 2003; Barnett et al. 2005). Such a negative system might look towards a comprehensive change in agriculture and food production. In contrast to current paradigm of growth, organic has the potential to add to a changed system. Organic as an alternative is legitimized through its diverse qualities of health, quality, ecological benefits, improved animal welfare, and more climate friendly qualities (Tanner 2000).

The organic, or low-input agrofood paradigm (Freyer 2007) focuses on small-scale farms, incorporates socio-cultural concerns, and is driven by ethical, ecological and social justice foundations (Blowfield 1999). This approach commonly includes attention to food security (Halberg et al. 2006a),[2] food sovereignty (e.g., Padel et al. 2007, p. 17; Pimbert 2009), and to regionalized food systems and food access (see Starke 2011).[3] It takes a systemic approach to how to feed the world, not restricted to the sole focus of increased production.

A wide range of empirical evidence demonstrates that organic and equivalent low-input agricultural systems can contribute significantly to overcoming the agro-ecological negative impacts of the high-input agriculture. There is scientific evidence that organic farming:

- Maintains the soils (e.g., Atkinson et al. 1995; Mäder et al. 2002);
- Sustains and increases biodiversity (e.g., McNeely and Scherr 2003; Hole et al. 2005);
- Reduces energy and water consumption (e.g., Stolze et al. 2000; Lotter et al. 2003; Ziesemer 2007; Posner et al. 2008; Seufert et al. 2012);

[2]Food security: Defined at the 1996 World Food Summit, food security: "[. . .] exists when all people, at all times, have physical, social, and economic access to sufficient, safe, and nutritious food that meets their dietary needs and food preferences for an active and healthy life" (Food Ethic Council 2012).

[3]Examples from the field see http://www.ecologyandfarming.com/

- Increases P-efficiency (crop's capacity to use phosphorous) (e.g., Martin et al. 2007; Onwonga et al. 2008; Dangour et al. 2009);
- Increases yields through agro-ecological strategies (Pretty and Hine 2001; Scialabba and Hattaam 2002; Pretty 2003, 2006);
- Contributes to resilience in the face of climate changes (e.g., seed and crop diversity; soil cover; organic manure) (Borron 2006; Scialabba and Müller-Lindenlauf 2010);
- Reduces the risks from the spread of epidemic diseases because of the dispersed production footprint and the bio-diverse practices (Van Bruggen 1995); and
- Fosters regionalized, and highly diverse and healthy food, because it is adapted to specific agro-ecological zones (Heaton 2001).

Yet many of these statements are disputed and counterarguments from both sides are continuously appearing. On the issue of yield, it is important to keep in mind that certified organic production is practiced on scarcely 1 % of the world's cultivated land (Willer and Kilcher 2011). Additionally, there is not much known about the performance of organic in comparison with conventional under same climate and soil conditions. Moreover, as Halberg et al. (2006a) note, there is little agreement about the different methods of estimating organic yield under different agro-ecological conditions (see also Connor 2008; de Ponti et al. 2012). Organic yields are generally estimated to be 20–30 % lower from those on non-organic farms (IFAD 2005; Pimentel et al. 2005; Badgley et al. 2007; Edwards 2007; Erb et al. 2009; Edwards et al. 2010). Furthermore, the interpretation of some trials might be either too optimistic (Badgley et al. 2007), or underestimated, or simply not identified (Trewavas 2001). Nevertheless, on-going research documents significant potential of organic practices to increase yields. In Austria, for example, the yield of the main arable crops (wheat, potato and maize) increased between 2007 and 2011 by approximately 30 % (Lebensministerium 2012). Furthermore, a series of promising sustainable and low-input systems, similar to the organic approach, in developing countries demonstrate a significant potential to increase yields without damaging either social systems (Scialabba and Hattaam 2002) or nature (Pretty et al. 2006).

In short, it is not known how much of the reported "lower organic yields" are a function of the limited agronomic research and the limited data, specifically with respect to developing countries. Organic farming has never benefitted from the same level of research and capacity building or funding and policy support as high-input agriculture (Niggli et al. 2008). As a result, the data from detailed large scale modeling to assess the possibility for organic production to "feed the world" is limited (Halberg et al. 2006a).

It must also be acknowledged that we lack convincing empirical evidence about the productivity of the organic paradigm to feed the world. Agronomic research on organic farming pales in comparison to research on conventional farming. There are no serious empirical, socio-economic or policy studies that identify alternative scenarios from the perspective of organic agriculture for any region of the world. The scenarios that have been proposed use only very selective samples of data from research plots where production results are based on optimal conditions (e.g., Connor 2008, p. 5), or models developed on simplified assumptions (e.g., Zanoli et al. 2000; Znaor et al. 2007).

While many question whether organic farming could produce the quantity of food that would be affordable for enough people "to feed the world", others do support this thesis. In short, both perspectives are based on overly simplified assumptions. Those who criticize the capacity of organic ignore the contributions of organic practices to production diversification, the avoidance of erosion, modified food consumption patterns and several social and economic factors that are of relevance to assess the potential of organic agriculture. Others ignore that in addition to techno-ecological issues, moving to organic is a question of changing power relations all along the food and farming chain. Furthermore, the idea of feeding the world needs a more differentiated view of farm size from one based on understanding the potential of smallholder farming because most farmers in the developing world live on small farms and are the key for food security (World Bank & Commission on Growth 2008).

5.2.3 *Feeding the World Beyond a Technological Perspective*

As is seen by the above discussion, debates continue to occur over whether organic products are healthier, tastier or environmentally friendlier, or how they contribute to resolving numerous social issues, whether it is affordable for the average consumer, or if it can seriously contribute to feed the world.[4] Moreover, assessments of organic production never question either the agronomic or political-economic and policy assumptions on which they are based. Consequently, these scenarios cannot be used to identify the potential of organic to feed the world. However, they are important milestones to better understand the capability of the organic system to contribute to food security.

In order to better understand organic's contribution to food security, it is not a question of choosing one model over another. Nor is a question of accepting some type of compromise, which is essentially the current situation. Furthermore, it is not a question of accumulating only more data to show that one model could be more successful in feeding the world. The ecological implications and the political economy of each model are so fundamentally different that any comparison would be like comparing apples to oranges. In addition, a wide and diverse number of factors influence any efforts to achieve food security. This includes: land tenure, credit availability, post-harvest food losses, labor, technical skills and education, the contribution of off-farm employment, gender roles, malaria and other health issues, the availability of mineral fertilizers, labor-saving technologies and, production or marketing subsidies.[5]

[4](see diverse discussions and assumptions on this issue, e.g., Lewin et al. 2004; Giovannucci 2005; Mittal 2006; Zundel and Kilcher 2007; Connor 2008; LaSalle et al. 2008; Pretty et al. 2010; Seufert et al. 2012).

[5](Kader 2004; Halberg et al. 2006b, p. 297; Khaledi et al. 2007; Gustavsson et al. 2011).

In summary, quantitative models do not help in understanding how to achieve food security. A broader approach, that, questions the assumptions behind the whole agrofood system, as well as the strategies and targets for change is required (Bartley 2007). From our perspective to better understand the impact of the organic approach is to question what kind of ethics motivates the different agricultural paradigms (Beus and Dunlap 1990). Organic farming is unique in its foundation on ethical principles. But these principles are rarely, if ever acknowledged in the debates over food security (e.g., Bartley 2007). Nevertheless, there are some remarkable contributions on ethical consumerism, food sovereignty, organic farming and the need for a new public ethic of care in the context of climate change (Morgan 2010) or within the context of peasants (Altieri and Toledo 2011). Thus, it becomes critical to consider the ethical dimensions of an organically driven approach to food security and to thinking about how to feed the world.

5.3 The IFOAM Principles and Food Security

This section focuses on the relationship between the IFOAM Principles and various dimensions of food supply, demand, access and the achievement of food security. We argue that the IFOAM Principles extend an entry-point for considering whether organic can feed the world (IFOAM 2012).

The IFOAM Principles offer an ethically founded guide to organic practices for consumption, production, processing, distribution and trade, through a normative ethical framework (see Chap. 2). The IFOAM Principles further represent an ethical framework that invites all actors from producers to consumers in the organic agrofood chain (IFOAM 2012). Additionally, in contrast to the mainstream discussion of "eco-intensification" (cf. Hunt et al. 2010) within conventional farming (often associated with high-inputs), organic eco-intensification, drawing parallels to the IFOAM framework, is sensitive to broader perspective. The organic eco-intensification includes the food chain as a whole aknowledging technical and ecological relevant practices, health strategies and food consumption patterns, as well as far reaching social and economic dimensions in a holistic sense (Breman et al. 2001; Arbenz 2011; Abouleish 2013). The Principles include feedback loops that integrate the impact of all activities on the whole agrofood system. This allows for the discovery and possible avoidance of unintended consequences of any activities in the organic agrofood chain on other human-nature systems. Furthermore, this explains why the IFOAM Principles cannot be applied piecemeal: the Principles are overlapping, and their message is always interrelated with the other Principles.

Organic actors do not simply practice or participate in the organic agrofood chain. These actors actively create and define the organic principles through their active and ongoing participation (IFOAM 2012). Thus, the development of the organic agrofood chain is not only an issue for farmers, processors, or certifiers, but also a responsibility of all actors, including consumers and traders. Seeing the organic agrofood system as a socio-ecological system, requires avoiding any harm to nature and any active defrauding of relationships. Together, these value-oriented characteristics of the IFOAM Principles should have consequences on food security. The following tables introduce how the IFOAM Principles can be applied to assessing the quantitative impact of organic on food supply, processing, distribution and consumption (see also Padel et al. 2007, pp. 15–22; IFOAM 2012, pp. 9–12) (Tables 5.1, 5.2, 5.3, and 5.4).

Table 5.1 The IFOAM Principles and food supply

Impact on food supply	Specific quantitative impact
Health	
Sustain and enhance soil health	Increase of productivity; risk reduction while droughts and floods
Avoid inputs that create health risks in production	Increase of high water quality
Conserve/enhance fertile soils and clean water	Increase of long term productivity
Assure animal health through preventive health care	Reduction of food losses through diseases or low quality or contaminations
Offer diverse, high quality, nutritious and healthy food	Increase of animal life-time achievement
Adopt appropriate technologies	
Ecology	
Protect an ecologically sound environment	Reduction of Input losses
Establish an efficient and ecologically oriented management of material, resources and energy	Increase of crop yields
Reuse uncontaminated inputs	Increase of food production
Increase species and genetic diversity; contributes to improved, long-term soil productivity and reduced production risks	Increase of food safety
Recycle and reinvest any organic matter, nutrients and energy, to enhance phosphorus availability	
Fairness	
Act fairly with respect to environmental and life opportunities, equity, respect, justice and stewardship	Facilitating best growth conditions, and avoiding any plant losses
Respect animal needs, behavior, and well being that contributes to animal health	Increase of animal production
Care	
Prevent risks by adopting appropriate technologies and rejecting unpredictable ones	Reduction of soil damages, e.g. soil compaction with positive impact on plant growth
Avoid any soil, plant, animal, air, water or food contamination	Reducing food losses over the long-term
Responsible product pricing to assure survival of small scale ecologically sound production units	Increase of accessibility of food for poor population

Source: following IFOAM. 2009. http://www.ifoam.org/about_ifoam/principles/index.html

Table 5.2 The IFOAM Principles and food processing

Impact on food processing	Specific quantitative impact
Health	
Reduce amount of processed food and increase use of fresh products and with that food quality	Reduction of food energy losses while food processing
Protect the health (quality) of products by excluding food additives	Increase of nutrient content
Protect the quality and quantity of products	Increase of the share of non-processed food reducing food waste
Assure healthy products with high nutrient content through the use of natural products	
Ecology	
Establish processing that includes full recycling	Increase of efficient use of food
Closing energy cycles	Increase of recycled nutrients
Exclude of food additives	Increase the efficient use of energy
	Increase of food safety
Fairness	
Assure fair relationships between producer and processor	Increase of consumable food
Foster the use of processing techniques that allow for a diversity of forms, sizes and qualities of plants and animal products	Reduction of food waste
Care	
Use processing technology that minimizes product loss without compromising food quality or quantity	Increase of consumable food

5.4 Food Security and the IFOAM Norms

In this section we introduce how the IFOAM Norms open the space for numerous opportunities for contributing to improved food security. The construct of IFOAM Norms (including the Principles, Standards and issues on certification) is the institution that is steering the status of values for the whole system – for the standards, control and certification.

The question we would like to explore now is: how are the IFOAM Principles expressed and embodied in the three key documents of the IFOAM Norms? The more the three documents embody the Principles, the easier it will be to understand their contribution to addressing concretely the quantitative dimension of food security in the organic agrofood chain.

The IFOAM Norms for production and processing define the basis for the concrete practices by organic movement participants in addition to serving as a framework for specific national and international (e.g., EU, USDA) guidelines for production, processing and trade (IFOAM 2012, pp. 6, 7). These three documents are (IFOAM 2012, p. 6):

- Common Objectives and Requirements of Organic Standards (COROS) – IFOAM Standards Requirements

Table 5.3 The IFOAM Principles and food distribution/marketing

Impact on food distribution/marketing	Specific quantitative impact
Health	
Assure storage and distribution conditions that protect the health and naturalness of products, e.g. through short supply chains	Reduction of food losses
Strictly define the health standards and criteria of products	Increase of the share of healthy food
Ecology	
Foster the efficient resource use of packaging attention to shorter supply chains	Reduction of food losses
Offer non-standardized products for a lower price instead discarding them	Increase of food usability, less food waste
Fairness	
Develop fair relations all along the agro-food supply chain	Increase of food avoiding stock market initiated food losses
Make food more accessible to lower income consumers	Reduce of not marketable high priced food
Account for real environmental and social costs	Increase of food sovereignty
Share "fair" costs and income distribution among all participants, including fair partnership with small-scale farmers and between farmers and consumers	
Care	
Protect health and well-being of current and future generations and the environment	Increase the efficient use of produced food
Decentralize food supply	Increase of food through less transportation risk
Foster availability of regional and seasonal	Increase of food through efficient use of regional and seasonal available food

- IFOAM Standards for Organic Production and Processing
- IFOAM Accreditation Requirements for Bodies Certifying Organic Production and Processing

The responsibility and mission of these documents are as follows (IFOAM 2012):

"The *COROS* articulates the broad objectives which the production rules in organic Standards and regulations commonly seek to achieve, and presents the common detailed requirements that relate to these various objectives. The COROS contains only requirements that were commonly found in organic Standards and Regulations globally (ibid, 13); The COROS is intended for use in international equivalence assessments of organic Standards and Regulations.... it is proposed as a template to guide governments and other stakeholders in conducting objective based equivalence assessments of two or more organic standards or regulations. In the context of the IFOAM Organic Guarantee System, it serves as the IFOAM Standards Requirements: the international reference against which all organic standards and regulations will be assessed against, for the purpose of inclusion in the IFOAM Family of Standards" (ibid, 14);

The *IFOAM Standard* (IS) is an internationally applicable organic standard developed by IFOAM. It is a good, practical interpretation of the IFOAM Standards Requirements

Table 5.4 The IFOAM Principles and food demand

Impact on food demand	Specific quantitative impact
Health	
Foster purchasing of healthy and diverse products	Increase of (bio)diverse production
	Increase of healthy environment
Ecology	
Link demand for meat-based diets to the carrying capacity of defined landscapes	Increase of soil fertility
	Decrease of water consumption for food production
	Increase of food energy
Fairness	
Assure payment for products that includes environmental and social costs	Increase of an ecologically and economically viable and sustainable production
Promote consumption guided by the awareness for farmers' conditions	Increase of food access and food security
Include concerns with equity, respect, justice and stewardship in food consumption practices	
Share limited resources fairly	
Care	
Orient demand with respect to products that are environmental friendly	Increase of the protection of natural resources
Encourage food consumption that contributes to preventive health care	Increase of efficient food consumption

> (Common Objectives and Requirements of Organic Standards), hence belongs to the IFOAM Family of Standards. The IFOAM Standard is an off-the-shelf standard which can be used by those wanting to outsource standard setting and maintenance and see the benefits of sharing the work with others and creating synergies on an international level (ibid, 26);
>
> The *IFOAM Accreditation Requirements* (IAR) establishes requirements for the conduct of organic certification by the certification body, including procedures and practices of the operator that the certification body must verify (ibid, 83)."

In order to become an IFOAM Accredited Certification Body (ACB), organizations must apply either the IFOAM Standards, or a standard that complies with the IFOAM Standards. The IFOAM Standards and the IFOAM Accreditation Requirements (IAR) are used by the International Organic Accreditation Service (IOAS) in the IFOAM accreditation process for organic certification bodies. The IOAS evaluates the standards (used by the certifier) against the IFOAM Standard and certification body performance against the IFOAM Accreditation Requirements.

A first relevant observation is that while the Principles themselves serve for the whole organic agrofood chain, the Norms only address the farmers and processors but not the traders and consumers. This means, that without their inclusion, due to their significant role in organic, there is no instrument that informs and identifies the potential the IFOAM Principles have in feeding the world. The relatively weak integration of consumers and traders in the IFOAM Norms and also in the Principles reduces the potential and options of the organic system to contribute as a whole to food security.

5.4.1 The IFOAM Standards Requirements

The statement of the IFOAM Standard Requirements cover a wide range of production and processing practices introduced in the IFOAM Principles. Several of these offer technical and ecological grounds achieving food security. Remarkably, only one paragraph refers to social aspects (IFOAM 2012, p. 19):

> "10. Fairness, respect and justice, equal opportunities and non-discrimination is afforded to employees and workers: Organic operations in countries where social legislation is not in place or not enforced have social policies in place. Such policies should be in accordance with the International Labor Organization's Declaration on Fundamental Principles and Rights at Work."

A commentary to the paragraph states: "this objective is commonly addressed in private standards although not usually in the scope of government organic standards (ibid, p. 19)." This commentary indicates that little importance is given to this issue. Although the IFOAM Standards Requirements directly address the IFOAM Principles of Health and Ecology, they do not similarly discuss or elaborate upon the Principle of Fairness and Care. As a result, there is no obligation or even recommendation to address social and economic justice.

5.4.2 IFOAM Standards

The IFOAM Standards are fundamental to the quality of organic agriculture and to its claim for sustainability. The Standards elaborate and confirm the contribution of organic to meeting practical ecological and health concerns similar to the IFOAM Standard Requirements. These include a wide range of technical and ecological concerns, such as how to fertilize, the exclusion of chemical pesticides or access and duration of animals to pasture. It also further covers a range of food safety relevant procedures (cf. Hansen et al. 2002).

With respect to social issues, the Standards deal exclusively with worker and labor rights. They identify fairness, respect and justice, equal opportunities and non-discrimination for employees and workers. This objective "is commonly addressed in private standards although not usually in the scope of government organic standards" (IFOAM 2012, p. 62). The social justice objective is defined as: "Social justice and social rights are an integral part of organic agriculture and processing" (Ibid, 62). More specifically, the Standards recommend that: "Operators should positively and actively encourage the collective organization of their employees or contracted smallholders."

International regulations follow the IFOAM Standards. The EU Regulation (EEC) 2092/91 builds significantly on the value elements of the IFOAM Principles Ecology and Health, but is not defined with respect to the Principles of Fairness and Care—specifically in relation to competition, transparency, the prohibition of GMOs, fairness, equity, respect, justice, animal welfare and future generations are not considered (Padel et al. 2007, pp. 22, 23; Padel et al. 2009). Lockie et al. (2006) note that most private organic standards as well, do not codify such values. Similar to the IFOAM Standard Requirements, social and economic justice issues are underrepresented. Also the IFOAM Standards do not address the broader meaning of the Principles of Fairness and Care. Thus the essence of the Principles of Fairness and Care gets lost in the IFOAM Standards. This not only weakens the impact of organic on food security, but also food sovereignty.

5.4.3 *The IFOAM Accreditation Requirements*

The third part of IFOAM Norms address the "IFOAM Accreditation Requirements for bodies certifying organic production and processing" (IFOAM 2012, p. 1). This deals with the requirements for certification bodies themselves, different types of certification and what we call the responsibility of the certifiers in the certification process. The issue of concern is how the control and certification procedures embody the IFOAM Principles ethics and in this way address the issues of food security.

The content of control and certification of production and processing practices refers to the fulfillment of the respective national or international regulations in line with IFOAM Standards. The control process assesses the conformity of diverse interpretations of regulations in the production and processing.

The control and certification process focus on "how" an actor in the organic agrofood chain fulfills specified technical standards. Therefore, control does not directly refer to the IFOAM Principles. It addresses ethical issues only indirectly if they are part of IFOAM Standards or related regulations (see Chap. 13). Consistent with the IFOAM Accreditation Requirements and Standards, Principles of Health and Ecology are central in the control process while Fairness and Care are not. In short, the absence of attention to these ethical concerns weakens the ability to apply the Principles in a concrete way to thinking about "feeding the world."

5.5 Discussion and Conclusions

In addressing if the IFOAM Principles provide an ethical framework that if applied in practice, would positively contribute to food security, we explored the relevance of the IFOAM Principles to food supply, processing, distribution/marketing and food demand. We further examined the three IFOAM Norms documents "IFOAM

Standards Requirements, IFOAM Standards and IFOAM Accreditation Requirements" too understand how far they refer to the Principles and further transport the identified potential for food security towards the organic agrofood chain.

Obviously, the IFOAM Principles provide a coherent ethical framework for setting priorities towards food security. The specific focus is on resource efficiency, long-term soil fertility, and health of soils, plants and animals. Through the Principles, especially Health and Ecology, individuals and collectives along the organic agrofood chain are invited to contribute to food security. The Principles of Fairness and Care address social and economic relevant issues affecting food access, food demand and specifically food sovereignty. Specifically these Principles broaden the perspective beyond a pure technical interpretation of food security.

The Principles offer a concrete ethical orientation to engage in food security for the producers, but less so for processors, distributors or consumers; they place special emphasis on the balance between humans and nature, instead of the permanent growth of wealth of individuals. They represent a systemic and holistic ethical approach that requires a deep "process of systemic change" (Pimbert 2009) by all actors and stakeholders in the organic agrofood system, when they enter the organic system.

The Principles offer several ethical formulations which have widespread consequences for the quantification of food supply, demand, processing, distribution/marketing and finally food security, if put into practice. In the IFOAM Norms ecological, health and technical criteria dominate. IFOAM Standard Requirements and Standards focus on providing the best quality of secure and healthy food while protecting the environment and natural resources. Together, the Principles of ecology and health are indicative of a positive quantitative influence on food security. However, ethics addressing non-technical values on social and economic justice that are highly relevant for food access and food sovereignty, are excluded. "The difficulty is that incorporating these wider concerns [social issues] into definitions of, and standards for organic farming is problematical. Standards are far more able to refer to prohibited inputs than to deal with precise criteria for the assessment of whether producers and processors are acting in a manner which is socially just or ecologically responsible" (Rigby and Cáceres 2001, p. 27). IFOAM Accreditation Requirements secure the compliance of IFOAM Standards and a fair control and certification process. From that particular perspective, there is no specific additional contribution to food security than that of securing the ecological and technical standards.

With respect to food security, certification and smallholder farmers, the argument is the more smallholders have access to knowledge and markets –, e.g., through Participatory Guarantee Systems (PGS), to become sovereign for their food the more they are enabled to contribute to food security (Zundel and Kilcher 2007). The PGS serves as an excellent example of bringing the Principles of Fairness and Care into practice. Not only is PGS oriented to the application of system knowledge and learning, it brings the consumer into the certification process, and thus offers a means for creating more trust and shared responsibility in the whole food system. PGS also strengthens farmers' production through low cost access to knowledge and information, supports the access to local markets and eventually international markets, and contributes to food access for consumers by strengthening local market development (see Sect. 13.2).

The relatively weak integration of consumers in the IFOAM Norms reduces the potential and options of the organic agrofood chain to contribute as a whole to the food security issue. An example can be made of high-income countries, in that if a regional, seasonal and less meat oriented production (Halberg et al. 2006b) is not "honored" through corresponding consumer food habits, the agro-ecological approach of organic is not efficient enough to feed the world (Erb et al. 2009), a statement which simply cannot be ignored.

The future relevance of the IFOAM Principles for food security and food sovereignty will depend on their awareness from the whole organic agrofood chain. The IFOAM Principles provide an ethical orientation to Health and Ecology that contribute to food security and are reflected in the IFOAM Standards and part of a certification system. To meet the goal of food security and food sovereignty and if the food issue is seen as a common societal challenge, there is need for more exposure of the IFOAM Principles to processors, traders and specifically consumers. Bringing social and economic justice into the center of the IFOAM agenda and specifically of the IFOAM Standards is a further precondition for organics serious contribution for feeding the world, specifically when it comes to food sovereignty.

References

Abouleish, H. 2013. SEKEM–Greening the desert. In *Linking science to society*, ed. M. Yahia, M. El-Faham, M. El-Wakil, and I. Serageldin, 31–77. Alexandria: Bibliotheca Alexandrina.

AGree. 2012. Facing the future: Critical challenges to food and agriculture. Retrieved from http://www.foodandagpolicy.org/sites/default/files/Facing_the_Future.pdf. Accessed 1 Jan 2013.

Aiking, H. 2011. Future protein supply. *Trends in Food Science & Technology* 22(2): 112–120.

Altieri, M.A., and V.M. Toledo. 2011. The agroecological revolution in Latin America: Rescuing nature, ensuring food sovereignty and empowering peasants. *Journal of Peasant Studies* 38(3): 587–612.

Arbenz, M. 2011. Creating power through statistical evidence: The organic alternative. In *The world of organic agriculture*, ed. H. Willer and L. Kilcher. Bonn/Frick: IFOAM/FIBL.

Atkinson, D., C.A. Watson, J.E. Hooker, and K.E. Black. 1995. Nutrient conservation in 16 organic crop production systems. In *Soil management in sustainable agriculture*, vol. 17, ed. H.F. Cook and H.C. Lee, 54–60. Ashford: Wye College Press.

Badgley, C., J. Moghtader, E. Quintero, E. Zakem, M.J. Chappell, K. Aviles-Vazquez, and I. Perfecto. 2007. Organic agriculture and the global food supply. *Renewable Agriculture and Food Systems* 22(2): 86–108.

Barham, E. 2002. Towards a theory of values-based labeling. *Agriculture and Human Values* 19: 349–360.

Barnett, C., P. Cloke, N. Clarke, and A. Malpass. 2005. Consuming ethics: Articulating the subjects and spaces of ethical consumption. *Antipode* 37(1): 23–45.

Barr, S. 2003. Strategies for sustainability: Citizens and responsible environmental behaviour. *Area* 35(2): 227–240.

Bartley, T. 2007. Institutional emergence in an era of globalization: The rise of transnational private regulation of labor and environmental conditions. *American Journal of Sociology* 113(2): 297–351.

Beus, C.E., and R.E. Dunlap. 1990. Conventional versus alternative agriculture: The paradigmatic roots of the debate. *Rural Sociology* 55(4): 590–616.

Blowfield, M. 1999. Ethical trade: A review of developments and issues. *Third World Quarterly* 20(4): 753–770.

Borron, S. 2006. *Building resilience for an unpredictable future: How organic agriculture can help farmers adapt to climate change*. Rome: Food and Agriculture Organization of the United Nations.

Breman, H., J. Groot, and H. van Keulen. 2001. Resource limitations in Sahelian agriculture. *Global Environmental Change* 11(1): 59–68.

Carlsson-Kanyama, A., and A.D. González. 2009. Potential contributions of food consumption patterns to climate change. *The American Journal of Clinical Nutrition* 89(5): 1704–1709.

Connor, D.J. 2008. Organic agriculture cannot feed the world. *Field Crops Research* 106(2): 187–190.

Cordell, D., J.-O. Drangert, and S. White. 2009. The story of phosphorus: Global food security and food for thought. *Global Environmental Change* 19(2): 292–305.

Cotula, L. 2009. *Land grab or development opportunity?: Agricultural investment and international land deals in Africa*. London/Rome: IIED/FAO/IFAD.

Dangour, A.D., S.K. Dodhia, A. Hayter, E. Allen, K. Lock, and R. Uauy. 2009. Nutritional quality of organic foods: A systematic review. *The American Journal of Clinical Nutrition* 90(3): 680–685.

Daniel, S. 2011. Land grabbing and potential implications for world food security. In *Sustainable agricultural development – Recent approaches in resources management and environmentally-balanced production enhancement*, ed. M. Behnassi, S. Shabbier, and J. D'Silva, 25–42. Dordrecht/Heidelberg/London/New York: Springer.

de Ponti, T., B. Rijk, and M.K. van Ittersum. 2012. The crop yield gap between organic and conventional agriculture. *Agricultural Systems* 108: 1–9.

Delgado, C.L. 2003. Rising consumption of meat and milk in developing countries has created a new food revolution. *The Journal of Nutrition* 133(11): 3907–3910.

Delgado, C.L. 2005. Rising demand for meat and milk in developing countries: Implications for grasslands-based livestock production. In *Grassland: A global resource*, ed. D.A. McGilloway, 29–39. Wageningen: Wageningen Academic Publishers.

Dietz, T., E.A. Rosa, and R. York. 2007. Driving the human ecological footprint. *Frontiers in Ecology and the Environment* 5(1): 13–18.

Drewnowski, A., and B.M. Popkin. 1997. The nutrition transition: New trends in the global diet. *Nutrition Reviews* 55(2): 31–43.

Duchin, F. 2005. Sustainable consumption of food: A framework for analyzing scenarios about changes in diets. *Journal of Industrial Ecology* 9(1–2): 99–114.

Edwards, S. 2007. The impact of compost use on crop yields in Tigray, Ethiopia. Paper presented at the Institute for Sustainable Development (ISD). In *Proceedings of the international conference on organic agriculture and food security*, Rome.

Edwards, S., T.B.G. Egziabher, and H. Araya. 2010. *Successes and challenges in ecological agriculture: Experiences from Tigray, Ethiopia*. Rome: Food and Agriculture Organization of the United Nations.

Ehrlich, P.R. 1968. *The population bomb: Population control or race to oblivion*. New York: Ballantine.

Ehrlich, P.R., and A.H. Ehrlich. 2013. Can a collapse of global civilization be avoided? *Proceedings of the Royal Society B: Biological Sciences* 280(1754).

Erb, K.H., H. Haberl, F. Krausmann, C. Lauk, C. Plutzar, J.K. Steinberger, C. Müller, A. Bondeau, K. Waha, and G. Pollack. 2009. Eating the planet. Feeding and fuelling the world sustainably, fairly and humanely–a scoping study. *Commissioned by Compassion in World Farming and*

Friends of the Earth UK. Institute of Social Ecology and PIK Potsdam. Vienna, Austria and Potsdam, Germany: Social Ecology Working Paper, 116.

Food Ethic Council. 2012. Food security. Retrieved from http://www.foodethicscouncil.org/topic/Food%2Bsecurity. Accessed 31 Jan 2012.

Freyer, B. 2007. Traditional and ecological farming systems in (sub) tropical countries–history, interactions and future perspectives. *Journal of Agriculture and Rural Development in the Tropics and Subtropics* 89: 1–30.

Fromartz, S. 2007. *Organic, Inc.: Natural foods and how they grew*. Orlando: Houghton Mifflin Harcourt.

Galloway, J.N., A.R. Townsend, J.W. Erisman, M. Bekunda, Z. Cai, J.R. Freney, and M.A. Sutton. 2008. Transformation of the nitrogen cycle: Recent trends, questions, and potential solutions. *Science* 320(5878): 889–892.

Gerbens-Leenes, P., and S. Nonhebel. 2002. Consumption patterns and their effects on land required for food. *Ecological Economics* 42(1): 185–199.

Giovannucci, D. 2005. *Organic agriculture and poverty reduction in Asia: China and India focus*. Rome: IFAD Office of Evaluation.

Godfray, H.C.J., J.R. Beddington, I.R. Crute, L. Haddad, D. Lawrence, J.F. Muir, and C. Toulmin. 2010. Food security: The challenge of feeding 9 billion people. *Science* 327(5967): 812–818.

Goolsby, D.A., and W.A. Battaglin. 2000. Nitrogen in the Mississippi Basin-estimating sources and predicting flux to the Gulf of Mexico. *USGS Fact Sheet*, 135–100.

Government Office for Science. 2011. The future of food and farming *Final project report*. London: The Government Office for Science.

Grievink, J.W. 2002. The changing face of the global food industry. Retrieved from http://webdomino1.oecd.org/comnet/agr/foodeco.nsf/viewHtml/index/$FILE/GrievinkPPT.pdf. Accessed 1 Oct 2005.

Gurian-Sherman, D. 2009. *Failure to yield: Evaluating the performance of genetically engineered crops*. Cambridge, MA: Union of Concerned Scientists.

Gustavsson, J., C. Cederberg, U. Sonesson, R. Van Otterdijk, and A. Meybeck. 2011. *Global food losses and food waste*. Rome: Food and Agriculture Organization of the United Nations.

Halberg, N., H.F. Alrøe, M. Trydeman Knudsen, and E.S. Kristensen. 2006a. *Global development of organic agriculture. Challenges and prospects*. Wallingford: CAB International.

Halberg, N., T.B. Sulser, H. Høgh-Jensen, M.W. Rosegrant, and M. Trydeman Knudsen. 2006b. The impact of organic farming on food security in a regional and global perspective. In *Global development of organic agriculture: Challenges and prospects*, ed. N. Halberg, H.F. Alrøe, M. Trydeman Knudsen, and E.S. Kristensen, 277–322. Wallingford: CAB International.

Hansen, B., H.F. Alrøe, E.S. Kristensen, and M. Wier. 2002. *Assessment of food safety in organic farming*, vol. 52. Tjele: DARCOF Working Papers.

Heaton, S. 2001. *Organic farming, food quality and human health: A review of the evidence*. Bristol: Soil Association United Kingdom.

Hendrickson, M.K., and H.S. James. 2005. The ethics of constrained choice: How the industrialization of agriculture impacts farming and farmer behavior. *Journal of Agricultural and Environmental Ethics* 18(3): 269–291.

Hole, D., A. Perkins, J. Wilson, I. Alexander, P. Grice, and A. Evans. 2005. Does organic farming benefit biodiversity? *Biological Conservation* 122(1): 113–130.

Humphrey, J., and O. Memedovic. 2006. *Global value chains in the agrifood sector*. Vienna: UNIDO.

Hunt, L., C. Rosin, J. Fairweather, H. Moller, and Others in ARGOS. 2009. *Doing the unthinkable: Is breadth of view important in bringing innovation to organic practices?* Paper presented at the organic sector conference, University of Waikato, Waikato.

Hunt, L., J. Benge, H. Campbell, J.R. Fairweather, G. Greer, D. Lucock, J. Manhire, H. Moller, J.R. Reid, C. Rosin, and C. Saunders. 2010. *Experiences of transdisciplinarity in research on agricultural sustainability*. Paper presented at the 9th European IFSA symposium, Vienna.

Hunt, L., J. Fairweather, C. Rosin, H. Campbell, D. Lucock, and G. Greer. 2011. *Doing the unthinkable: Linking farmer's breadth of view and adaptive propensity to the achievement of social, environmental and economic outcomes*. Paper presented at the 18th International Farm

Management Association Congress 'Thriving in a global market: Innovation, co-operation and leadership', Canterbury.
IFAD. 2005. Organic agriculture and poverty reduction in Asia: China and India focus, vol. Report No. 1664. Rome: IFAD.
IFOAM. 2009. The principles of organic agriculture. Retrieved from http://www.ifoam.org/about_ifoam/principles/index.html. Accessed 1 Sept 2011.
IFOAM. 2012. *The IFOAM norms for organic production and processing*. Retrieved from http://www.ifoam.org/en/ifoam-norms. Accessed 1 Jan 2013.
Imhoff, M.L., L. Bounoua, T. Ricketts, C. Loucks, R. Harriss, and W.T. Lawrence. 2004. Global patterns in human consumption of net primary production. *Nature* 429(6994): 870–873.
Kachika, T. 2010. Land grabbing in Africa: A review of the impacts and the possible policy responses *Pan Africa Programme of Oxfam International*. Oxford: Oxfam International.
Kader, A.A. 2004. *Increasing food availability by reducing postharvest losses of fresh produce.* Paper presented at the V international postharvest symposium 682, Verona.
Khaledi, M., R. Gray, S. Weseen, and E. Sawyer. 2007. *Assessing the barriers to conversion to organic farming: An institutional analysis*. Saskatoon: Department of Agricultural Economics, University of Saskatchewan.
Kirchmann, H., L. Bergström, T. Kätterer, O. Andrén, and R. Andersson. 2008. Can organic crop production feed the world? In *Organic crop production–Ambitions and limitations*, ed. H. Kirchmann and L. Bergström, 39–72. Dordrecht: Springer.
LaSalle, T., P. Hepperly, and A. Diop. 2008. The organic green revolution. 16. Retrieved from http://classic.ifoam.org/growing_organic/1_arguments_for_oa/food_security/pdfs/RodaleInst_TheOrganicGreenRevolution.pdf. Accessed 1 Jan 2013.
Lebensministerium. 2012. *Grüner Bericht*, vol. 53. Vienna: Die Republik Österreich, vertreten durch den Bundesminister für Land- und Forstwirtschaft, Umwelt und Wasserwirtschaft.
Lewin, B., D. Giovannucci, and P. Varangis. 2004. *Coffee markets: New paradigms in global supply and demand*, World bank agriculture and rural development discussion paper(3). Washington, DC: Agriculture and Rural Development.
Lockie, S., K. Lyons, G. Lawrence, and D. Halpin. 2006. *Going organic: Mobilizing networks for environmentally responsible food production*. Wallingford: CAB International.
Lotter, D., R. Seidel, and W. Liebhardt. 2003. The performance of organic and conventional cropping systems in an extreme climate year. *American Journal of Alternative Agriculture* 18(3): 146–154.
Lunneryd, D. 2003. *Unique decision making with focus on information use: The case of converting to organic milk production.* Dissertation, Swedish University of Agricultural Sciences Uppsala, Department of Economics.
Mäder, P., A. Fliessbach, D. Dubois, L. Gunst, P. Fried, and U. Niggli. 2002. Soil fertility and biodiversity in organic farming. *Science* 296(5573): 1694–1697.
Malhi, Y., J.T. Roberts, R.A. Betts, T.J. Killeen, W. Li, and C.A. Nobre. 2008. Climate change, deforestation, and the fate of the Amazon. *Science* 319(5860): 169–172.
Mancino, L. 2005. Going with the grain: Consumers responding to new dietary guidelines. *Amber Waves* 3(5): 4–5.
Martin, R.C., D.H. Lynch, B. Frick, and P. van Straaten. 2007. Phosphorus status on Canadian organic farms. *Journal of the Science of Food and Agriculture* 87(15): 2737–2740.
Martínez-González, M.A., E. Fernández-Jarne, M. Serrano-Martínez, A. Marti, J.A. Martinez, and J.M. Martín-Moreno. 2002. Mediterranean diet and reduction in the risk of a first acute myocardial infarction: An operational healthy dietary score. *European Journal of Nutrition* 41(4): 153–160.
McAlpine, C.A., A. Etter, P.M. Fearnside, L. Seabrook, and W.F. Laurance. 2009. Increasing world consumption of beef as a driver of regional and global change: A call for policy action based on evidence from Queensland (Australia), Colombia and Brazil. *Global Environmental Change* 19(1): 21–33.
McIntyre, B.D., H.R. Herren, J. Wakhungu, and R.T. Watson (eds.). 2009. *IAASTD international assessment of agricultural knowledge, science and technology for development: Global report.* Washington DC: Island Press.

McMichael, P. 2009. Food sovereignty, social reproduction and the agrarian question. In *Peasants and globalization: Political economy, rural transformation and the agrarian question*, vol. 2, ed. A.H. Akram-Lodhi and C. Kay, 288–313. New York: Routledge.

McMichael, P. 2012. The land grab and corporate food regime restructuring. *Journal of Peasant Studies* 39(3–4): 681–701.

McMichael, A.J., J.W. Powles, C.D. Butler, and R. Uauy. 2007. Food, livestock production, energy, climate change, and health. *The Lancet* 370(9594): 1253–1263.

McNeely, J.A., and S.J. Scherr. 2003. *Ecoagriculture: Strategies to feed the world and save wild biodiversity*. Washington, DC: Island Press.

Meat & Wool New Zealand. 2010. *Compendium of New Zealand farm facts*. Wellington: Wellington Meat & Wool New Zealand Economic Service.

Miller, P., and N. Rose. 1997. Mobilizing the consumer: Assembling the subject of consumption. *Theory, Culture, and Society* 14(1): 1–36.

Mittal, A. 2006. Food security: Empty promises of technological solutions. *Development* 49(4): 33–38.

Morgan, K. 2010. Local and green, global and fair: The ethical foodscape and the politics of care. *Environment and Planning A* 42(8): 1852.

Nellemann, C. 2009. *The environmental food crisis: The Environment's role in averting future food crises: A UNEP rapid response assessment*. Arendal: UNEP.

Niggli, U., A. Slabe, O. Schmid, N. Halberg, and M. Schlüter. 2008. *Vision for an organic food and farming research agenda 2025. Organic knowledge for the future*. Brüssel/Bonn: IFOAM/ISOFAR.

Oliveira, G. 2013. Land regularization in Brazil and the global land grab. *Development and Change* 44(2): 261–283.

Onwonga, R., J. Lelei, B. Freyer, J. Friedel, S. Mwonga, and P. Wandhawa. 2008. Low cost technologies for enhancing N and P availability and maize (Zea mays L.) performance on acid soils. *World Journal of Agricultural Sciences* 4: 862–873.

Paarlberg, R. 2009. The ethics of modern agriculture. *Society* 46(1): 4–8.

Padel, S. 2002. *Conversion to organic milk production: The change process and farmers' information needs*. Dissertation, University of Wales, Aberystwyth.

Padel, S., H. Röcklinsberg, H. Verhoog, H. Fjelsted Alrøe, J. De Wit, C. Kjeldsen, and O. Schmid. 2007. *Balancing and integrating basic values in the development of organic regulations and standards: Proposal for a procedure using case studies of conflicting areas*. Aberyswyth/Tjele: Danish Research Centre for organic Farming (DARCOF).

Padel, S., H. Röcklinsberg, and O. Schmid. 2009. The implementation of organic principles and values in the European regulation for organic food. *Food Policy* 34(3): 245–251.

Pimbert, M.P. 2009. *Towards food sovereignty*. London: IIED.

Pimentel, D., P. Hepperly, J. Hanson, D. Douds, and R. Seidel. 2005. Environmental, energetic, and economic comparisons of organic and conventional farming systems. *BioScience* 55(7): 573–582.

Pinstrup-Andersen, P. 2009. Food security: Definition and measurement. *Food Security* 1(1): 5–7.

Pinstrup-Andersen, P., R. Pandya-Lorch, and M.W. Rosegrant. 1999. World food prospects: Critical issues for the early twenty-first century *Food policy report*. Washington, DC: International Food Policy Research Institute.

Pollan, M. 2006. *The omnivore's dilemma: A natural history of four meals*. New York: Penguin.

Posner, J.L., J.O. Baldock, and J.L. Hedtcke. 2008. Organic and conventional production systems in the Wisconsin integrated cropping systems trials: I. Productivity 1990–2002. *Agronomy Journal* 100(2): 253–260.

Pretty, J. 2003. Social capital and the collective management of resources. *Science* 302(5652): 1912–1914.

Pretty, J. 2006. Agroecological approaches to agricultural development *Background Paper for the World Development Report 2008*, 1–37. Santiago: RIMSIP.

Pretty, J. 2008. Agricultural sustainability: Concepts, principles and evidence. *Philosophical Transactions of the Royal Society, B: Biological Sciences* 363(1491): 447–465.

Pretty, J., and R. Hine. 2001. *Reducing food poverty with sustainable agriculture: A summary of new evidence*. Colchester: University of Essex.

Pretty, J., C. Brett, D. Gee, R. Hine, C. Mason, J. Morison, and G. Van der Bijl. 2000. An assessment of the total external costs of UK agriculture. *Agricultural Systems* 65(2): 113–136.

Pretty, J., A. Noble, D. Bossio, J. Dixon, R. Hine, F. Penning de Vries, and J. Morison. 2006. Resource-conserving agriculture increases yields in developing countries. *Environmental Science & Technology* 40(4): 1114–1119.

Pretty, J., W.J. Sutherland, J. Ashby, J. Auburn, D. Baulcombe, M. Bell, and J. Burke. 2010. The top 100 questions of importance to the future of global agriculture. *International Journal of Agricultural Sustainability* 8(4): 219–236.

Renard, M.C. 2003. Fair trade: Quality, market and conventions. *Journal of Rural Studies* 19(1): 87–96.

Rigby, D., and D. Cáceres. 2001. Organic farming and the sustainability of agricultural systems. *Agricultural Systems* 68(1): 21–40.

Rockström, J., M. Lannerstad, and M. Falkenmark. 2007. Assessing the water challenge of a new green revolution in developing countries. *Proceedings of the National Academy of Sciences* 104(15): 6253–6260.

Rosset, P. 2003. Food sovereignty: Global rallying cry of farmer movements. *Food First Backgrounder* 9(4): 1–4.

Runge, C.F., B. Senauer, P.G. Pardey, and M.W. Rosegrant. 2003. *Ending hunger in our lifetime: Food security and globalization*. Baltimore: Johns Hopkins University Press.

Scialabba, N., and C. Hattaam. 2002. *Organic agriculture, environment and food security*, vol. 4. Rome: Food and Agriculture Organization of the United Nations.

Scialabba, N., and M. Müller-Lindenlauf. 2010. Organic agriculture and climate change. *Renewable Agriculture and Food Systems* 25(02): 158–169.

Searchinger, T., R. Heimlich, R.A. Houghton, F. Dong, A. Elobeid, J. Fabiosa, and T.-H. Yu. 2008. Use of US croplands for biofuels increases greenhouse gases through emissions from land-use change. *Science* 319(5867): 1238–1240.

Setboonsarng, S. 2006. *Organic agriculture, poverty reduction, and the millennium development goals.* Paper presented at the international workshop on sufficiency economy, poverty reduction, and the MDGs – Organized under the umbrella of the Exposition of Sufficiency Economy for Sustainable Development, Bangkok, Thailand.

Seufert, V., N. Ramankutty, and J.A. Foley. 2012. Comparing the yields of organic and conventional agriculture. *Nature* 485(7397): 229–232.

Smil, V. 2002. Worldwide transformation of diets, burdens of meat production and opportunities for novel food proteins. *Enzyme and Microbial Technology* 30(3): 305–311.

Speedy, A.W. 2003. Global production and consumption of animal source foods. *The Journal of Nutrition* 133(11): 4048S–4053S.

Starke, L. (ed.). 2011. *State of the world. Innovations that nourish the planet*. Washington, DC: The Worldwatch Institute.

Stoate, C., N. Boatman, R. Borralho, C. Carvalho, Gd. Snoo, and P. Eden. 2001. Ecological impacts of arable intensification in Europe. *Journal of Environmental Management* 63(4): 337–365.

Stolze, M., A. Piorr, A. Häring, and S. Dabbert. 2000. The environmental impacts of organic farming in Europe. In *Organic farming in Europe: Economics and policy*, vol. 6, ed. S. Dabbert. Stuttgart-Hohenheim: University of Hohenheim.

Tanner, B. 2000. Independent assessment by TPC bodies. *Food Control* 11: 415–417.

Tegtmeier, E.M., and M.D. Duffy. 2004. External costs of agricultural production in the United States. *International Journal of Agricultural Sustainability* 2(1): 1–20.

Thompson, R. 2008. Implications of increasing food prices for agricultural and energy policy. Retrieved from http://www.chicagofed.org/digital_assets/others/events/2008/agricultural_markets_and_food_price_inflation/presentation_implications_increasing_food_prices.pdf Accessed 1 Jan 2013.

Tilman, D., K.G. Cassman, P.A. Matson, R. Naylor, and S. Polasky. 2002. Agricultural sustainability and intensive production practices. *Nature* 418(6898): 671–677.

Trewavas, A. 2001. Urban myths of organic farming. *Nature* 410(6827): 409–410.
Trigo, E.J., and E.J. Cap. 2004. The impact of the introduction of transgenic crops in Argentinean agriculture. *AgBioForum – Journal of Agrobiotechnology Management & Economics* 6(1).
Trostle, R. 2008. *Global agricultural supply and demand: Factors contributing to the recent increase in food commodity prices*. Washington, DC: United States Department of Agriculture, Economic Research Service.
Tscharntke, T., Y. Clough, T.C. Wanger, L. Jackson, I. Motzke, I. Perfecto, and A. Whitbread. 2012. Global food security, biodiversity conservation and the future of agricultural intensification. *Biological Conservation* 151: 53–59.
United States Department of Agriculture. 2009, 12/08/2012. Healthy Incentives Pilot. Retrieved from http://www.fns.usda.gov/SNAP/HIP/qa-s.htm. Accessed 31 Jan 2012.
Van Bruggen, A.H.C. 1995. Plant disease severity in high-input compared to reduced-input and organic farming systems. *Plant Disease* 79(10): 976–984.
van Solinge, T.B. 2010. Deforestation crimes and conflicts in the Amazon. *Critical Criminology* 18(4): 263–277.
Volpi, G. 2007. Climate mitigation, deforestation and human development in Brazil *Human development report*, vol. 2008. New York: UNDP.
Wahl, P. 2009. *Food speculation: The main factor of the price bubble in 2008*. Berlin: WEED Weltwirtschaft, Ökologie & Entwicklung.
Walters, R. 2006. Crime, bio-agriculture and the exploitation of hunger. *British Journal of Criminology* 46(1): 26–45.
Wang, Y., M.A. Beydoun, L. Liang, B. Caballero, and S.K. Kumanyika. 2008. Will all Americans become overweight or obese? Estimating the progression and cost of the US obesity epidemic. *Obesity* 16(10): 2323–2330.
Willer, H., and L. Kilcher. 2011. *The world of organic agriculture: Statistics and emerging trends 2011*. Bonn/Frick: IFOAM/FIBL.
Willett, W.C., F. Sacks, A. Trichopoulou, G. Drescher, A. Ferro-Luzzi, E. Helsing, and D. Trichopoulos. 1995. Mediterranean diet pyramid: A cultural model for healthy eating. *The American Journal of Clinical Nutrition* 61(6): 1402S–1406S.
Winter, M., and M. Lobley. 2009. Conclusions: The emerging contours of the new land debate. In *What is land for?: "The food, fuel and climate change debate"*, ed. M. Winter and M. Lobley. London/Sterling: Earthscan.
World Bank, & Commission on Growth. 2008. *The growth report: Strategies for sustained growth and inclusive development*. Washington, DC: World Bank.
Zacune, J. 2011. Who benefits from GM crops? An industry built on myths. Retrieved from http://www.google.at/url?sa=t&rct=j&q=&esrc=s&source=web&cd=1&ved=0CCwQFjAA&url=http%3A%2F%2Fwww.foei.org%2Fen%2Fresources%2Fpublications%2Fpdfs%2F2011%2Fwho-benefits-from-gm-crops-2011&ei=D7qTUsGxCIGD4ASc_oCgCA&usg=AFQjCNH-xUCuLCwUJsFIaidMg58UHXRU_g&sig2=LGomhK-YPrXb7RBQuD5B-Q&bvm=bv.57127890,d.bGE. Accessed 1 Jan 2013.
Zanoli, R., D. Gambelli, and D. Vairo. 2000. Organic farming in Europe by 2010: Scenarios for the future. In *Organic farming in Europe: Economics and policy*, vol. 8, ed. S. Dabbert. Stuttgart-Hohenheim: University of Hohenheim.
Ziesemer, J. 2007. *Energy use in organic food systems*. Rome: Food and Agriculture Organization of the United Nations.
Znaor, D., J. Pretty, S. Todorovic, and J. Morison. 2007. *Impact of organic agriculture on the environment and economic performance of Croatia*. Paper presented at the international conference on organic agriculture and food security, Rome.
Zundel, C., and L. Kilcher. 2007. Organic agriculture and food availability. Retrieved from http://orgprints.org/10753/1/zundel-kilcher-2007-food-availability.pdf. Accessed 1 Jan 2013.

Part III
Markets and Consumers

Overview and Observations

The next set of chapters examines organic ethical perspectives through market and consumer relationships. Two chapters specifically deal with labeling, or how values are expressed in product labels used in the EU and the US. The final chapter addresses market and consumer perspectives in the Israeli organic movement and how globalization affects the movement.

Ginevra Adamoli's 'Who came first? The egg or the carton?' An analysis of organic packaging for Italian retail chain Esselunga' discusses the implications of labeling of organic egg cartons, designed by one of the largest Italian supermarket chains, for the citizen-consumer. The hybrid citizen-consumer (Johnston 2008) is a concept implying the social practice of voting with your Euro or Dollar. It also describes a concept in which citizenship is revived through consumerism and expressed in 'life-politics'. This concept can interestingly enough, satisfy dual ideologies of citizenship (collective responsibility) and consumerism (self-interest). The author describes how the case of egg carton labels appears to respond to both the individual and the societal/collective view of the consumer.

These competing ideologies about the consumer have led to questions about consumer agency and specifically consumer power within the organic market. In this case, the implications for the citizen-consumer shopping organically are based both on the involvement of the state in regulating packaging as well as green marketing strategies. Neither action is necessarily that informative for the average consumer. The wealth of information that is put on the package, and the labels are less than transparent since they assume that the consumer knows what they mean.

What attracts and motivates consumers to buy organic, and which corresponding IFOAM Principles apply to consumer values and behavior? Kristin Getter et al., in 'Increasing Demand for Pasture-Based Dairy: What Attributes and Images Do Consumers Want?' studied consumer values. They created and tested promotional messages with potential consumers to investigate current perceptions and attitudes about milk and its attributes. This research was motivated by earlier studies showing that many farmers lack effective promotional messages for milk products, thus creating a significant barrier to marketing.

This study focused on Michigan consumers and what they value most about milk products. The three most valued attributes of milk products were identified as: organic, local, and grass-fed, which were linked primarily to beliefs in improved human health; animal welfare; and, environmental benefit. These values are clearly connected to the IFOAM Principles, even though the authors did not discuss the relationship.

Getter et al. provide an example of how self-interested reasoning prevails over altruism in the profile of consumers. This could also be described as a model of hybrid consumer-citizenship. Beyond individual consumer perceptions of, and interests in organic, local, and grass-fed dairy for personal health, consumers also valued the positive impact on animal welfare as well as the environmental benefits of pasturing. This strengthens the observation that self-interested behavior does not necessarily preclude a collective ecological or social benefit. In other words, consumers share values with the farmers that are fundamental in the IFOAM Principles (see Schösler et al. 2012).

Both chapters beg questions about the societal influence of the organic ethical principles. Finding a way to present the principles as an ethical framework for consumers and producers alike could unite otherwise fragmented discussions such as these.

Rafi Grosglik's 'Post-National Organic and the Field of Organic Food in Israel' acknowledges the centrality of globalization, not consumer values, in organic marketing in Israel. He describes the evolution of organic food in Israel to reveal the global, cultural and economic conditions that define the organic marketplace in this country. Two trends characterize this process. The first involves the 'commodification of organic food' and is expressed through the widespread export of organic products, as well as the appearance of upscale organic supermarkets. The second includes the acceptance of anti-global practices, many of which are also global phenomenal, such as the spread of community-supported agriculture.

Grosglik's differentiation does not offer a complete picture of current market trends. But we feel that distinguishing between four existing market types helps illustrate how diverse types link differently to the IFOAM principles. In the regionally oriented market type, values play an important role in the relations between farmers and consumers; in contrast, the national market is characterized by anonymous farmer consumer relationships. The global organic market is divided into food chains that follow conventional market approaches and that are characterized by fair trade related value-based relationships between farmers and consumers. The national and international conventionally oriented market type excludes the IFOAM principles of fairness and care, but might meet the principles of health and ecology principles, with some limitations (e.g., food miles). The regional and international fair trade oriented approaches promise to meet the principle of fairness and care in addition to health and ecology. However, the dynamics of globalization also foster possibilities for broadening the organic social movement by means of communicating ideas, information and examples of local organizing such as farmers' markets, CSAs and cooperatives. In these instances, IFOAM's role for integrating ethics into both 'branches' (regional/international) of organic becomes increasingly relevant.

As Grosglik argue, the multiplicity of ideals, images and representations that are associated with organic food in Israel, turns organic into a carrier of different (and even contradictory) post-modernist positions. Israeli organic food has created a new consumer-cultural discourse that is fundamentally different from that of previous agricultural discourse. This discourse connects with the process of the conventionalization and commodification of organic through which organic becomes a post-Fordist commodity. At issue are the ways in which IFOAM can address the ethics of the multiple organics.

Each of the chapters address the relationship between markets and consumers, and the values brought to this relationship. As Adamoli observes, the values of the consumers are often reflected in the purchasing choices. Her study helps to illustrate the decisions of organic, hybrid citizen-consumers. In Michigan, particular images invoked customer's preferences and values that are in line with the principles of health, ecology and to some extent fairness; but they do not evoke ideas related to care. The Israeli case illustrates an additional dimension in which different consumer values are seen in the context of global commodification.

Whatever might be the preferred values of consumers, organic is bound to multiple values. When looked at from the perspective of the IFOAM principles of Health, Ecology, Fairness and Care, one sees how the individual values intertwine. This means that even if the main motivation for a consumer is self-interest for personal health (Hughner et al. 2007), the consumer can also support values related to animal welfare and environmental sustainability (i.e., an altruistic value orientation instead of an individualistic one). But, in general these chapters illustrate that self-interest has proven to be much more influential globally in motivating consumer purchasing of organic products than altruistic values and reasoning (i.e. animal welfare, environmental reasons, supporting farmers, fair prices etc.).[1]

The comparison of the three chapters allows us to highlight the significance and meaning of the importance of emotion in building trust between producer and consumer. In Adamoli's survey the codes and labels should impart trust. But, without information about the meaning of the codes and labels, any trust that is created is superficial or emotion-based, and not the result of empirical information. Moreover, it does not generate or foster any social connections. Getter et al. indicate that consumers trust local milk because they can see farmer practices. In this case the consumer is socially connected with the farmer. However, often overlooked is the fact that knowing or seeing the farmer does not automatically mean that the production follows the IFOAM Principles. Being connected with the farmer is social and emotional. In contrast, in the case of fair trade oriented market frameworks discussed by Grosglik, written rules document the characteristics of production in

[1](McEachern and Mcclean 2002; Lockie et al. 2004; Yiridoe et al. 2005; see Aertsens et al. 2009).

which the consumer trusts without knowing the farmer. There might also be an emotional component and the purchase of fair trade products is socially oriented, but in this case not based on a face-to-face relationship.

References

Aertsens, J., W. Verbeke, K. Mondelaers, and G. Van Huylenbroeck. 2009. Personal determinants of organic food consumption: A review. *British Food Journal* 111(10): 1140–1167.

Hughner, R.S., P. McDonagh, A. Prothero, C.J. Shultz, and J. Stanton. 2007. Who are organic food consumers? A compilation and review of why people purchase organic food. *Journal of Consumer Behaviour* 6(2–3): 94–110.

Johnston, J. 2008. The citizen-consumer hybrid: Ideological tensions and the case of Whole Foods Market. *Theory and Society* 37(3): 229–270.

Lockie, S., K. Lyons, G. Lawrence, and J. Grice. 2004. Choosing organics: A path analysis of factors underlying the selection of organic food among Australian consumers. *Appetite* 43(2): 135–146.

McEachern, M.G., and P. Mcclean. 2002. Organic purchasing motivations and attitudes: Are they ethical? *International Journal of Consumer Studies* 26(2): 85–92.

Schösler, H., J. De Boer, and J.J. Boersema. 2013. The organic food philosophy: A qualitative exploration of the practices, values, and beliefs of Dutch organic consumers within a cultural–historical frame. *Journal of Agricultural and Environmental Ethics* 26(2): 439–460.

Yiridoe, E.K., S. Bonti-Ankomah, and R.C. Martin. 2005. Comparison of consumer perceptions and preference toward organic versus conventionally produced foods: A review and update of the literature. *Renewable Agriculture and Food Systems* 20(4): 193–205.

Chapter 6
Who Came First? The Egg or the Carton? An Analysis of Organic Packaging for Italian Retail Chain Esselunga

Ginevra Adamoli

In light of the growing organic products and the rise of the hybrid citizen-consumer, defined by Johnston (2008) "as a concept (that) implies a social practice – "voting with your dollar" – that can satisfy competing ideologies of consumerism (an idea rooted in individual self-interest) and citizenship (an ideal rooted in collective responsibility to a social and ecological commons)" (p. 229), calls for more studies into green marketing on the connection between consumers agency and organic packages of food products is needed. This chapter examines organic advertising of the organic egg carton by *Esselunga*, one of the largest Italian chain supermarkets. Visually the package of the eggs reflects nostalgia for family owned farms, while also providing the consumer with technical legal jargon associated to organic farming and marketing regulations in Italy in alignment also with European Union regulations. These images and text serve to appease both the self and societal interest of the buyer. Through a qualitative content analysis of the package, this chapter argued that these forms of textual and visual communication are consistent with 'compliance marketing' and 'green selling' (King 1984; Peattie and Crane 2005), rather than serving to empower the buyer through valuable information. The paper discussed the implications of the findings for the citizen-consumer.

G. Adamoli (✉)
School of Communication, Florida State University, 4100 University Center, bLD C, Tallahassee, FL 32306, USA
e-mail: gincoolette@gmail.com

B. Freyer, J. Bingen (eds.), *Re-Thinking Organic Food and Farming in a Changing World*, The International Library of Environmental, Agricultural and Food Ethics 22, DOI 10.1007/978-94-017-9190-8_6

6.1 Background: Consumer Citizen Within Green Marketing

Healthy, natural, organic, sustainable, ethical, authentic, tasty, and humanitarian are common words used to label organic food, providing a recurrent persuasive marketing discourse for consumers of natural products. The choice of organic discourse whether in the form of text or image has twofold functions. One is to help consumers in making a product selection among the many competitors (Caswell and Padberg 1992; Hughner et al. 2007; Brent and McMullen 2008) and the other is to empower the buyer (Holder 1991; Johnston 2008). In theory, the distinction between the two is that while the first aims to maximize the profit of a company, eliminating competition, the other serves to educate the buyer on the benefits and/or harms of the product, allowing a consumer to exercise agency (Holder 1991; Winson 2004; Brent and McMullen 2008). The first function of organic discourse is usually associated with corporate marketing practices, while the second can be associated with the involvement of governmental agencies in protecting people's interest (e.g., food policies) (Holder 1991; Zinkhan and Carlson 1995; Libery and Kneafsey 1998; Winson 2004; Peattie and Crane 2005). However, in many cases, the state has failed to protect people's interest, favoring entrepreneurs who are entering the organic market (Lockie 2009). According to Lockie (2009), "in the domain of food consumption, the role of the state has shifted somewhat from protecting the public... to helping "consumers" and "entrepreneurs" make the correct choices by providing them with technical assistance and information" (p. 195) (Draper and Green 2002). Food labels on packages are an example of the adopted role by state that aims at protecting consumers, while at the same time favoring companies (Nestle 2002). As Caswell and Padberg (1992) note,

> labels are designed for their impact on the whole food marketing system rather than simply as a consumer information [...] Food labels and media advertising are closely linked because firms coordinate label and advertising messages to produce a consistent product image. (pp. 463–465)

Nestle in her book *Food Politics*, explains further the close relationships between food firms and politics. She calls attention at interlocks between governmental agencies and corporations by providing an overview of the food industry and how this latter has influenced every aspect of food style, including food-labeling policy. According to the author (2002),

> food package labels, (said Dr. Kessler,) were the result of politics, not science, and had become "so opaque or confusing that only consumers with hermeneutic abilities of a Talmudic scholar can peel back the encoded layers of meaning. That is because labels spring not from disinterested scientific reasoning but from lobbying, negotiation, and compromise" (p. 249).

There are two important points addressed in the above quote. First, there seems to be an overabundance of information that appears on food packaging, which can lead to confusing the buyer. Marketing text is alternated with technical terms imposed by governmental agencies (e.g., nutritional labels). This latter form of discourse aims to appease legal litigations between food firms and the government, rather than informing consumers. Second, there is the underlined assumption, between governmental agencies and corporations, that an economic market drives individual lifestyle and not vice versa. For example, in 2009 former Monsanto lobbyist, Michael R. Taylor, was appointed by the Obama's administration as FDA's deputy commissioner for food safety (Carney 2010) – a clear example of the revolving door. Taylor, before serving as the lobbyist for Monsanto, was an executive assistant for the FDA, who contributed to the FDA approval of Monsanto's artificial growth hormone (Smart Publications 2010). This hormone was found to be carcinogenic. The case led to the infamous lawsuit of New World Communication of Tampa v. Jane Akre, 2003.[1] It is only now, in 2010 that the "United States Federal Court overturned a ban on labeling milk as artificial hormone free" (Pugliese 2010, p. 5).

As a result, even the organic market becomes subject to the same capitalistic maneuvers as conventional markets, where products are commoditized to maximize the spirit of the term 'organic,' but in many cases do not fulfill the actions the phrases connote (Polonsky 1994; Guthman 2002; Sassatelli 2006; Johnston 2008; Lockie 2009). In this way, the organic market that aims to make a difference in society, counteracting corporate practices, seems to fall short on its cause. As a result, concepts like the citizen-consumer hybrid, which "implies a social practice that can satisfy competing ideologies of consumerism," the self and the social (Johnston 2008, p. 232) pose contradictions over consumer agency, leading scholars to raise questions about consumers' power within the organic market. It is important to note, in alignment with work by Lockie (2009) that there is a common understanding, which places the government to create favorable conditions for both businesses and consumers to exercise their rights. The tension rises when governmental policies such as organic food labeling and/or other organic regulations leave the consumer confused when attempting to shop consciously (Polonsky 1994; Peattie and Crane 2005; Hughner et al. 2007).

During the early 2000s, scholarship on the citizen consumer revolved around the idea of an active buyer both at the societal and political level. The act of 'vote with your dollars' supported the belief that consumers carried agency through their action of shopping (Dickinson and Carsky 2005; Schor 2007). Thus, by consciously choosing how to invest one's own money, alternative food production, distribution, and consumption emerged (Baker 2004; Lockie 2009). Activities that epitomized on the active shopper included community gardens, farmers' markets, co-ops, buying green-products, boycotting of brands, recycling, and reducing consumption (Latta 2007; Lockie 2009). With regard to this new type of citizenship within a discourse of consumerism Scammell (2000) states that "citizenship is not dead, or dying, but found in new places, in life-politics [...] and in consumption" (p. 351). Scammell (2000) argues for scholars to move from the pre-established notion that political involvement rests in production; rather she calls for an understanding of how consumption is becoming a way for consumers to become politically involved. She

[1] For more information about the case please refer to New World Communications of Tampa, INC, d/b/a WTVT-TV v. Jane Akre, 866 So. 2d 1231, (Fla. Dist. Ct. App. 2003).

critiques Adorno's notion that the buyer or viewer of a certain product is nonetheless a corporate worker – consumer. By pointing out the results of globalization in terms of consumers pressing corporations to adapt 'socially responsible' programs, Scammell (2000) sees a politicization of the consumer, which leads to a consumer who is demanding, concerned and socially and politically active. As she notes,

> by drawing attention to their capacity to escape state regulation, they (corporations) inadvertently highlight their own responsibility for good or ill. They are no longer disguised as an almost nonpolitical fact of life, as they were in the welfare democracies, where the state is the focus of all politics. In the process they politicize consumption. (p. 353)

Scammell (2000) argues that activism at the beginning of the twenty-first century has changed in response to deregulations of the 1980s and 1990s, which led to a privatization, internationalization, and concentration of corporate power. As a result of the "corporate hijacking of political power" Klein (2001, p. 340), citizens used their consumer power to pressure corporations to be more environmental, socially and politically involved. Example of consumers' power relied on brands' boycotts (e.g. Taco Bell and the Coalition of Immokalee Workers (CIW), Monsanto's products), and a continuous battle to implement food labeling, for example, for genetically modified organisms[2] (Adamoli 2012).

While Scammell (2000) seems optimistic about the concept of the citizen-consumer, recently published articles that have explored the citizen-consumer model of environmental issues such as organic farming and marketing, have concentrated more on contradictions of coexisting ideologies within capitalistic practices (Smith 1998; Guthman 2003; Schröder and McEachern 2004; Sassatelli 2006; Johnston 2008; Lockie 2009), thus pointing at limitations of the citizen-consumer model due to the commodification of food and the conventionalization of organic foods (Smith 1998; DeLind 2002; Johnston 2008; Lockie 2009).

According to these scholars, despite recognition that consumers play a major role in decision-making and hence have agency, corporations can co-opt this agency (DuPuis 2000; Guthman 2003; Pollan 2006). A common assumption is that even with the emergence of a new socially responsible buyer, food companies have adapted new marketing strategies to fulfill their corporate agenda (Smith 1998; Johnston 2008). Johnston (2008) leaves the reader by noting that the contradiction between consumerism and citizenship is problematic to analyze and to resolve. As he writes,

> this articulation of the citizen-consumer ... engenders profound contradictions that severely limit its transformative potential. However, it is important to emphasize that discourses are never homogenous. Possibilities for a more balanced citizenship-focused hybrid may be found in different modes of food provisioning, particularly when they are framed by non-profit organizations more able to de-center the idea of consumer choice in the service of ideals like social justice, solidarity, and sustainability (p. 263).

Livingstone et al. (2007) discuss the complexity of the rhetoric and discourse of the citizen-consumers in UK regulatory apparatus. They identify advantages

[2]For more information on the Coalition of Immokalee Workers (CIW), please visit www.ciw-online.org/.

and disadvantages for activists and citizens to position themselves as citizen-consumers. The advantage of defining citizens in terms of consumers lies in its economic-political influence. Livingstone and colleagues point at the importance of discourse in advancing the interests of the 'citizen-consumer' for the media and communications environment. In their interview, the Chairman of Voice of the Listener and Viewer, notes that,

> It is much easier to regulate consumer issues, which are basically economic issues and redress and fair representation and so on than citizenship issues which involve social, cultural, democratic issues which are far more difficult to quantify and measure (Livingstone et al. 2007, p. 72).

This statement illustrates that when working within a neoliberal regulatory market, the most efficient way to empower citizens is through shopping. Livingstone et al. (2007) note the outcome of this emphasis on consumers, rather than citizens:

> The outcome is a conception of the citizen as a vulnerable minority, leaving the majority to express their citizen interest primarily through their active role as consumers in the marketplace. But this is a conception that critics would question, because it does not offer citizens a route to represent themselves directly, and because it concentrates the citizen interest on the vulnerable few rather than the public as a whole (p. 85).

On one hand, the citizen is encouraged to shop (whether locally, or to boycott certain brands) to assert his/her political voice. On the other hand, the environment in which this concept originates lies on an economic agenda of market regulations, problematizing the concept of consumers and agency.

6.2 Green Marketing and Organic Foods

As the consumer-citizen established its position in society, green marketing emerged during the late 1980s early 1990s and has promulgated into a variety of definitions ranging from environmental to ecological marketing at a point that there is not one definition that connotes the term. A broad concept of green marketing encompasses consumers' goods, services and industrial goods (Polonsky 1994; Peattie and Crane 2005), including tourism, food, automobile, etc. The term relies on marketing products that are safe for the environment and for the health of the population. Thus, these elements are based on the processes of production, distribution, and consumption (Polonsky 1994; Peattie and Crane 2005), from packing goods, to waste disposal, to modifying advertisement to appeal to green consumers. According to *Marketing for Food* (2011), "They are all diverse names (terms referring to green marketing) but they respond to a common objective: satisfying the needs of a consumer concerned with the environment whilst simultaneously favoring sustainable development" ("Green Marketing" 2012, pp. 2–5).

In late 1980s and early 1990s, corporations responded to an increase in consumers' demand for green products (Hume 1991; Peattie and Crane 2005), characterized by a willingness to pay more (Mintel 1991; Coddington 1993; Peattie and Crane 2005). According to Peattie and Crane (2005) during the 1990s, "the volume of green print ads grew by 430 %, and that of green TV ads by 367 %" (p. 358) and companies like McDonalds adopted changes across its operations, including the alteration of food packages with the elimination of chlorofluorocarbons from manufacturing processes (Hume 1991). Yet, much of the talk around green initiatives during these early years focused on the environmental impacts of corporate practices (e.g., pollution, waste disposal). Limited change was made in terms of conventional (i.e., non-organic) foods, even though concerns from the population kept rising and grass root movements were emerging (Schlosser 2001; Nestle 2002; Pollan 2006). For example, in the United States the issue of organic certification and food labels had a peak in 1998 after consumers rallied to protect organic agricultural farming from the "U.S. Department of Agriculture's controversial proposed national regulations for organic food" (Organic Consumers Association 2011, pp. 19–20; Adamoli 2012). The rise of food-safety scandals of the 1980s and 1990s pushed scholars and journalists to be more involved, by publishing exposé pieces including *Fast Food Nation* by Eric Schlosser (2001), *The Omnivore Dilemma* by Michael Pollan (2006), and *Food Politics* by Marion Nestle (2002), in an attempt to inform the public.

With an increase on food concerns and emerging trends in the sector of organic farming and food-labeling on the global level (e.g., ban on GMOs cultivations in Europe, mandatory GMOs food labeling in European countries, organic food movements), green marketing strategies adopted by firms were soon evaluated by corporations as failing to increase revenue (Lawerence 1991; Ottman 1992; Peattie and Crane 2005). King (1984) attributed the failure of green marketing from the 1990s to four business issues. This chapter focuses on two of them: 'green selling,' and 'compliance marketing'.

Much of the failure of green marketing is associated to the exploitations of consumers through advertisement in the form of misleading or deceptive green messages without actually changing the company's products (Mendleson and Polonsky 1995; Libery and Kneafsey 1998; Peattie and Crane 2005; Hughner et al. 2007). The intention of informing potential buyers about the environmental impacts of certain products was outweighed by a desire to sell at a higher cost (Libery and Kneafsey 1998; Peattie and Crane 2005) due to the fact that people had indicated willingness to pay more for green products (Canavari et al. 2002; Krystallis and Chryssohoidis 2005; Hughner et al. 2007). As Peattie and Crane (2005) note, "Facile, meaningless, and unproven green claims were slapped on unchanged products in failed attempts to boost sales, leading to mounting consumer cynicism and suspicion, and concerns about a potential consumer backlash . . . Some firms have realized that their claims lacked independent authentication" (p. 361). A recent example is the lawsuit against the company Naked Juice, which advertises its products as natural and GMOs free when in actuality they contain synthetic and GMOs ingredients (Goldberg 2012). Green selling denoted businesses' practices to generate sales adapting a green discourse without providing changes in production or distribution of the item(s) "in order to take advantage of any environmental concerns of consumers" (Peattie and Crane 2005, p. 361). This business model resulted in consumers' cynicism, distrust for green products and companies, and a request by citizens to impose federal and state regulations both at the marketing level as well as at the production level (Schlossberg 1993; Polonsky 1994; Triplett 1994; Mendleson and Polonsky 1995; Nestle 2002; Peattie and Crane 2005).

The intervention from governmental entities in green corporate practices did not resolve the issue of producing and promoting alternative products and solutions to environmental/food issues. Peattie and Crane (2005) define 'compliance marketing' businesses' practices "whose environmental initiatives do not go beyond responding to regulation" (p. 363). Thus, even with the involvement of state or federal authorities firms were still reluctance in drastically changing operations of production, distribution, and consumption. This is not to say that the implementation of food-labeling laws did not bring positive outcomes for the consumer, the environment, and the company. Libery and Kneafsey (1998) discussed the conceptualization of accreditation and labeling schemes in Europe, pointing at how regulatory programs in food labeling for non-industrialized foods can add credibility and authenticity to the product and the company, thus favoring locally and family owned firms. In recent years, even companies, including Wal-Mart and McDonalds have been able to improve their environmental credibility and green image by complying to laws, revamping green marketing (Polonsky 1994; Mendleson and Polonsky 1995; Smith 1998; Peattie and Crane 2005). A recent example is the partnership with Michelle Obama and Wal-Mart in fighting obesity. The retail store aims to offer more 'healthy foods' at a lower price by 2015, including opening new stores in deserted food areas (Wilgoren and Mui 2011). This intervention of the state and public officials like Michelle Obama, in influencing the credibility of green initiatives poses a question of consumer-citizenship linked to the marketing and labeling system. Does additional information on packages of green products reflect consumers' power or a mixture of compliance marketing and green selling?

6.3 Organic Marketing & Food Labels in Italy

Generally speaking, food labeling is defined as "direct consumer information, with the federal government intervening in the two-party relationship between seller and buyer to remedy information imperfections and failures" (Caswell and Padberg 1992, p. 462). From a company's prospective, food labels (e.g., healthy claims, slogans that attract the buyer) should have limited regulation (Caswell and Padberg 1992) because they should help entrepreneurs to enter the market and have a positive impact on the economy (Peattie and Crane 2005). Food labels for the seller mean freedom of marketing a product with limited attention to environmental and health consequences. To address the potential harm of false claims many countries and unions, including the European Union and the United States, require firms to follow specific procedures when marketing food products, in addition to guidelines for the production and distribution of goods (Libery and Kneafsey 1998; European Commission 2007, 2008; Gold 2008). For example, the U.S. Food and Drug Administration (FDA) requires that all health claims must fall within the FDA guidelines (U.S. Food and Drug Administration 2010a, b).[3] The FDA food labels are intended to improve the American diet (Caswell and Padberg 1992). In Italy, food-labeling laws must be in alignment with the European Union. In case "EU law may be incomplete or absent, the law of each member state applies" (Foreign Agricultural Service 2009), thus each European country shall have the liberty to extend EU food labeling regulations. In Australia, the Australian Trade Practices Commission's (TPC) has designed laws to control green marketing claims. Similarly, in the United States the US Federal Trade Commission has issued guidelines for firms to follow when making green marketing claims, whether for dishwasher soaps or green automobiles (Peattie and Crane 2005; "Part 260 -Guides for the use of environmental marketing claims" 2012). In general, governmental regulations aim to protect citizens, providing them with the opportunities to evaluate the safety of the product(s) for sale. In relation to environmental marketing, these laws in accordance with Peattie and Crane (2005) are designed to protect citizens by reducing "production of harmful good or by-products; modify consumer and industry's use and/or consumption of harmful goods; or ensure that all types of consumers have the ability to evaluate the environmental composition of goods" (p. 4). In other terms, green labels should identify the processes of production, distribution, and consumption of goods in such a way for consumers to make a well-informed decision.

With special attention to organic marketing of foods, companies must adhere to a variety of regulations ranging from organic farming/agricultural laws to organic marketing of products to food labeling rules, which apply to all type of food items.[4] For example, food information on packages in Italy must be displayed in the form of labels, packages and information must be presented following specific guidelines and products must display nutritional labeling specifications (European Commission – Agriculture and Rural Development 2009). In particular, Italian labeling regulations for organic produces rely on three elements: (1) the clarification of information, (2) the transparency of organic procedures through text and images, and (3) stimulating the adaptation of organic processes in the food market:

> To provide clarity to the consumers in the entire market it is necessary to make mandatory the use of the UE logo for all organic food produces that are prepackaged and have been produced in the European Community [. . .] In addition, with the aim of informing consumers of the transparency of the market and to stimulate the use of organic ingredients,

---

[3]FDA guidelines only apply to processed food. A company is required to provide product information only when health claims are made, but the agency has never issued formal rules about the term. It is the manufacture's responsibility to assure that health claims meet the FDA guidelines. In case a health claim is made the agency evaluates the credibility of the claim based on the totality of publicly available scientific evidence.

[4]For more information on U.S. and European organic agricultural laws please refer to the "Federal Organic Food Production Act of (OFPA)" (1990), the EC 834/2007 and EC 889/2008 (European Commission 2007, 2008). See also *Agricoltura e Sviluppo Rurale* at http://ec.europa.eu/agriculture/organic/consumer-confidence/logo-labelling_it.

> one shall consent also, based on specific conditions, to insert in the list of ingredients references to the organic production (European Commission 2007 189/2-189/3; author trans.; 2008).[5]

Organic products must display the organic logo, which certifies that the item has been subject to organic procedures and has been inspected by organic certified agents. Visual information of the logo for each country and detailed information on organic production and labeling of organic products with regard to organic production can be found in the European Council Regulation reports, EC 834/2007 and EC 889/2008. For example, the first two pages of the EC 889/2008 provides an overview of the emergence of an organic market and the need to protect this new economic asset, assuring fair competition to the seller, buyer and consumer. Labeling of organic products is seen as beneficiary for all parties including the environment. Emphasis is given on environmental issues (contamination); rather than solely on the economy or individual. Following this introduction, the document lists the products to which the regulation applies. These products include: unprocessed crops, livestock and unprocessed livestock, feed, processed crop and livestock products. The rest of the document defines products that do not fall within the organic law, including GMO products and derives. For example, a chicken that was fed GMO corn should not be considered organic.

Labeling specifications for organic products including origins of provenance from the feeding, to process and distribution are also included in the manuscript. Organic products must contain at least 95 % or organic ingredients or derive from ingredients that are 95 % organic. The product should not contained or been exposed to chemicals or ionizing radiation. The Act also includes a section on seeding including seeding procedures and information about supplier of seed. Inspection details, imports from other countries, clarification on utilizing renewable natural resources as a way to develop and maintain sustainable agriculture are also items addressed in the document. The last section describes and defines free range and housing for livestock including mammals, poultry, and apiaries, as well as disease prevention for animals and plants. The law also specifies general conditions for all products including information about soil, fertilizer, animal excrements, compost, and feed. The document ends with visual documentation of organic logo for European countries that each product will display.

In terms of information required to appear on packaging of organic produces the EU indicates the following: Organic logo, "the name or code number of the certification body that has certified the organic producer" as well as the first two digits of the country of origin: "The indication of the place where the agricultural raw materials of which the products is composed have been farmed, as referred to in Article 24(1)(c) of Regulation (EC) 834/2007" (European Commission 2007,

[5]"Per dare chiarezza ai consumatori in tutto il mercato comunitario, occorre rendere obbligatorio il logo UE per tutti i prodotti alimentari biologici in imballaggio preconfezionato ottenuti nella Comunità... Inoltre, ai fini dell'informazione dei consumatori, della trasparenza del mercato e per stimolare l'uso di ingredienti biologici, si dovrebbe anche consentire, a determinate condizioni, di inserire nell'elenco degli ingredienti riferimenti alla produzione biologica."

p. 23); a list of ingredients, including the percentage of organic ingredients if the product is not 100 % organic (CCPB 2012). In addition, the following information is mandatory: (1) the name or business name and address of the manufacturer or packager, date of production and expiration and label of the European Community (European Commission 2007; Minutoli 2012 [Interview]). The labels indicating the above information must be displayed in two colors, either green/white or black/white (CCPB 2012, p. 19). In case of feeds, the label must specify feed material and percentage of feed material. Additional information must be displayed in accordance to labeling laws for all types of foods. For example, for eggs (organic and non-organic) labels must indicate the quantity expressed in weight or number of eggs. For organic eggs, it is mandatory to include information on the package and/or each egg about the type of farming (e.g., organic, cage, free-cage) as well as codes for the location where the eggs were hatched and how the chickens were raised. According to the Decree DM MiPAAF n. 91436 of 04.08.

> It is possible to identify organic eggs those products packaged according to the modalities of the organic production, ultimately stamped on the shell at the time of origin or manufacturing. The package will carry references about the type of organic production, properties of the product, and location of processing. ("Decree DM MiPAFF 91436 – Produzioni Animali Biologiche," 2000, p. 14 author translation).[6]

All commercial eggs including organic eggs, must show data to allow traceability: country or state of production, town hall of production, province of production, name and place where the chicken was raised, and date of expiration ("La Carta d'Identita' delle Uova" 2011).

In addition to specific labeling rules that apply to organic products, marketing for food packaging must respect certain dimensions applicable to text and graphic. For example, according to Nuciari (2011), "*a 80 cm*/Mandatory captions on labels shall have typographical characters not inferior to 1.2 mm (taking as a referent point the letter "x" not capitalized), or 0.9 mm if the packaging presents a surface inferior to 80 cm" (pp. 60–63). These rigid rules leaves little space for companies to market their product in such a way to detach their products from other similar ones. Marketers in the area of Italian foods stress the importance of telling a story through the visual and text impact of packaging. Logos and mascots are recommended to use to evoke memories and make a memorable impression in consumers' mind, helping them to sort through items while shopping. According to a marketing report for farmers and food producers, "Logos summarize in a visual image and/or in a brief text key elements that one wishes the buyer to remember, seek, and easily identified." (ISMEA 2008, p. 45). These images or text must appear on a package that has been designed to capture the attention of the consumer from competing items (ISMEA 2008, p. 47). This is done with colors, selected fonts, and graphic that must conform to regulations.

---

[6]"Sono identificabili come uova biologiche i prodotti confezionati secondo le modalità del disciplinare di produzione biologica, eventualmente timbrate sul guscio all'origine o alla lavorazione. La confezione riporterà il riferimento al metodo di produzione biologico, il riferimento dell'unità ed il riferimento del centro di confezionamento."

While companies have capitalized on green consumerism, applying green marketing for purpose of return of investment, the state in Europe and in particular in Italy has responded by aggressively implementing regulatory laws for food labeling of organic products, with the aim of protecting consumers from false 'natural' claims. King (1984) has addressed the intervention of governmental entities in marketing of goods as 'compliance marketing,' concluding that laws do not necessarily address the issue of providing alternative green solutions to consumers, rather corporations comply with what has been asked, without going beyond. The following case study of organic eggs from the Italian retail chain supermarket *Esselunga* reflects on the concepts of 'green selling' and 'compliance marketing,' leading to a discussion on the implications for the consumer-citizen. Does information on food packages help the buyer?

6.4 The Case Study and Methodology

The case study of the egg carton from the Italian supermarket *Esselunga* was chosen for two main reasons. First, according to research on the organic market in Italy, the retail chain *Esselunga* offers the highest volume in assortment for organic products compared to other mass-produced supermarkets in Italy (Santucci 2009; Asioli et al. 2011). Schaak and Willer (2010) reported that large retail chain food markets have increased their selection of organic products, at times launching their own organic line, as in the case of *Esselunga.* Second, the chain retailer was chosen for its similarities in terms of practices, size and distribution to other international large retailers such as Tesco, Waitrose, Big Bear, Kroger, Winn Dixie, and Giant Eagle located in the UK and U.S. respectively. Revenue for *Esselunga* in the year 2009 was estimated to be \$5.7 billion, while net income for the company was \$202.6 million. The company owns 130 supermarkets mostly in North Italy and has 142,000 employees ("Esselunga's History" 2010). According to *Hoovers*, Kroger, despite having 334,000 workers and 3,619 stores, has similar annual revenue (\$76,733 billion) to the Italian retail chain ("Krogers Revenue" 2010). Annual net income for the company is \$70 million. Both firms offer a range of organic products between 100 and 200 while producing and distributing their articles ("Kroger" 2010; Esselunga 2010). Thus, future studies could be conducted comparing organic markets across retailers in different countries. As a result, the egg carton was chosen for its ubiquity and commonality in people's diets across supermarkets and countries. Eggs are a universal staple food.

The data for the study consisted of textual and visual material from *Esselunga* official website and its organic items collected at selected stores. This material was then situated within a theoretical framework of consumer power and marketing. Items were collected through visits to the store as well as visits to the web site over a 1-month period. The material was then analyzed visually to document whether the information on the package informs the consumer or reflects compliance marketing and green selling. Qualitative content analysis was selected because, according to Neuman (1997) this technique is a non-intrusive method to derive significant findings. He notes that, "the 'content' refers to words, meanings, pictures, symbols, ideas, themes, or any message that can be communicated. The 'text' is anything written, visual, or spoken that serves as a medium for communication" (pp. 272–273). The following items were collected and analyzed in the study: (1) textual material gathered from the website of the company about organic products, sustainability initiatives, partnerships to save endangered species and recycling programs; (2) product packaging of organic products including egg cartons; and (3) informational food material found in the store to inform customers about products. These items were collected at different stores, in different cities and regions in Italy, including the regions of Sardinia, Tuscany, Lazio, and Emilia Romagna.

6.5 How Esselunga Packages Organic Products

In recent years a proliferation of foods that appear to bring health and safety benefits has rapidly created a need for retail chains such as *Esselunga* to embrace the organic movement and develop marketing campaigns designed to meet the consumer's demand and new food information (Libery and Kneafsey 1998; Richter et al. 2000; Brunori et al. 2008). The role of *Esselunga* in organic marketing, illustrates its commitment to organic food or original approach to marketing organic food (Richter et al. 2000, p. 1). The company website states its commitment to ecological causes, including details on several initiatives supported by the company and a section devoted exclusively to defining food labeling and 'organic' according to the European and Italian government standards (Esselunga 2010). In this way, the company defines itself as a socially responsible company utilizing materials that have a low impact on the environment (e.g., green buildings, usage of recycle packages, educative programs for customers, partnerships with humanitarian and animal rights organizations, organic and eco-friendly line) and serving as a channel to advance the cause of the environmental movement by empowering the consumer to choose products that make a difference. The store, in fact, offers more than 200 organic products ranging from fresh goods (vegetable and fruit) to more convenience products such as ketchup, pizza, pre-cooked and frozen goods, and soda (Richter et al. 2000).

To allow the customer to choose among other competing products, organic items are allocated among conventional goods, hence in the sauce section organic and non-organic tomato jars are placed next to each other. This distribution policy follows a particular communication strategy. In order to not only empower the customer but also to retain customers "who will revert to conventional products" (Richter et al. 2000, p. 4) organic items must differentiate themselves and pop out. This differentiation is done with packaging and food labeling appealing to the emotive element of the consumer (Asioli et al. 2011). In this way, the store capitalizes on its organic commitment, by alluding to the improvement of the consumer and the environment. As Smith (1998) points out, what can happen is that "individuals are lured into these uncontrollable consumer orgies by marketing masterminds" (p. 8). The emotive element is a key factor in the organic market as retailers must justify the price difference (organic products are more expensive) to a more responsible consumer (Richter et al. 2000; Asioli et al. 2011). According to a survey on consumer satisfaction conducted by Consumers' Forum, an independent Italian research entity, 73 % of the Italian population demanded more green products, and 70.4 % indicated the desire for companies to include more useful food label including information about the companies providing the product ("Consumers' Forum" 2010). As consumers become more socially active while shopping, retail stores like *Esselunga* must adopt a rhetoric that attracts customers to buy socially responsible products (Corbett 2006). To do so, clear signage, providing more information, is installed. As Richter et al. (2000) write, "many organic products have the labels of both the retailer and the farming or certification association in order to add credibility and to make the organic assortment appear more professional" (p. 4).

Using a juxtaposition of text and images the notion of social and personal action, which allures consumers concerned with hedonism, pleasure and achievement, serves as a driving force to buy the product (Naspetti and Zanoli 2004, p. 2). The ochre color of the carton is gentle and natural and does not obfuscate the message that the product carries – organic. The only images on the box that use same tone colors as the card-box are that of a farm with two chickens on the right side. On the opposite side a large egg is shown. These pictures compose a referent system of signifiers that combined with text create a set of meaning fulfilling the expectation of the consumer (Guthman 2002). In other words, text and images appearing on the package convey specific meaning to each consumer, alluding to social and personal empowerment. The consumer identifies the signifier (text or image) with a set of cultural meanings. In the case of the card-box the soft colors, the picture of the egg and chicken and the farm in the background allude to the idea of a safe and natural environment, almost if the chicken was raised by a single farmer on a vast land. In terms of discourse, the text is more prevalent than the images, and it is only by combining the two together that the picture goes beyond the referent meaning. Information about the product (from type of farming, location where it was produced, contact information, due date, category of the product) covers the box on all sides, placing 'green' discourse as a key element for the success of the sale. 'Traceable code,' 'organic,' 'bio,' 'fresh,' 'category A,' 'toll-free number,' 'no GMO organism,' are just few of the words appearing on the box. On the side, 12 lines explain the characteristic of the organic product. Using a bright red background specifications about the product origins are expressed alluding customers concerned with GMO that the product they are buying does not derive from genetically modified organisms or by-product. The rhetoric of organic campaign continues as one opens the box. On the top of the inside box, a legend is provided to the consumer. The image shows an egg. A code made out of digits and letters is stamped on the egg to reflect both EU and Italian food regulation for all eggs both conventional and organic. The legend explains how to read each digit. An example of the code is here provided: 0IT104RA386 ENTRO 09/06. Each egg contained in the carton has the following information printed in blue. 0 refers to the typology of farming, meaning that the chicken "scratched about, freely and outdoor, eating chicken feed deprived of fat and animal flour obtained by organic farming, for which it is not provided the usage of chemical mixture of synthesis for the feeding of the animals, weed and pesticide and the battle against parasites in field and after the harvesting" (Esselunga 2010). IT stands for the country of origins (Italy); 014 (Ravenna) refers to municipality of the city where the eggs were produced; RA (Ravenna) is the province of origin; and 386 refers to the location where the egg was hatched. Unfortunately no information on the location was found based on the three digits, suggesting limitations of the consumer-citizens due to technical jargon. *Entro* suggests the day/month/year by which to consume the product.

6.6 Discussion and Conclusion

On the surface, the abundant information suggests that "information can be brought together in different ways to accommodate different points of view" (Smith 1998, p. 5). On one hand, the consumer is empowered to choose a product that will bring pleasure (self-interest) and will sustain a social cause bringing "the ordinary details of life to become sites of political struggle" (Smith 1998, p. 5). On the other, the retailer will have successfully endorsed her/his advertising campaign in compliance with governmental food regulations. It is only through discourse and the use of prose and images that both parties are self-satisfied. As Smith (1998) notes, "it is discourse that gives actions and behaviors," (p. 10) the tools to interpret a message in accordance with a pre-determined ideology. The responsible customer seems more "pleasure oriented" (Richter et al. 2000, p. 7), driven by self-actualization, inner harmony and happiness (Naspetti and Zanoli 2004) by the variety of information provided. However, further analysis of the egg carton revealed a problematic picture for the consumer-citizen.

The level of details in the labels deserves recognition, as well as caution. The identification code on the cart box only provides digits and letters. Even if the text explains how to read those numbers (e.g., IT = country of production; 014 = code of municipality), it fails to inform the customer that 014 stands for Ravenna. This information is available in legal documents not accessible at supermarkets, but can be located after extensive research on the web. Moreover, it was noted during the analysis that even after extensive research there were still codes not identifiable by the researcher, pointing at a gap between state, people, and food packaging. These limitations of decoding caused by technical jargon do not facilitate consumer power. The jargon of politics does not translate into the language of the mass; rather it only serves to reinstate a distinction and to maintain a gap between the elite and the masses. As Williamson (1978) states, "to fill in gaps we must know what to fill in, to decipher and solve problems we must know the rules of the game" (p. 99). The consumer is left to express his/her agency through the creation of trust via food labels.

These legal terms present on the organic egg carton while they limit consumers from understanding the production and distribution of the food they are buying, they help building brand credibility. Compliance marketing merges green selling with the aim of influencing buyers' decisions. Williamson explains the way advertising produces knowledge. For the author, words and images establish a legitimization of a certain product that is automatically regarded as truthful by its consumers because of "a pre-existing bodies of knowledge allow(ing) reference to take the place of description, connotation of denotation, in ads" (Williamson 1978, p. 100). Hence, complying with food regulations by including text that underlines the nature of the product creates a sense of authenticity and legitimization for *Esselunga*. Legal terms and information are juxtaposed with images of green selling, in this way associating marketing to food labeling regulations. For example, when a consumer sees an image of chickens running in an infinite field reinforced by the text "free range," this person associates the package as adhering to state and federal organic and food regulations, suggesting that the brand and company are green. In other words, the sign has become the signifier, "a system of meanings, a referent system, is used in its entirety to give significance to the product" (Williamson 1978, p. 106). Text and images chosen by marketers in compliance with state and federal laws aim to persuade customers to act in a certain way. As a result, when the consumer believes she/he is choosing, the choice has already been pre-packed by the retailer as well as the state.

The implications for the consumer-citizen when shopping organically is based both on the intervention from the state in regulating packaging of goods as well as green marketing strategies to sell a product. Such combination does not necessarily empower the buyer. On the contrary, the case of *Esselunga* illustrates that confusion can emerge from the overload of technical terms associated to the packaging of organic eggs. In this way, the consumer is left to act passively, meaning that the individual must assume the information provided is adherent with his/her socially and political ideologies. In regards to green selling as the practice of developing marketing strategies to generate profit for a company, it must be noted that companies are faced with the challenge of complying with laws imposed by the government. Italian regulatory laws must be in compliance with the European Union. Eggs must contain information for traceability purposes and must meet organic standards, limiting food firms from covering packages with deceptive green slogans. However, this paper also noted that complying to state regulations can be an asset for companies like *Esselunga* that are tying to infiltrate the organic market, by adding credibility to their brand and company. As a result, the intervention of the government becomes to be known as 'complying marketing,' whereas information on goods that must be included due to organic laws becomes a marketing strategy to sell at a higher price. More studies on the role of food labels on organic products is needed to evaluate from the point of view of the consumers whether this overabundance of information increases the willingness to buy green products. A point, however, from this study is clear, the information provided is too technical and complex to decode, thus suggesting a dependable consumer agency.

References

Adamoli, G. 2012. *Social media and social movements: A critical analysis of audience's use of facebook to advocate Food Activism Offline*. Phd, Florida State University Tallahassee.

Asioli, D., M. Canavari, A. Castellini, T. de Magistris, F. Gottardi, P. Lombardi, and R. Spadoni. 2011. The role of sensory attributes in marketing organic food: Findings from a qualitative study on Italian consumers. *Journal of Food Distribution Research* 42: 16–21.

Baker, L.E. 2004. Tending cultural landscapes and food citizenship in Torontos's community gardens. *Geographical Review* 94(3): 305–325.

Brent, B., and T. McMullen. 2008. Visual communication to children in the supermarket context: Health protective or exploitive? *Agriculture and Human Values* 25(3): 333–348.

Brunori, G., R. Cerruti, S. Medeot, and A. Rossi. 2008. Looking for alternatives: The construction of the organic beef chain in Mugello, Tuscany. *International Journal of Agricultural Resources, Governance and Ecology* 7(1): 126–143.

Canavari, M., G.M. Bazzani, R. Spadoni, and D. Regazzi. 2002. Food safety and organic fruit demand in Italy: A survey. *British Food Journal* 104(3/4/5): 220–232.

Carney, T. 2010. *Obama makes a mockery of his own Lobbyist Ban*. Retrieved from http://www.washingtonexaminer.com/politics/Obama-makes-a-mockery-of-his-own-lobbyist-ban-83385832.html#ixzz12jOESbst. Accessed 3 Feb 2010.

Caswell, J.A., and D.I. Padberg. 1992. Toward a more comprehensive theory of food labels. *American Journal of Agricultural Economics* 74(2): 460–468.

CCPB. 2012. *Normative per L'etichettatura dei Prodotti Biologici e Tipologie di Applicazione*. Retrieved from http://www.ccpb.it/. Accessed 8 July 2010.

Coddington, W. 1993. *Environmental marketing: Positive strategies for researching green consumer*. New York: McGraw-Hill.

Consumers' Forum. 2010. Retrieved from http://www.consumersforum.it/sito/eventi/eventi.php?id=25. Accessed 8 July 2010.

Corbett, J.B. 2006. *Communicating nature: How we create and understand environmental messages*. Washington, DC: Island Press.

Decree DM MiPAFF 91436 – Produzioni Animali Biologiche. 2000. (2000).

DeLind, L.B. 2002. Place, work, and civic agriculture: Common fields for cultivation. *Agriculture and Human Values* 19(3): 217–224.

Dickinson, H., and M.K. Carsky. 2005. The consumer as economic voter. In *The ethical consumer*, ed. R. Harrison, T. Newholm, and D. Shaw. Thousand Oaks: Sage Publications.

Draper, A., and J. Green. 2002. Food safety and consumers: Constructions of choice and risk. *Social Policy & Administration* 36(6): 610–625.

DuPuis, M. 2000. Not in My body: rBGH and the rise of organic milk. *Agriculture and Human Values* 17(3): 285–295.

Esselunga. 2010. *Esselunga*. Retrieved from www.esselunga.it. Accessed 2010.

Esselunga's History. 2010. Retrieved from http://investing.businessweek.com/research/stocks/private/snapshot.asp?privcapId=5467388. Accessed 6 July 2010.

European Commission. 2007. Council Regulation (EC) No 834/2007, No 834/2007 C.F.R. (2007).

European Commission. 2008. Council Regulation (EC) No 889/2008, No 834/2007 C.F.R. (2008).

European Commission – Agriculture and Rural Development. 2009. *Logo ed Etichettatura*. Retrieved from http://ec.europa.eu/agriculture/organic/consumer-confidence/logo-labelling_it. Accessed 12 Nov 2013.

Federal Organic Food Production Act of (OFPA). 1990.

Foreign Agricultural Service. 2009. Italy food and agricultural import regulations and standards narrative. Retrieved from http://gain.fas.usda.gov/recent%20gain%20publications/food%20and%20agricultural%20import%20regulations%20and%20standards%20-%20narrative_rome_italy_7-30-2009.pdf. Accessed 23 Sept 2012.

Gold, V. 2008. *Guide to U.S. organic marketing laws and regulations*. Retrieved from http://www.nal.usda.gov/afsic/pubs/OAP/OAPGuide1.shtml. Accessed 4 May 2012.

Goldberg, M. 2012. *Guides to organic laws*. Retrieved from http://livingmaxwell.com/update-on-the-naked-juice-kashi-class-action-lawsuits. Accessed 4 May 2012.

Green Marketing. 2012. Retrieved from http://www.marketing4food.com/green-marketing-and-organic-or-eco-friendly-marketing-in-food/?lang=en. Accessed 4 May 2012.

Guthman, J. 2002. Commodified meanings, meaningful commodities: Re-thinking production-consumption links through the organic system of provision. *Sociologia Ruralis* 42(4): 295–311.

Guthman, J. 2003. Fast food/organic food: Reflexive tastes and the making of 'yuppie chow'. *Social & Cultural Geography* 4(1): 45–58.

Holder, J. 1991. Regulating green advertising in the motor car industry. *Journal of Law and Society* 18(3): 323–346.

Hughner, R.S., P. McDonagh, A. Prothero, C.J. Shultz, and J. Stanton. 2007. Who are organic food consumers? A compilation and review of why people purchase organic food. *Journal of Consumer Behaviour* 6(2–3): 94–110.

Hume, S. 1991. *McDonald's: Case Study Advertising Age* 62(5): 32.

ISMEA. 2008. *Manuale di marketing per la vendita diretta di prodotti agricoli*. Retrieved from http://www.ismea.it/flex/cm/pages/ServeBLOB.php/L/IT/IDPagina/6843. Accessed 1 Jan 2012.

Johnston, J. 2008. The citizen-consumer hybrid: Ideological tensions and the case of Whole Foods Market. *Theory and Society* 37(3): 229–270.

King, S. 1984. Has marketing failed, or was it never really tried? *Journal of Marketing Management* 1: 1–19.

Klein, N. 2001. *No logo: No space no choice*. Flamingo: No Jobs.

Kroger. 2010. Retrieved from www.kroger.com. Accessed 4 May 2010.

Krogers Revenue. 2010. Retrieved from http://www.lexisnexis.com/hottopics/lnacademic/?. Accessed 16 July 2010.

Krystallis, A., and G. Chryssohoidis. 2005. Consumers' willingness to pay for organic food: Factors that affect it and variation per organic product type. *British Food Journal* 107(5): 320–343.

La Carta d'Identita' delle Uova. 2011. Retrieved from http://www.codacons.piemonte.it/Alimentazione/Etichettatura%20uova.htm. Accessed 3 Feb 2010.

Latta, P.A. 2007. Locating democratic politics in ecological citizenship. *Environmental Politics* 16(3): 377–393.

Lawerence, J. 1991. From hero to villain. *Advertising Age* 62(5): 12–13.

Libery, B., and M. Kneafsey. 1998. Product and place promoting quality products and services in the lagging rural regions of the European union. *European Urban and Regional Studies* 5(4): 329–341.

Livingstone, S., P. Lunt, and L. Miller. 2007. Citizens, consumers and the citizen-consumer: Articulating the citizen interest in media and communications regulation. *Discourse & Communication* 1(1): 63–89.

Lockie, S. 2009. Responsibility and agency within alternative food networks: Assembling the "citizen consumer". *Agriculture and Human Values* 26(3): 193–201.

Mendleson, N., and M.J. Polonsky. 1995. Using strategic alliances to develop credible green marketing. *Journal of Consumer Marketing* 12(2): 4–18.

Mintel. 1991. *The green consumer report*. London: Mintel.

Naspetti, S., and R. Zanoli. 2004. Do consumers care about where they buy organic products? A means-end study with evidence from Italian data. In *Marketing trends for organic food in the 21st century MAICH, Greece*, ed. G. Baourakis. Singapore: World Scientific Publishing.

Nestle, M. 2002. *Food politics: How the food industry influences nutrition and health*. Berkeley: University of California Press.

Neuman, W.L. 1997. *Social research methods: Qualitative and quantitative approaches*. Needham Heights: Allyn & Bacon.

Nuciari, P. 2011. *Nuovo Regolamento Europeo sull'Etichettatura degli Alimenti*. Retrieved from http://pieronuciari.it/2011/09/18/il-nuovo-regolamento-europeo-sull\T1\textquoterightetichettatura-degli-alimenti. Accessed 1 Jan 2012.

Organic Consumers Association. 2011. *About the OCA: Who we are and what we're doing*. Retrieved from http://www.organicconsumers.org/aboutus.cfm. Accessed 8 June 2011.
Ottman, J. 1992. Brand 'stewards' will guide environmental planning. *Marketing News* 26(8): 15.
Part 260 -Guides for the use of environmental marketing claims. 2012. Retrieved from http://www.ftc.gov/bcp/grnrule/guides980427.htm. Accessed 23 Aug 2012.
Peattie, K., and A. Crane. 2005. Green marketing: Legend, myth, farce or prophesy? *Qualitative Market Research: An International Journal* 8(4): 357–370.
Pollan, M. 2006. *The omnivore's dilemma: A natural history of four meals*. New York: Penguin.
Polonsky, M.J. 1994. An introduction to green marketing. *Electronic Green Journal* 1(2): 1–10.
Pugliese, G. 2010. *Federal court ditches ban on labeling milk*. Retrieved from http://www.organicauthority.com/foodie-buzz/federal-court-ditches-ban-on-labeling-milk.html. Accessed 1 Jan 2012.
Richter, T., O. Schmid, P. Damary, P. van den Berge, U. Meier, and D. Halpin. 2000. *International study of retail chains with regard to their activities in the marketing of organic products*. Paper presented at the International Seminar Organics in the Supermarket, Frick.
Santucci, F.M. 2009. *I circuiti commerciali dei prodotti biologici*. Ancona: AGRIREGIONIEUROPA.
Sassatelli, R. 2006. Virtue, responsibility and consumer choice. In *Consuming cultures, global perspective*, ed. J. Brewer and F. Trentmann, 219–250. Oxford: Berg Publisher.
Scammell, M. 2000. The internet and civic engagement: The age of the citizen-consumer. *Political Communication* 17(4): 351–355.
Schaak, D., and H. Willer. 2010. Development of the organic market in Europe. In *The world of organic agriculture – Statistics and emerging trends 2010*, ed. H. Willer and L. Kilcher, 141–144. Frick: IFOAM/FiBL.
Schlossberg, H. 1993. Effect of FTC green guidelines still doubtful for some marketers. *Marketing News* 27(3): 1.
Schlosser, E. 2001. *Fast food nation: The dark side of the all-American meal*. Boston: Houghton Mifflin.
Schor, J. 2007. *Consumer-Topia: Envisioning a new culture of consuming*. Paper presented at the American Sociological Association Annual Meeting, New York.
Schröder, M.J., and M.G. McEachern. 2004. Consumer value conflicts surrounding ethical food purchase decisions: A focus on animal welfare. *International Journal of Consumer Studies* 28(2): 168–177.
Smart Publications. 2010. *The FDA and Monsanto: Strange bedfellows*. Retrieved from http://www.smart-publications.com/articles/view/lies-and-deception-how-the-fda-does-not-protect-your-best-interests/. Accessed 4 May 2012.
Smith, T.M. 1998. *The myth of green marketing: Tending our goats at the edge of apocalypse*. Toronto: University of Toronto Press.
Triplett, T. 1994. Three firms abandon green marketing claims. *Marketing News* 28(15): 10.
U.S. Food and Drug Administration. 2010a. *Labeling & nutrition*. Retrieved from http://www.fda.gov/Food/LabelingNutrition/default.htm. Accessed 2010.
U.S. Food and Drug Administration. 2010b. *Claims*. Retrieved from http://www.fda.gov/Food/GuidanceComplianceRegulatoryInformation/GuidanceDocuments/FoodLabelingNutrition/FoodLabelingGuide/ucm064908.htm. Accessed 2010.
Wilgoren, D., and Y.G. Mui. 2011. *With praise from Michelle Obama, Wal-Mart announces healthy food campaign*. Retrieved from http://www.washingtonpost.com/wp-dyn/content/article/2011/01/20/AR2011012001581.html. January 20. Accessed 20 Jan 2011.
Williamson, J. 1978. *Decoding advertisements: Ideology and meaning in advertising*. London: Marion Boyers.
Winson, A. 2004. Bringing political economy into the debate on the obesity epidemic. *Agriculture and Human Values* 21(4): 299–312.
Zinkhan, G.M., and L. Carlson. 1995. Green advertising and the reluctant consumer. *Journal of Advertising* 24(2): 1–6.

Chapter 7
Increasing Demand for Pasture-Based Dairy: What Attributes and Images Do Consumers Want?

Kristin L. Getter, Bridget K. Behe, Philip H. Howard, David S. Conner, and Lia M. Spaniolo

7.1 Introduction

Pasture-based dairy farming in Michigan has been identified as a component of a possible solution to the dual problems of aging farmer population and disappearance of small- and medium-scale farms, as this model is particularly appropriate for younger and beginning farmers (Conner et al. 2007). Previous research suggests a host of ancillary environmental and social benefits accompanying the use of this model such as reduced soil erosion and phosphorus losses, increased soil carbon sequestration potential, improved quality of life at the farm level (including increased income per cow and less capital investment) and community level (including lower poverty and unemployment rates and greater civic engagement) as compared to traditional systems (DiGiacomo et al. 2001; Rotz et al. 2002; Conant et al. 2003; Bishop et al. 2005; Taylor and Foltz 2006; Conner et al. 2007). Due to its lower start-up costs and greater profitability per production and animal unit, pasture-based dairy can provide a viable family income on a family farm scale (Dartt et al.

K.L. Getter (✉) • B.K. Behe
Department of Horticulture, Michigan State University, 1066 Bogue Street, Room A240-E, East Lansing, MI 48824, USA
e-mail: getterk@msu.edu; behe@msu.edu

P.H. Howard • L.M. Spaniolo
Department of Community Sustainability, Michigan State University, 480 Wilson Road, Room 316, East Lansing, MI 48824, USA
e-mail: howardp@msu.edu; spaniol9@msu.edu

D.S. Conner
Community, Development and Applied Economics, University of Vermont, Morrill Hall Rm 205H, Burlington, VT 05405, USA
e-mail: David.Conner@uvm.edu

B. Freyer, J. Bingen (eds.), *Re-Thinking Organic Food and Farming in a Changing World*, The International Library of Environmental, Agricultural and Food Ethics 22, DOI 10.1007/978-94-017-9190-8_7

1999; Nott 2002; Kriegl and Frank 2004; Conner et al. 2007). Product differentiation based on consumer demand for specific attributes is a well-established strategy for small- and medium-scale farmers who lack the volume to provide family viable incomes given the tight margins of commodity markets (Lancaster 1966; Porter 1985; Duffy 1998). Yet, according to a recent study, most (80 %) pasture-based dairy farmers in Michigan sold their products to commodity pools, including Michigan Milk Producers Association, rather than as differentiated products (Haan et al. 2011). This presents a probable missed opportunity, as research in Michigan finds that consumers value pasture-based dairy products, are likely to buy them if available and the majority (87 %) are willing to pay a premium price (Conner and Oppenheim 2008). Demand for these products is based on perceived benefits, including improved human health, and animal welfare, and reduced environmental impacts, which are also rated as important by large majorities of consumers in the United States (Howard 2006; Conner et al. 2008; Howard and Allen 2010).

Consumers in one recent study, for example, stated they were willing to pay a $1.00 per gallon (median) premium for pasture-raised milk (Conner and Oppenheim 2008). Multiplied by one-half (reflecting median) of the state's population (about five million) (U.S. Census Bureau 2010) and per capita milk consumption (23.9 gal per year according to USDA (Economic Research Service 2012), these findings imply a net gain of more than $117 million to the state's dairy industry if half the state's milk consumption could be converted to pasture-raised milk, made available to consumers and identified as such. While this number is admittedly an upper bound, it demonstrates the potential gains if this demand is met.

One key message of this study was the current lack of effective promotional messages to communicate desired product attributes to potential pasture-based dairy consumers (Conner and Oppenheim 2008). The goal of this research was to fill this gap by using focus groups to create and test promotional messages, while also investigating the current perceptions and attitudes of dairy consumers. The results will be incorporated into a subsequent online survey and may ultimately be used to expand the demand for pasture-based dairy. The goal was to generate and/or identify key themes that resonate best with milk consumers to help connect them with the important attributes of the product, leading to increased profitable sales.

Consumers participating in this study were recruited from retail food co-operatives and natural food stores in Michigan, since the growth of organic eco-labeled products took root in these niche markets and are likely places for alternative eco-labeled products, like pasture-raised, to also gain support (Organic Trade Association 2010). Similar motivations and values are shared between natural food store shoppers and food-co-op members, although inherent differences also exist. Food co-operative members who frequently purchase organic foods tend to also be concerned for environmental health (Goldman and Clancy 1991; Wilkins and Hillers 1994). Altruistic and relational values are reflected in retail food co-operatives' ideological features of organizational participation and democratic decision-making (Brown 1985) despite these structure's potential hindrances to economic viability (Cotterill 1983). Natural food retailers do not share the ideological structure of cooperatives but have still contributed to the growth of the alternative eco-product's success (Dimitri and Greene 2002). Consumers who frequent markets, like natural health and food retailers, have a "highly adaptable micro-cultural frame of reference for understanding the etiology of illness, the deep sources of wellness [...] a wide range of ecological and social issues and the preferred trajectory of their lives" (Thompson and Troester 2002). As a result of these concerns, natural food and food co-op consumers are among those most likely to provide early support for niche products that seek to address such issues.

7.2 Methods

Two researchers trained in focus group methodology conducted six focus groups dispersed geographically in Lower Michigan in May and June, 2010. Two focus groups were held at each of three cities: East Lansing, MI; Traverse City, MI; Ann Arbor, MI. These cities were selected to represent a wide geographic area of Lower Michigan. Each focus group lasted less than 90 min and participants were paid $30 for their participation.

In each city, researchers recruited participants for one focus group from food co-operative store consumers and the other focus group from natural food store consumers (Table 7.1). However, in one city (Traverse City, MI), there was minimal interest from the natural food store consumers to participate in the focus group, so researchers recruited participants for both focus groups from the local food co-operative in the same city.

Recruitment occurred 4–7 days before each focus group, by a combination of personal and electronic means. For personal recruitment, researchers set up a table with a poster inviting participation in the research near the entrance of the store. Customers of the store were approached by the researchers asking them for their participation. Customers were invited to participate if they were 18 years or older and a consumer of cow's milk, although occasional users were not prohibited, recruiters encouraged participation from more than occasional users. At other times

Table 7.1 Dates and locations of six focus groups held in Lower Michigan

Focus group number	Date	Location	Type of store participants recruited[a]	Type of recruitment method	Recruitment date and time of day
1	5/25/10	East Lansing, MI	N	In Person	5/22/10 mid-morning
2	5/26/10	East Lansing, MI	C	E-mail listserv	5/22/10 afternoon
3	6/09/10	Traverse City, MI	C	In Person	6/2/10 evening
4	6/10/10	Traverse City, MI	N[b]	In Person	6/2/10 evening
5	6/21/10	Ann Arbor, MI	N	In Person	6/17/10 late morning
6	6/28/10	Ann Arbor, MI	C	Facebook	6/21/10 afternoon

[a]*N* natural food store, *C* food co-operative

[b]In Traverse City, MI, there was low customer volume and consumer interest in this research at the natural food store. As a result, some food co-operative shoppers were also invited to this focus group

customers approached the table voluntarily if they were interested in the poster developed specifically to recruit consumers who drank milk for a study on milk attributes. For electronic recruiting, researchers sent out an e-mail to a store's customer listserv or used a Facebook announcement from the store, both which invited participation in the study. Potential participants were instructed to contact the researchers via email. The researchers then replied to their interest by setting up an appointment to the focus group. Recruiting was performed during a variety of days (weekdays, weekends) and times (morning, afternoon, evening) in an effort to capture the full variability of demographics in potential shoppers.

Each focus group session was audio recorded and followed the same format and sequence of question stimuli. Using crayons, participants were first asked to "Please draw a milk label that represents attributes that you are looking for while purchasing milk". Crayons were specifically used to give the impression that this was a fun project and the artwork itself was not being judged, but rather to encourage creative generation of important points or attributes as the main goal. The second portion of the focus group consisted of answering open-ended oral questions which focused on the reasons they purchase their current milk product and opinions of alternative milk products. Questions included 'What is the most important thing you look for when you buy milk? Why?', 'What are your opinions about pasture-raised or organic milk?' Prior to the latter question, we defined organic milk according to the USDA definition (Agricultural Marketing Service 2009) and pasture-raised milk (or pasture-grazed or simply pastured milk) as milk coming from cows that spend most of their time outdoors in pasture (rather than in a barn) grazing on grasses and forage for a significant amount of their food, rather than eating mixed feed rations.

The third portion of the focus group included showing the participants five potential label images that were developed by a local graphic-artist (Fig. 7.1). Each potential image contained two versions shown next to each other on a piece of paper, one with a black and white cow (Holstein) and one with a brown cow (Jersey). Participants were asked for their reaction and input regarding these potential labels. Images were shown randomly at each focus group. Questions asked about each image included 'What words come to mind when you see this image?', 'What do you like or not like about this image?', 'Would you be interested in buying a product with this label? Why or why not?' After all images had been examined, participants were asked which label they liked best and why. Finally, consumers were asked to complete a written survey that queried demographic information (such as age, race, gender, income level, and education).

The audio tapes were transcribed verbatim by one researcher who was present at all six focus group sessions. This same researcher then read and coded the transcripts using standard procedures (Morgan and Krueger 1998). Emergent themes were developed from the coding. The same researcher who transcribed the focus groups also visually examined each of the participant hand-drawn images and generated a list all components of the drawings in both the pictures represented and the wording on the labels. Statistical analysis (SAS 9.2, SAS Institute, Cary, NC) was used to generate means and standard deviations from the written questionnaires (demographic information).

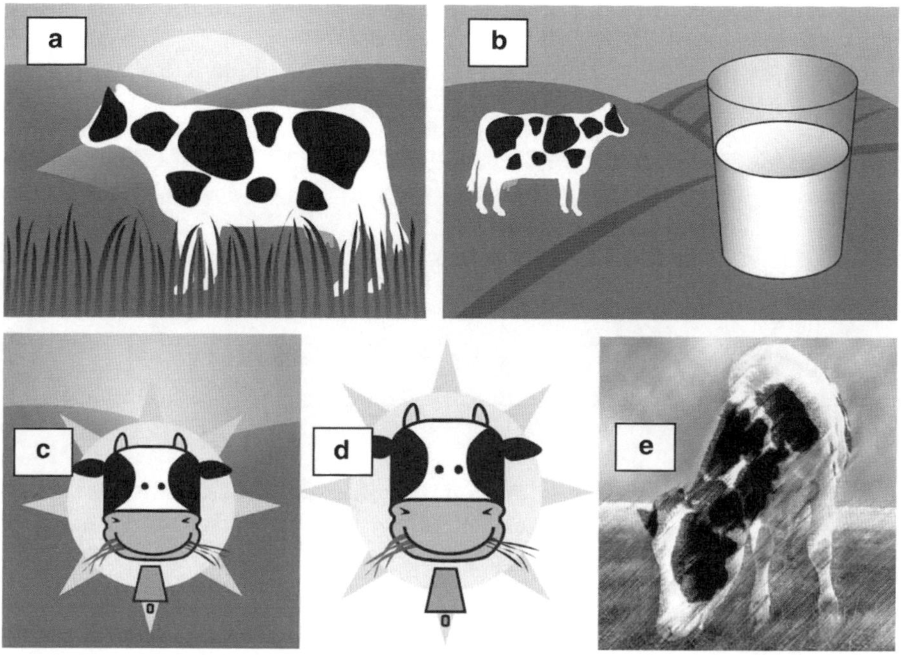

Fig. 7.1 Potential marketing label images developed by a graphic artist. Images were shown to participants in random order. For the first two focus groups, each image was shown twice (for a total of ten images); once as represented here and once with the same image, but substituting a brown cow. Due to a strong dislike of the brown cow images, only the images here were shown in the last four focus groups

7.3 Results and Discussion

7.3.1 Demographics

A total of 55 people participated in the six focus groups (Table 7.2). From the self-administered written survey, demographics of the participants indicated that the majority of participants were Caucasian (87 %) and female (75 %), the average age was 43 years, and the most frequent family gross income category for 2009 was $20,000–$60,000. In addition, all participants had received at least a high school education and 75 % had at least a college baccalaureate degree. Ethnicity, average age, and percent female are all higher than Michigan population demographic statistics which reports these averages as 80.2 % Caucasian, 35.5 years, and 51 % female, respectively. Education levels reported here are also higher than the Michigan average of 83.4 % of the population earning high school degrees or higher and only 21.8 % of the population holding bachelor degrees or higher. However, incomes reported here are similar to the state median household income of $61,617 (U.S. Census Bureau 2010).

Table 7.2 Demographic characteristics of participants in six focus groups in Lower Michigan

	Focus Group Number						
Characteristic	1	2	3	4	5	6	Total
N	11	11	7	14	8	4	55
Age (SD)	38.3 (13.6)	43.9 (11.1)	53.8 (16.7)	40.8 (14.9)	39.8 (11.8)	48.5 (12.9)	43.0 (13.9)
Gender							
M	63.6	18.2	0.0	14.3	25.0	25.0	15.5
F	36.4	81.8	100.0	85.7	75.0	75.0	74.5
Total Gross Family Income in 2009 (%)							
Less than $20,000	18.2	30.0	20.0	21.4	25.0	66.7	25.5
$20,000–$60,000	27.3	10.0	20.0	57.1	50.0	33.3	35.3
$60,000–$100,000	36.3	40.0	40.0	14.3	12.5	0.0	25.5
$100,000+	18.2	20.0	20.0	7.1	12.5	0.0	13.7
Education (%)							
Completed High School	9.0	36.4	28.6	0.0	12.5	25.0	16.7
Completed Associates Degree	0.0	9.1	0.0	15.4	12.5	0.0	7.4
Completed Bachelors Degree	45.5	45.5	71.4	76.9	37.5	50.0	55.6
Completed Graduate Degree	45.5	9.1	0.0	7.7	37.5	25.0	20.3
Ethnicity (%)							
African-American	9.1	0.0	0.0	7.1	12.5	0.0	5.6
Asian	9.1	0.0	0.0	0.0	0.0	0.0	1.8
Caucasian	63.6	90.0	100.0	92.9	87.5	100.0	87.0
Hispanic	9.1	10.0	0.0	0.0	0.0	0.0	3.7
Native American	9.1	0.0	0.0	0.0	0.0	0.0	1.8
Average Gallons/week of milk purchased (SD)							
Whole milk	0.58 (0.38)	0.69 (0.48)	1.7 (2.0)	0.7 (0.3)	0.6 (0.3)	0.8 (0.4)	0.8 (0.7)
Reduced Fat	0.7 (0.3)	0.9 (0.3)	1.0 (0.5)	0.8 (0.3)	0.5 (na)	1.0 (na)	0.8 (0.3)
Skim	0.75 (0.3)	0.5 (na)	0.5 (na)	1.1 (0.8)	0.7 (0.3)	1.0 (na)	0.8 (0.4)

Participants routinely purchased nearly 1 gal per week of milk products, which is significantly higher than the U.S. average of 0.4 gal of plain fluid milk consumption per week (University of Wisconsin Dairy Marketing and Risk Management Program 2005). However, we did recruit only people who routinely purchase milk products, excluding those who did not. In addition, milk purchases were self reported as evenly split between whole milk, reduced fat milk (1 % or 2 %) and skim milk, which is also slightly different than the national average whereby nearly half of plain milk purchased is reduced fat milk (University of Wisconsin Dairy Marketing and Risk Management Program 2005).

7.3.2 Participant Generated Images

A total of 52 images (30 from natural food store customers and 22 from cooperative food store customers) were collected from participant's hand-drawn milk labels that represented the attributes that were important to them when making a purchasing decision about milk. The most common themes in hand-drawn images included cows in pasture, blue skies, and sunshine (Table 7.3), which were similar to the themes that the professional graphic artist chose for potential pasture-based milk labeling (Fig. 7.1). Wording on these same labels indicated that participants generally wanted organic, hormone free, local, grass-fed milk products (Table 7.4). Differences in images and oral questions (below) between the two recruiting sites were tested, but were not significantly different.

Table 7.3 Themes from participant-generated images across six focus groups (n = 52)

Image	Number of times image appeared (% of n)
Pasture	23 (44 %)
Single cow	20 (38 %)
Sunshine	19 (36 %)
No picture	13 (25 %)
Blue sky	9 (17 %)
Multiple cows	9 (17 %)
Fence	6 (11 %)
Glass bottle	6 (11 %)
Barn	4 (8 %)
Trees	3 (6 %)
Chewing grass	2 (4 %)
Flowers	2 (4 %)
Glass of milk	1 (2 %)
Label like cow blotches	1 (2 %)
Processing plant	1 (2 %)
Udder	1 (2 %)

Table 7.4 Wording found in participant-generated labels across six focus groups (n = 52)

Wording	Number of times wording appeared (% of n)
Organic	32 (62 %)
No hormones	24 (46 %)
Local	15 (29 %)
Grass fed	11 (21 %)
Fresh product	10 (19 %)
Free range	7 (13 %)
No antibiotics	7 (13 %)
Happy	6 (12 %)
Healthy	6 (12 %)
Expiration date	5 (10 %)
Natural	4 (8 %)
Pasteurization	4 (8 %)
Family owned	3 (6 %)
No pesticides	3 (6 %)
Non-GMO	3 (6 %)
Pure, clean	3 (6 %)
Grass	2 (4 %)
Homogenized	2 (4 %)
Sunshine	2 (4 %)
Animal welfare	1 (2 %)
Fresh air	1 (2 %)
Humane certified	1 (2 %)
Lactose free	1 (2 %)
No commercial feed	1 (2 %)
No factory farm	1 (2 %)
Price	1 (2 %)
Raw milk	1 (2 %)
Taste	1 (2 %)
Wholesome	1 (2 %)

7.3.3 Oral Questions

Most important attributes of milk. Prior to any mention of pasture-raised or organic milk by researchers, participants were asked what the most important attributes they were looking for when buying milk. Participants' responses followed several themes, but overwhelmingly most sought organic products. Some specifically wanted the product to be certified (USDA or another certifying agency), while other participants said milk didn't need to be certified but they wanted it to be pure (free of additives).

The majority felt that these attributes resulted in a healthier and tastier product to consume compared to conventional milk. This viewpoint seems to reflect a growing body of research that organic milk, as well as pastured milk, often have higher conjugated linoleic acid (CLA) profile than conventionally raised milk cows (Bloksma et al. 2008; Butler et al. 2009) which may be beneficial to human health in their anti-carcinogenic and anti-atherogenic properties (Bauman and Lock 2006). However, some authors argue that this CLA difference is so small and that organic milks are higher in non-desirable fatty acids (saturated and trans-fatty, which are associated with coronary heart disease) that overall these "specialty labeled milks are similar in nutritional quality, and wholesomeness to their conventional counterparts" (O'Donnell et al. 2010). Research conclusions are also mixed as to whether organic milk tastes better than conventionally raised milk products. Some studies have found no significant difference in flavor between conventional and organic milk, even though organic was described as creamier and consisting of a more grassy taste than conventional milk (Bloksma et al. 2008). Others have found flavor differences between pasture-raised cows that used organic practices (although not certified organic) and conventional total mixed ration fed cow's milk (Croissant et al. 2007).

Participant's also expressed concern about the welfare of the animals and environmental concerns (e.g., free range). Many wanted to know that the cows were treated ethically. One participant commented that "It's just that you want to feel like you're doing something good with the products that you consume, you want to feel like you are supporting something that doesn't do damage to the environment or do horrible things to animals." In fact, they felt the most ethical treatment of the cows was to pasture them as much as possible, a viewpoint in-line with research which found improved animal welfare with increasing time at pasture (Hernandez-Mendo et al. 2007; Olmos et al. 2009). Participants indicated that locally produced milk was also important because they felt a local producer would be held more accountable for the quality of the product. In addition, they believed less fuel would be used due to the close proximity of producer to consumer, thus contributing to the sustainability of the product.

Other concerns included milk price and safety. These safety topics ranged, but were mostly confined to additives and GMOs. Although the price of milk was a concern for many participants in this study, it was rarely the very first attribute mentioned as being most important in their milk purchase. In addition, when asked by the moderator which product they would purchase if price were equal between conventional milk and organic milk, participants all indicated they would buy organic milk. For safety, most said they wanted pasteurized milk, although discussions about raw milk (milk that has not been pasteurized or homogenized) occurred in all but one focus group. In Michigan, all milk sold is required to be pasteurized, but raw milk is still available directly from the farm by purchasing a cow-share. Of those participants who currently drink raw milk, they indicated that safety was not an issue because they visit and see the farm and how milk was handled and processed. One participant who consumes raw milk said "(Raw milk) is way more nutritious and we know the people that are producing it and so feel safe, because we know where the source is." Others were concerned about getting sick from consuming non-pasteurized milk. One participant said, "I didn't want to rain on anyone's parade, but tuberculosis passes from cows to us if the farmers don't keep things cleaned, so you have to trust the farmer with your health. I mean tuberculosis will affect you."

Overall, these participants indicated that how the cow was raised was the most important milk attribute, which is different than other research. For instance, Pirog (2004) and Shelquist (2002) both found that freshness and taste were the most important attributes of milk for consumers. Differences between this present study and the others mentioned here are likely due to demographic differences, as both studies did not exclude those people who do not regularly purchase milk, while this present one did.

Pastured milk. Following the initial question about general milk attributes that are important to them, participants were then asked to discuss their opinions in-depth regarding pastured milk. In the first two focus groups, the term 'pasture-raised' or 'pasture-grazed' milk was used to describe cows feeding in pasture, but these terms were confusing to many participants, sounding too similar to pasteurized milk. This is contrary to findings of another study in the Midwest that reported the term 'pasture raised' to be favored by focus group participants (Shelquist 2002). However, that study was not specifically aimed at milk, but rather all meat, poultry, and diary purchases. Because of the confusion that arose between the terms "pasteurized" and "pasture raised", the term 'pastured' milk was used exclusively in the last four focus groups. This reduced confusion in subsequent focus groups and focused more time on milk attribute discussion.

Overwhelmingly, specific opinions about pastured milk included concern for the animal. Participants felt that there was a moral issue of how the animals are treated. One participant indicated "it seems like if we are going to ask them to feed us, we then have an obligation to steward them properly and ethically and morally." Another consumer felt that pasturing cows is "a more pure and natural way for the cow and for the end product. It naturally has to follow that it would be better for us." This sentiment is consistent with a previous study on pasture-raised animal products (Shelquist 2002).

While it may be more expensive, which was a concern for a few participants, many felt that the increased price is worth it. Not only does pastured milk taste better, but they felt it is better for human health as well as the environment, too. One person stated, "to my mind, you are going to pay for it whether you pay for it up front when you buy the milk or on the back end when you go to the doctor." Others agreed in that they felt the environmental degradation that is potentially caused by some conventional producers of milk is not reflected in traditional milk pricing. These findings are similar to Pirog (2004) who found that 60 % of conventional milk buyers knew of potential human health benefits and more humane treatment of cows in pastured dairy systems.

Organic milk. Participants were also asked to discuss their opinions of organic milk. The same three themes (better for human health, better for the animal's health, and better for the environment) were found when discussing organic milk specifically. One participant stated, "The organic thing to me is important – it is in the farming practices – it is in how we are treating the land, it's the chemicals that we are infusing the environment with, its pesticides and fertilizers, etc., etc. I know we can do without this and I know it will take a long time to get there, because we have a lot of correcting to do, but I think it is something we should be looking for and striving for."

Some people were more concerned about the cows receiving no antibiotics and growth hormones than if the cow is pastured due to human health reasons, while other participants were equally concerned about the treatment of the animal. Many assumed organic practices meant humane animal treatment and there was general surprise that organic milk wasn't necessarily pastured milk. One participant said, "I couldn't believe that an organic farm wouldn't pasture their cows. When I realized that, I went 'what?'"

Based on these comments, participants in this study clearly thought organic milk meant that the cows were also pastured and most then wanted to make sure that their organic milk was also pastured milk. While the USDA has recently implemented new grazing rules for their organic certification program, existing certified organic milk producers have until 17 June 2011 to comply. The new grazing rule mandates that animals must graze during the grazing season, which is defined locally and must be at least 120 days per year. During this time, animals must have at least 30 % of their dry matter intake from pasture (Agricultural Marketing Service 2010).

Even though most of these consumers indicated that they regularly purchase organic milk, there was still much discussion about whether the USDA organic certification could be trusted. One participant stated, "How do you define or do you even know what the regulations are and do you trust the USDA in their oversight of the farm versus your own eyes (seeing the farm)?" Most agreed that the program is fallible because the government is in charge of enforcing the standards and because "mega-corporations" had the power to block regulations. From an economic viewpoint, one participant commented that "If I knew I could trust the organic label, then I would be more likely to buy it. If you are paying twice as much (for the product) and you're not sure (if the organic standards are being followed), then you won't buy it." These comments are consistent with a recent audit of the National Organic Program that found weaknesses in the programs oversight and enforcement of regulations (U.S. Department of Agriculture Office of Inspector General 2010). In fact, there were several instances in this report where producers were falsely marketing their products as organic when they knowingly did not follow the organic standards.

Due to this issue, many participants indicated that they would prefer to buy their milk from local farmers that have organic practices (regardless of whether they are certified organic or not), because they could see how the farmer produced the product. One participant said, "There is a huge disconnect between humans and food consumption – we need to see how farms are run. Even if the farmers aren't certified organic, I want to buy from a farmer who has the organic mindset – not just in it for the money." Even though it is not feasible to observe farmer operations 24-h a day, participant's still felt that a local farmer would be more trustworthy.

One consumer said, "At some point you have to trust your farmer [. . .] and that is why I like to buy local products. It seems to me even if you don't know them, other people do and word-of-mouth, and it seems to me they would be more responsible to their public/consumers than a conglomerate shipping across the country."

7.3.4 Responses to Graphic Artist Images

Because two versions (one with a black and white cow (Holstein) and one with a brown cow (Jersey)) of each of the five images were shown during the first two focus groups, most of the discussion involved which cow species better represented their ideal alternative label milk product. Overwhelmingly, unless the attendee knew something about cows, participants identified with the black and white cow, which reflected a typical Holstein breed, and did not like the brown cow, which depicted a more typical Jersey cow. Consequently, in order to get feedback on the image itself, in subsequent focus groups the brown cow image was dropped and only the black and white cow image remained.

Of the five potential label images presented, the participants voted Fig. 7.1e, a realistic depiction of a cow in the act of grazing on pasture, as their favorite, receiving 19 votes of 38. They liked the fact that the image was both artistic and realistic and conveyed exactly what they were looking for: a cow grazing on grass outside. Some people didn't like the blurry effect and said they would prefer a hand-drawn or a real picture of a cow grazing at pasture like this one. Figure 7.1a, an iconic cow on grass with a sun in the background, was also well-liked, receiving 14 votes as the favorite image. They described this picture as being happy, serene, or peaceful, as well as being realistic and serious. One participant said that "I would expect to see this type of label on an organic pasture-based product." Another said, "I could put myself in this picture." Others felt it implied a more 'natural' product. Negative comments on this image were similar to the previous image in that people didn't like the 'clip-art' style and would prefer a hand-drawn or real picture.

Participants were divided in their liking or disliking for Fig. 7.1c, d, both of which displayed a logo-type iconic cow's head in front of a sun either displayed in front of a pasture or on a white background, respectively. While only voted as being a favorite by two and three people respectively, they still described these figures as sunny, bright, and cheerful. Many people especially liked the simplicity of Fig. 7.1d, while most found that Fig. 7.1c was too busy. Negative responses to both images included too cartoonish or childish and unrealistic. When asked if they would buy this product, the response was generally neutral.

Figure 7.1b, a small iconic cow on grass next to a large glass of milk, was overwhelmingly disliked, receiving zero votes as the favorite image. Participant's generally felt that the scale of the cow to the glass of milk was 'ridiculous'. This was the only image where participants said they would likely not purchase a milk product with that image on it. All other images were either favorable or neutral in future purchasing decisions.

There were two other themes that developed during discussion of these potential marketing labels. First, several people said they wanted to see a story on the label, such as how the milk was produced or a history or philosophy about the farm. One person said, "All of these images are so generic. If a farm wanted to put out their product, I mean, I would want to know something about their farm." Others agreed, saying that the product should have a picture of their farm on it that says 'this is where the milk comes from'. A second theme developed in five of the six focus groups, whereby the participants preferred glass-bottled milk. They felt that the glass bottle alone with the product in it is enough marketing in and of itself; no labeling is required, because milk is such a simple product and the extra advertising is unnecessary. They also felt that glass bottled milk gives the impression of an old-fashioned farmer and has a wholesome connotation. One participant said, "It's that image of 1950s milk delivered right to your door." Other participants also cited that glass was better because of human health concerns with leaching in the traditional plastic milk containers and that the taste of the milk is maintained in a glass container. These findings suggest that future research should evaluate more than the milk label, but rather the entire package.

7.4 Conclusions

These food co-operative and natural food store consumers who participated in these six focus groups all expressed an interest in pastured milk. This was based on beliefs that non-conventional methods for raising milking cows are better for human health, better for the animal's health, and better for the environment. While price of the product was a concern for most participants in this study, it was rarely the very first attribute mentioned as being important in their choosing of a milk product. Many participants felt that locally produced milk was also very desirable because they could see how the farmer produced the product. Opinions on potential milk labels for these alternative milk products varied, but most preferred an image that was both artistic and realistic and portrayed a cow actively feeding at pasture. Because there was also discussion about not just the milk label, but also the entire milk container, future research should evaluate the entire milk package.

Results led researchers to conclude that several items should be included in subsequent research to expand the market for pastured milk. The package needs to be examined as it has much to do with shelf life, product quality preservation, as well as the attitudinal and perceptual impact on consumers. The product label likely needs to include some information on the differences and benefits of pastured milk over organic milk. Consumers who participated in this study were typically unaware and surprised by the fact that organic did not automatically equal pastured. Interest about the farm and/or farmer was also important and should be included in subsequent package/label evaluations. This study provided important insight into the themes and concerns of milk drinkers as researchers seek effective communication points to help improve profitable sales of pastured milk.

The results of this study are limited to the participants in our study and are not generalizable to other populations. However, given the geographic scope of the studies and efforts to recruit from two types of stores, the results increase our understanding of the breadth of attitudes and beliefs held by likely consumers of these products.

Future directions of work include efforts to create and administer a survey based on the emergent themes of this study, to better understand frequency and interaction of these beliefs and impact on consumer behaviors in a larger sample. This survey is planned for early 2011 and will further increase understanding of consumer demand for milk products and inform marketing efforts.

References

Agricultural Marketing Service. 2009. National Organic Program. Retrieved from http://www.ams.usda.gov/AMSv1.0/nop. Accessed 2010.

Agricultural Marketing Service. 2010. *National Organic Program 2010 – Access to Pasture (Livestock)*, vol. 2010. Washington, DC: United States Department of Agriculture/Agricultural Marketing Service.

Bauman, D., and A. Lock. 2006. Conjugated linoleic acid: Biosynthesis and nutritional significance. In *Advanced dairy chemistry, vol. 2 lipids*, ed. J.L. Lojpersberger and J.A. Bloomfield, 93–136. New York: Springer.

Bishop, P.L., W.D. Hively, J.R. Stedinger, M.R. Rafferty, J.L. Lojpersberger, and J.A. Bloomfield. 2005. Multivariate analysis of paired watershed data to evaluate agricultural best management practice effects on stream water phosphorus. *Journal of Environmental Quality* 34(3): 1087–1101.

Bloksma, J., R. Adriaansen-Tennekes, M. Huber, L.P. van de Vijver, T. Baars, and J. de Wit. 2008. Comparison of organic and conventional raw milk quality in the Netherlands. *Biological Agriculture & Horticulture* 26(1): 69–83.

Brown, L.H. 1985. Democracy in organizations: Membership participation and organizational characteristics in US retail food co-operatives. *Organization Studies* 6(4): 313–334.

Butler, G., M. Collomb, B. Rehberger, R. Sanderson, M. Eyre, and C. Leifert. 2009. Conjugated linoleic acid isomer concentrations in milk from high-and low-input management dairy systems. *Journal of the Science of Food and Agriculture* 89(4): 697–705.

Conant, R.T., J. Six, and K. Paustian. 2003. Land use effects on soil carbon fractions in the southeastern United States. I. Management-intensive versus extensive grazing. *Biology and Fertility of Soils* 38(6): 386–392.

Conner, D.S., and D. Oppenheim. 2008. Demand for pasture-raised livestock products: Results from Michigan retail surveys. *Journal of Agribusiness* 26(1): 1–20.

Conner, D.S., M. Heller, S. Cocciarelli, and M. Hamm. 2007. *Opportunities in grazing dairy farms: Assessing future options*, Occasional white paper. East Lansing: CS Mott Group for Sustainable Food Systems at Michigan University.

Conner, D.S., V. Campbell-Arvai, and M.W. Hamm. 2008. Value in the values: Pasture-raised livestock products offer opportunities for reconnecting producers and consumers. *Renewable Agriculture and Food Systems* 23(01): 62–69.

Cotterill, R. 1983. Retail food cooperatives: Testing the "small is beautiful" hypothesis. *American Journal of Agricultural Economics* 65(1): 125–130.

Croissant, A.E., S. Washburn, L. Dean, and M. Drake. 2007. Chemical properties and consumer perception of fluid milk from conventional and pasture-based production systems. *Journal of Dairy Science* 90(11): 4942–4953.

Dartt, B., J. Lloyd, B. Radke, J. Black, and J. Kaneene. 1999. A comparison of profitability and economic efficiencies between management-intensive grazing and conventionally managed dairies in Michigan. *Journal of Dairy Science* 82(11): 2412–2420.

DiGiacomo, G., C.J. Iremonger, L. Kemp, C. van Schaik, and H. Murray. 2001. *Sustainable farming systems: Demonstrating environmental and economic performance*. St. Paul: Minnesota Institute for Sustainable Agriculture.

Dimitri, C., and C. Greene. 2002. *Recent growth patterns in the US organic foods market*, Agriculture Information Bulletin, 777. Washington, DC: U.S. Department of Agriculture.

Duffy, M. 1998. *How small farms compete*. Paper presented at the Agricultural outlook forum.

Economic Research Service. 2012. Organic production: Data sets overview. Retrieved from http://www.ers.usda.gov/data-products/organic-production.aspx. Accessed 1 March 2012.

Goldman, B.J., and K.L. Clancy. 1991. A survey of organic produce purchases and related attitudes of food cooperative shoppers. *American Journal of Alternative Agriculture* 6(2): 89–96.

Haan, M., D.S. Conner, and P. Taylor. 2011. Pasture-based dairying in Michigan: Farmer practices and needs. *Journal of Extension* 49(3).

Hernandez-Mendo, O., M. Von Keyserlingk, D. Veira, and D. Weary. 2007. Effects of pasture on lameness in dairy cows. *Journal of Dairy Science* 90(3): 1209–1214.

Howard, P. 2006. Central Coast consumers want more food-related information, from safety to ethics. *California Agriculture* 60(1): 14–19.

Howard, P., and P. Allen. 2010. Beyond organic and fair trade? An analysis of ecolabel preferences in the United States. *Rural Sociology* 75(2): 244–269.

Kriegl, T., and G. Frank. 2004. *An eight year economic look at Wisconsin dairy systems*. Madison: University of Wisconsin Center for Dairy Profitability.

Lancaster, K.J. 1966. A new approach to consumer theory. *The Journal of Political Economy* 74(2): 132–157.

Morgan, D.L., and R.A. Krueger. 1998. *Focus group kit*. Thousands Oaks: Sage Publications.

Nott, S. 2002. *Dairy grazing finances in 4 great lakes states 2000*. East Lansing: Michigan State University, Department of Agricultural, Food, and Resource Economics.

O'Donnell, A., K. Spatny, J. Vicini, and D. Bauman. 2010. Survey of the fatty acid composition of retail milk differing in label claims based on production management practices. *Journal of Dairy Science* 93(5): 1918–1925.

Olmos, G., L. Boyle, A. Hanlon, J. Patton, J.J. Murphy, and J.F. Mee. 2009. Hoof disorders, locomotion ability and lying times of cubicle-housed compared to pasture-based dairy cows. *Livestock Science* 125(2): 199–207.

Organic Trade Association. 2010. U.S. organic industry overview. Retrieved from http://www.ota.com/pics/documents/2010OrganicIndustrySurveySummary.pdf. Accessed 12 Oct 2012.

Pirog, R. 2004. *Consumer perceptions of pasture-raised beef and dairy products*, Leopold Center, Iowa State University. Ames: Leopold Center, Iowa State University.

Porter, M. 1985. *The competitive advantage: Creating and sustaining superior performance*. New York: Free Press.

Rotz, C.A., A.N. Sharpley, L.D. Satter, W.J. Gburek, and M.A. Sanderson. 2002. Production and feeding strategies for phosphorus management on dairy farms. *Journal of Dairy Science* 85(11): 3142–3153.

Shelquist, K. 2002. *Pasture raised products message and strategy; Consumer focus group study*. Ames: Political Farmers of Iowa.

Taylor, J., and J.D. Foltz. 2006. *Grazing in the dairy state: Pasture use in the Wisconsin dairy industry, 1993–2003*. Madison: University of Wisconsin.

Thompson, C.J., and M. Troester. 2002. Consumer value systems in the age of postmodern fragmentation: The case of the natural health microculture. *Journal of Consumer Research* 28(4): 550–571.

U.S. Census Bureau. 2010. State and county quick facts. Retrieved from http://quickfacts.census.gov/qfd/states/26000.html. Accessed 1 Jan 2013.

U.S. Department of Agriculture Office of Inspector General. 2010. *Oversight of the national organic program*, vol. 2010. Washington, DC: United States Department of Agriculture/Agricultural Marketing Service.

University of Wisconsin Dairy Marketing and Risk Management Program. 2005. Per capita total beverage milk consumption. Retrieved from http://future.aae.wisc.edu/data/annual_values/by_area/2165?tab=sales. Accessed 1 Jan 2013.

Wilkins, J.L., and V.N. Hillers. 1994. Influences of pesticide residue and environmental concerns on organic food preference among food cooperative members and non-members in Washington State. *Journal of Nutrition Education* 26(1): 26–33.

Chapter 8
Post-national Organic: Globalization and the Field of Organic Food in Israel

Rafi Grosglik

8.1 Introduction

In 1993 McDonald's opened its first branch in Israel. The chain soon became a clear culinary signifier of cultural globalization in Israel (Ram 2005). Seventeen years later, *Guy Maroz*, a journalist specializing in the political-documentary genre conducted a "sensational" television investigative report, accompanied by an experiment: for a month he ate only organic food for the purpose of examining its effect on his health. "*The idea was taken from Morgan Spurlock, the man from the movie 'Super Size Me', who ate only McDonald's*" said the journalist. At the end of the experiment, after continuously consuming organic foods, *Maroz* complained that he gained 6 lb of body weight and poured ewers of scorn against the "*organic food trend that conquered the Western world*".[1] This media event demonstrates that these days, organic food has become a symbol of Westernization and an object of condemnation – exactly the way McDonald's was in the 1990s.

The literature dealing with the social aspects of organic agriculture is not blind to the concept "globalization". Some see the appearance of organic agriculture as the most vocal opponent of industrial agriculture in the global era (Knight 2010, p. 203) and respectively, organic foodstuffs are perceived as the complete opposite of industrialized foodstuffs. Some argue that with its emergence, in the first half of the twentieth century, organic agriculture promoted concepts of production and consumption

[1]Taken from: (Maroz 2010 [in Hebrew]).

R. Grosglik (✉)
Department of Sociology and Anthropology, Ben-Gurion University of the Negev, Beer-Sheva 84105, Israel
e-mail: rafig@post.bgu.ac.il

B. Freyer, J. Bingen (eds.), *Re-Thinking Organic Food and Farming in a Changing World*, The International Library of Environmental, Agricultural and Food Ethics 22, DOI 10.1007/978-94-017-9190-8_8

based on mutual trust, local knowledge, species diversity and social justice. These concepts were derived from total resistance to industrialized food production based on efficiency, competitiveness and standardization; and against food systems that operate on a global scale (Arce and Marsden 1993; Murdoch et al. 2000).

On the opposite side, others claim that the field of organic food changed in the last few years from a network of local producers and consumers to a global institutionalized and industrialized system, and became the food of social elite (Buck et al. 1997; Guthman 1998, 2004a). In the United-States, for example, organic food attained the scornful moniker "Yuppie-chow" (Guthman 2003). Accordingly, various studies point to the accelerated new global economy of the last three decades as the cause for transforming organic agriculture from an "alternative" social movement to an agricultural industry with characteristics similar to conventional agriculture. In these studies of what is known as the "conventionalization thesis" (Guthman 2004a) it is argued that organic agriculture has been included, integrated and even "hijacked" (Engler 2012) into the dominant forms of the "conventional" and global agribusiness. This thesis points to the fallacy of the image that was cultivated by the organic food industry as embodying an alternative cultural and economic philosophy (Marsden and Arce 1995; Buck et al. 1997; Coombes and Campbell 1998; Guthman 1998), and suggests that the organic food sector became structurally assimilated into the prevailing global-industrialized food systems (Guthman 2004b).

These two theoretical perspectives ("*organic food production oriented to locality and driven by farmer-consumer partnership*" and "*organic food as driven by industrial organized production*") can be seen in the evolution of organic food and in globalization as separate processes. But is the realm of organic food, essentially, incompatible to globalization – and the inverse – are globalization and the processes of organic food production interrelated?

Exploring the Israeli field of organic food reveals that its' emergence, in the mid 1980s, was actually a function of processes occurring in the global macro-social level, and worked, essentially, to strengthen economic and cultural globalization in Israel. Furthermore, an increase of demand for organic food by Israeli consumers appeared from 2000 onward and a variety of production and distribution methods were developed. I will argue that these new methods embody different symbolic and materialistic aspects of globalization.

These arguments are based on an empirical study which included content analysis of 36 leaflets and protocols published by the *Israeli Bio-Organic Agriculture Organization* (1989–2006), supplementary data collected from popular media reports dealing with organic food in Israel and in-depth interviews with key agents and actors in the Israeli organic food field.

8.2 Organic Food in Israel as a Global-Cultural-Artifact

During the last two decades organic agriculture has gone through an extensive transformation and changed from an array of separate and local food systems to an institutionalized global system (Raynolds 2004). Global trading volume related to organic agriculture was estimated in 2007 at approximately 46 billion dollars and a growth of about 5 billion per year (Willer and Kilcher 2009, p. 20). Organic food is the nutritional-commercial category with the fastest growth rates in the global food industry (Raynolds 2004). The growth in production and marketing led to the expansion of the range of organic food products and agriculture (Raynolds 2004, p. 732). Likewise, changes occurred in methods of distribution and marketing. On the one hand the distribution of organic food through transnational commercial retail chains expanded greatly,[2] while on the other hand there has been increased activity by social networks and grass roots organizations (Castells 2003, pp. 186–189 [1997]) working to transfer information from a global level and to increase organic agriculture activity at the local level. Thus, it seems that organic food expresses the two main axes of "glocalization" (Robertson 1995): it is a product of global culture, driven by the force of the accelerated neo-liberal and post-Fordist production system, but also an outcome of the longing for local experience and a resistance to the cultural homogenization of industrial modernization.

How do these contradicting aspects of organic food spread and "put down roots" in places beyond its native land (North America and Western Europe)? How are the international organic standards, marketing practices and organizational configurations embedded in different cultures and what are the ways in which organic is "translated" to the Israeli political and cultural context? The appearance and evolvement of deliberately politicized organic practices, as it appears in the Israeli field of organic food, sheds light on these questions.

Since the 1970s a fundamental change has been taking place in the social structure and culture in Israel. This change is expressed, first, in the disintegration of the national political and cultural center (Zionism), with the transition from economic public-governance culture to economic private-business culture and the transformation of Israeli society to an affluent–consumer society. Second, it is possible to point to the formation of two polarized points of identity: the post-Zionist identity – which aspires to globalization and connection to global networks; and, a neo-Zionist identity – which promotes religious national locality (Ram 2005, pp. 27–29). Exploring the emergence and evolution of organic food from a socio-historical perspective demonstrates how organic food reflects, and even takes part in this "politics of identities".

Following is an analysis of the material and cultural dimensions related to organic food in Israel. First, a historical description of the emergence of Israeli organic agriculture (mid 1980s) is discussed. Second, the fragmentation in the organic sector (2000s onward) that led to different configurations of production and consumption is outlined. In this regard, I refer to the developments in organic food production as organized in a distinct space of social action. This space can be seen as a "field"

---

[2]In the United States, the major retail chain stores specializing in marketing organic food are Whole Foods Market and Wild Oats. Alongside them, the conventional retail chains market over a third of the organic food sold in the United States (Raynolds 2004). It should be noted that in most retail chains in the West, organic food products are sold at a higher price, in the range of 20–40 %, than comparable products that are non-organic (FAO/ITC/CTA 2001).

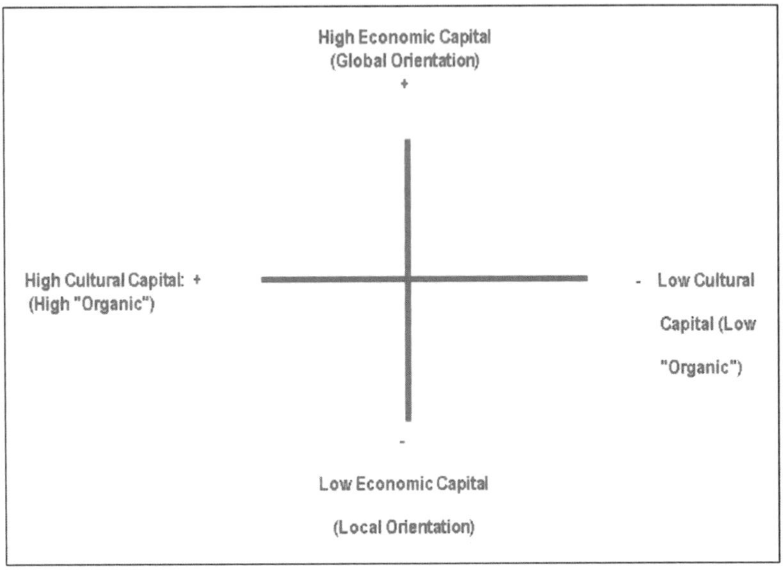

Fig. 8.1 Evaluation axis of the new configurations contained in the Israeli field of organic food

(Bourdieu 1993, 2005), or a space in which actors and agents struggle over the allocation of material resources, prestige and appreciation. This perspective suggests a typology of several cases (categories) representing the constant institutional actors operating in this field. These categories are organized according to different perceptions of the idea of organic inherent in their operation (and therefore they may be differentiated on the basis of degree of "organicness" attributed to them [i.e. organic cultural capital]). In addition, these categories are distinct from each other due to the economic capital and global orientation which they hold (see Fig. 8.1). Finally, in the discussion of these observations, I will claim that the global socio-economic conditions are the ones responsible for the fragmentation of this field and that the new organic configurations reflect major social changes occurring in contemporary Israeli society and culture.

8.3 The Genesis of Israeli Organic Agriculture

It is well known that agriculture was central in the Jewish settlement of land of Israel since the late nineteenth century. Agriculture was incorporated into the official national ideology of Israel – Zionism. This ethos led to widespread social support in resource allocation for agriculture, making agricultural production efficient (Weitz 1969, p. 165) and encouraging surplus produce for export (Weitz 1969, p. 169). Since 1967, agriculture was also incorporated into the colonial project of Jewish settlements in the occupied Palestinian territories (Svirsky 2004). In those years, Israeli agriculture relied on conventional agricultural practices, encouraging widespread use of chemical fertilizers and pesticides (Tal 2006, p. 69).

Beginning in 1985, a crisis occurred in Israeli agriculture. There were three principal reasons for this crisis: (1) the dissolution of the national-Zionist ethos led to a devaluation of the reputation of the agricultural sector. (2) A reduction of the share of agricultural produce in the private consumption expenditure was tied to the increase in per capita income[3] (3) Trends of liberalization in the agricultural sector reduced the profitability of local sales. These and other factors led to the growth of agricultural industrialization and a growing dependence on the export market (Yustman 2001, p. 581). Investments in agricultural research and development for new export industries strengthened this trend.

Organic farming, which was not practiced in Israel, seemed feasible only to a handful of people and it was certainly not considered as part of an export strategy. *Mario Levi* was among the first to be interested in organic farming, and his personal story reflects the genesis of organic agricultural in Israel.

Levi, an 87 year-old religious Jewish farmer,[4] immigrated to Israel about a decade before the founding of Israel. He joined the religious kibbutz (Israeli communal settlements) "*Sdeh Eliyahu*" in order to "*inhabit and settle the Land of Israel*" (Taken from an interview with Levi, 13 March, 2011).

Until the 1960s he exported peppers to Europe. This export business flourished until pests caused considerable damage to the crops. *Levi* and his colleagues frequently used pesticides, but to no avail, and the pepper industry was dropped. The sense of failure made him question the effectiveness of conventional agriculture and the economic loss forced him to seek other agricultural activities. During this period, *Levi* was sent by the Israeli Agriculture Ministry to a course on bio-organic agriculture in Switzerland, as part of a government research project to find new agricultural niches. Upon his return to Israel *Levi* decided to lead the establishment of organic agriculture in Israel. First, he turned to his friends in the Kibbutz and suggested that organic agriculture practices should be used to rebuild the exports lost to pests.

When the first attempts to grow crops succeeded, he contacted "*Agrexco*", a company owned by the government, to engage in exporting organic agricultural products. At that time, this company was the only channel linking Israeli agriculture to the global markets. At the same time, *Levi* joined the International Federation of Organic Agriculture (*IFOAM*) and founded the Israeli Organization of Organic Agriculture. The organization brought together a handful of farmer-owned factory farms for growing organic produce intended for export. The organization operated in full cooperation with *Agrexco*. This connection between an organic grower's

[3]In accordance with Engel's law (named after the statistician Ernst Engel), as income rises, the proportion of income spent on food falls, even if actual expenditure on food rises (Zimmerman 1932).

[4]Typical to religious Zionists, Levi holds an ideology that combines Zionism and Jewish religious faith. Typically as well, he strongly supports Zionist efforts to build a Jewish state in the land of Israel.

Table 8.1 Increase in export volume of organic produce in Israel

	Export volume	Financial turnover
1993	6,400 t	40 million NIS[a]
2008	80,000 t	1 billion NIS[a]

Sources: (Eytan 1993, p. 8 [in Hebrew]; Meirav 2009 [in Hebrew])
[a]NIS New Israelian Shekels

organization and a conventional trading company is the basis for the creation and growth of the field of organic food in Israel (see Table 8.1).

Levi's political outlook (which sees agriculture as an integral part of the Zionist national task); his being a part of conventional agricultural establishment; his professional moves that matched the formal national encouragement to find new export agricultural branches (in order to cope with a crisis in Israeli conventional agriculture) – all of these testify that the appearance of organic food in Israel was not related to any counter-culture ideology. It did not emerge from any ideological resistance to the conventional agricultural establishment. On the contrary – during almost two decades the organic supply for Israeli consumers was minimal and the attitude to organic agriculture was as a replacement for effected export branches. Thus, organic agriculture was preserved as a sector that has the potential to strengthen the national-agricultural project and as a means of integration into the global capitalist economy.

Furthermore, during this period the Israeli Organization of Organic Agriculture established many organic farms in Jewish settlements in the Palestinian territories under Israeli military control. This act loaded organic agriculture with neo-national local meanings (i.e. the neo-Zionist perspective of Jewish rights to the land of Israel, including Palestinian territories).

Thus, the dialectics which is typical to the glocalization era can be discerned in this early stage for the operation of the organic field: on the one hand – openness to the global economy through export of organic food, but on the other hand – using organic agriculture to promote local-national projects. In any case, these two processes were integrated with national and conventional agribusiness (and certainly do not represent opposition to it). As such, this early stage represents performance that might be identified with low organic cultural capital and local-national orientation.

8.4 From Agricultural Production to Cultural Production

In the last two decades the interest shown by Israelis in eating as a form of entertainment and in food as an expression of lifestyle is growing. As a result, the proliferation of unique and refined foods – ethnic foods, gourmet food, homemade and slow foods, artisan foods and organic food – has become prominent in the new Israeli culinary repertoire (Table 8.2).

Table 8.2 The growth in consumption of organic produce in Israel

Organic food consumers in Israel – 1993	Organic food consumers in Israel – 2005
1,000[a]	60,000[b]

Sources: [a](Zali 1993, p. 3 [in Hebrew]); [b](Mazori 2005 [in Hebrew])

Until the early 2000s, organic food consumption was demarcated by class and identified mainly with the upper classes, which had the ability to pay the relatively high priced organic food.[5] During this time most of organic food marketing in Israel was done through exclusive stores that specialize in health, medical and lifestyle products ("*Beit Teva*"- "*nature* stores"). Since the early 2000s a variety of marketing methods were developed and enabled more extensive access to organic food. This differentiation is rooted in the bourgeois revolution started in Israel, of which the empowerment of the material culture and adoption of global consumption patterns in daily life of Israelis is central (Ram 2005, pp. 37–46).

The four cases presented below represent a possible typology for the forms of organic production-consumption-distribution which became instituted in light of the rising of a "mosaic" of global and local identities and postmodern styles in Israel.

8.4.1 "Harduf": The Conventionalization of Anthroposophy

A well-known cultural-social category, ascribed in the public discourse to eating organic food, includes the participants of the 'new age' culture. They include in their private and public lives alternative activities to the main stream, and hold on to the perception of mutual ecology, which sees the world as an integral holistic entity (Ruah-Midbar 2006). The image of organic food as "ecological" and "alternative" is appropriate for this approach and therefore very common in so-called new age settlements. One of these settlements is Kibbutz *Harduf*.

Kibbutz *Harduf* was founded by a group of young people, second generation Zionist settlers in Israel, who while visiting Europe in the 1980s, were captivated by the anthroposophical doctrine and sought to establish a cooperative community following this doctrine.

Within the framework of the anthroposophy doctrine, the kibbutz founders established a branch of organic agricultural production, for the welfare of the kibbutz members, as a source of livelihood. Over the years this organic agricultural undertaking gained success on a national scale. Today *Harduf* is the most recognized retail brand of commercial organic food in Israel. This occurred largely because in 2002 Israel's largest food corporation, *Tnuva,* acquired the kibbutz

[5]Organic food products in Israel are 20–25 % more expensive than parallel products that are not organic (according to a survey conducted by Panels Institute for The Marker, published on 8 July 2010). See http://shivuk.themarker.com/news/index.dot?id=49889 [in Hebrew].

factories and the rights to use the brand.[6] *Harduf* products (and other organic food products, some imported) are distributed nationally under this brand.

The irony is that kibbutz *Harduf* is located in the *Jezreel Valley*, a place that is perceived as a symbol of Zionism and socialism. The name of this place (*Harduf* of the Jezreel Valley), in which – according to the Jewish historical narrative- the cooperative agricultural settlement in the State of Israel began in the start of the twentieth century, has now become a clear representation of the capitalization and industrialization of Israeli organic agriculture.

Thus, in its early stages *Harduf*, acquired high "organicness" and low economic capital. However, since 2002 *Harduf* has become a clear representation of capitalization and industrialization of Israeli organic agriculture.

8.4.2 "Eden Teva Market" – The "IKEA" of Organic

For the past 5 years several supermarket chains in Israel have draped themselves with images of organic, natural, healthy, ecological and other similar images. The most prominent among them is the chain "*Eden Teva Market*", designed according the model of the American supermarket chain "Whole Foods Market". The first branch was established in 2003 and today the network has 20 branches.

The chain store owner claims that all types of existing organic food products in Israel can be found in their stores. But the abundance in the stores does not end with local goods. A tour in *Eden Teva Market* is accompanied by a feeling of visiting a "cultural global market", in which there are dozens of kinds of beans, spices and dried fruits imported from around the globe, a stand for herbal tea and all kinds of coffee – also imported, food products from the Far East and even a sushi-bar.

Moreover, the chain store owner adopted administration and marketing practices that characterize the "American way of business management": low pricing, increased competition, a growing supply of organic products and an existence of "sales", "customer club" and special discounts. For these reasons *Eden Teva Market* clearly symbolizes post-Fordisation[7] and the conventionalization of the organic food in Israel. Post-Fordisation is also reflected clearly in the jingle accompanying the chain's advertising campaign:

> In Eden Teva Market you will find a huge selection ensuring that everyone can find just the right products for him. Like any other supermarket ... only healthy!

Since 2007 the company's sales increases 30 % annually. Some see the success of the chain positively and attribute it great importance in stimulating the organic

[6]Tnuva was controlled by the global venture capital firm "Apax Partners", and since May 2014 – is controlled by the multinational company "Bright food Group".

[7]Post- Fordism is the dominant system of current capitalist mode of production, identified by flexible production (Lash and Urry 1987).

market in Israel. But there are some who denounce the chain, including journalists who criticize it, similar to Michael Pollan's (2006) critique of the *Whole Foods Market* (*Eden Teva Market's* spiritual mother from the United States): "*This is a monster that does not stop planting branches across the country and has already been termed the 'IKEA' of the organic*" (Taken from : Lavi, A (13 December 2008) "Not only organic". *NRG-online* (Daily newspaper online). http://www.nrg.co.il/online/1/ART1/824/656.html.

Thus, *Eden Teva Market*, quite similar to the late period of *Harduf*, represents the conventionalization of Israeli organic and acquires high global orientation and low organic cultural capital.

8.4.3 "Orbanic" Market

Reducing the gap between the volume of produce grown for export and the produce for the domestic market (export 92 % + 8 % domestic market)[8] is one of the objectives that *Israeli Organization of Organic Agriculture* set for itself. Therefore, one of the latest initiatives was the establishment of a market selling solely organic agricultural products in the main urban center of Israel: Tel-Aviv.

And so, "*Hatachana*" plaza *("the station"* plaza*)* – an old train station compound, which went through gentrification process, became a site for entertainment and upscale shopping in May 2010. On this site, designed in urban-up to date-luxurious style, a "purely organic farmers market" was established: the "*Orbanic"* market.

From the beginning of its establishment, the market was attributed cosmopolitan images as a strategy to attract consumers. For example, the market's chosen name "*Orbanic"* – a combination of words organic and urban – symbolizes the founders' intention to provide a sense of updated-global urbanism. There are 40 well-designed and pleasant stalls in the market plaza. Prominently displayed on each stall is a certificate testifying that the food products (which are aesthetically arranged) are "organic certified". In addition to the food products, visitors to the market are offered participation in a "consumer recreation experience": One can join Tai-Chi practice, workshops on matters of health and ecology, and listen to "world music" from speakers placed in the center of the market plaza.

While wandering the market, and while listening in on the conversations taking place in it, I noticed that consumers and vendors alike respond to an imagined imperative that encourages them to engage in "locality": Consumers often asked about the source of the crops, the vendors responded willingly and spoke about their lives in the rural areas. Thus, during a conversation, the organic foods were loaded with a local image. But paradoxically, this local image in fact strengthens

---

[8]Taken from Sikoller (2009 [in Hebrew]).

the global significance of the market. For example, following is a text written on a web site that deals with lifestyle issues:

> *Orbanic* market– to buy local: Equipped with my dynamic shopping cart I go shopping, like in a small village in Italy or Greece. Fresh produce in small stalls.[9]

This description illustrates how the consumption experience in the market is perceived: hedonistic, trendy, young, vibrant and even with the feel of being a tourist. But contrary to the global orientation and to the cosmopolitan and urban atmosphere, most of the stalls are run by organic growers from rural and peripheral areas in Israel. This is the reason that the market acquired some degree of organic cultural capital.

8.4.4 Chobiza – Israeli CSA

The first Israeli Community-Supported Agriculture (CSA) is located in the village of *Ben-Nun*, not far from Jerusalem. In 2002, the founder of the CSA, a young woman in her late 30s, returned to Israel from an extended stay in California. There she was involved in community gardens and decided to try to establish an alternative agricultural model in Israel. In her words: "I wanted to engage in agriculture but also to see a human face". She rented some farmland on which she started the "*Chobiza* Farm". The chosen name "*Chobiza*", taken from Arabic, is intended to describe the plant *malva*, also known as "*Arab Bread*". This plant won a place of honor in the pantheon of Israeli Jewish heroism because the residents of Jerusalem had nothing to eat but *Chobiza* during the siege and starvation of the 1948 war – at least according to legend. But contrary to the nationalism and rootedness implied by the farm's name, what attracts the customers is a model that offers a worthy alternative to industrialized agriculture in different places in the Western world:

> A lot of our customers know what CSA is. There are many families of American Jews ... there are families who were overseas on a mission or from their hi-tech job where they first met it (Taken from an Interview with Chobiza founder November 21, 2010).

But the Israeli model is different from other CSAs in America or Europe. CSAs are based, by definition, on formal organization and an ongoing relationship between growers and consumers. This is reflected by the consumers depositing money with the growers during the establishment stages, taking part in the risks that may arise, participate in decision-making and take an active role in the farm (Cone and Myhre 2000; Stanford 2006). According to the Israeli model, the customers are actually subscribers who receive a weekly vegetable box and pay a monthly fee for the amount of vegetables consumed. Although subscribers give up the option of choosing the vegetables included in the box and express their support of the farmer who is free to grow seasonal crops without being subject to demand; they

---

[9]Taken from (Zvi 2010 [in Hebrew]).

may terminate their subscription or renew it at will. They did not contribute to the establishment of the farm, do not share in its management, and few of them take an active part in the growing process.

That being the case, one can ask if this is a partnership. Where is the community activity? To the founder of the CSA the answers seem obvious: a sense of cooperation is provided through an active website and the virtual weekly newsletter that she publishes. Thus, the relationship between farmers and customers is carried out, as befits a global information era, not through a physical meeting, but through a virtual meeting and the exchange of textual information. Global networking technology, virtual in character, is what re-establishes, in practice, the sense of partnership between growers and consumers.

Chobiza Farm currently provides a weekly box of vegetables to 450 families. The number of people asking to join is growing. But the founder of *Chobiza* does not want to widen the circle of customers:

> We were contacted from all kinds of farmers' markets who proposed to advertise us and we refused. We have no intention to grow. We do not want to gain customers. We are more interested in finding a solution to the problem of gourd flies. We are farmers. Not merchants (Taken from an Interview with Chobiza founder November 21, 2010).

Thus, the Israeli CSA represent an institutional actor which it's habitus includes high organic cultural capital and local orientation, although global culture and technology are those renewing this locality.

8.5 Discussion

This article demonstrates the centralism of global cultural and economic conditions within the emergence and evolvement of the field of organic food in Israel. The influence of global neo-liberalization on organic food, as it appears from these case studies, is not restricted solely by conventionalization and structural changes that occur in the production process itself. Global cultural and economic processes are connected to the appearance of organic food and to the interaction between the fields of organic food production and conventional food production. These processes play a major role in the symbolic dimensions and political representations occurring in the Israeli field of organic food.

The multiplicity of ideals, images and representations associated with organic food in Israel turns it into a carrier of different (and even contradictory) post modernist aspects: local (*Chobiza; Harduf* between 1982 and 2002), national (the beginning of organic agriculture in *Sdeh Eliyahu*), Global/American (*Eden Teva Market*), hyper-consumerism (*Eden Teva Market; Harduf* 2002 onward) and counter-consumerism (*Chobiza*), lifestyle, self-care and (symbolic) care for community relationships (*Orbanic* market).

The numerous representations of organic food are not unique to Israel, but the modes of identity and identification connected to it are derived from two main particular sociological aspects, which occurred in the appearance and development of organic food in the Israeli context: First, as noted in the description of the first stages of the Israeli organic field, the appearance of organic food in Israel is not a derivative of subversive counter-culture. The development of the Israeli organic agriculture was based on an economic-instrumental rationale, and not ideological. In other words: the collaboration with conventional agricultural export companies, creating an organic market for export and minimal investment in local organic market development – all these indicate that the import of the "organic" idea, and its implementation in Israel, stemmed from constraints and interests, created as a result of the effects of globalization in the field of conventional agriculture. Thus, paradoxically, organic agriculture strengthened conventional agriculture while the latter was coping with global and local economic and cultural crises.

Second, the organization of the Israeli field of organic food, from 2000 and on, is based on a new consumer-cultural discourse, fundamentally different from the previous agricultural discourse. This discourse, which relies on the globalization of food and global consumption culture, promotes the commodification of organic food, that is to say, changing the agricultural product to a post-Fordistic commodity, characterized by production flexibility and expanding commercialization. These new forms of organic production embody diverse (and even polarized) symbols and meanings and are produced by several institutional actors. The position of these actors in the field is presented in the following figure (Fig. 8.2).

The first axis (horizontal in the following figure) represents an estimate of the degree of cultural capital attributed the actor. The positive side of the axis indicates

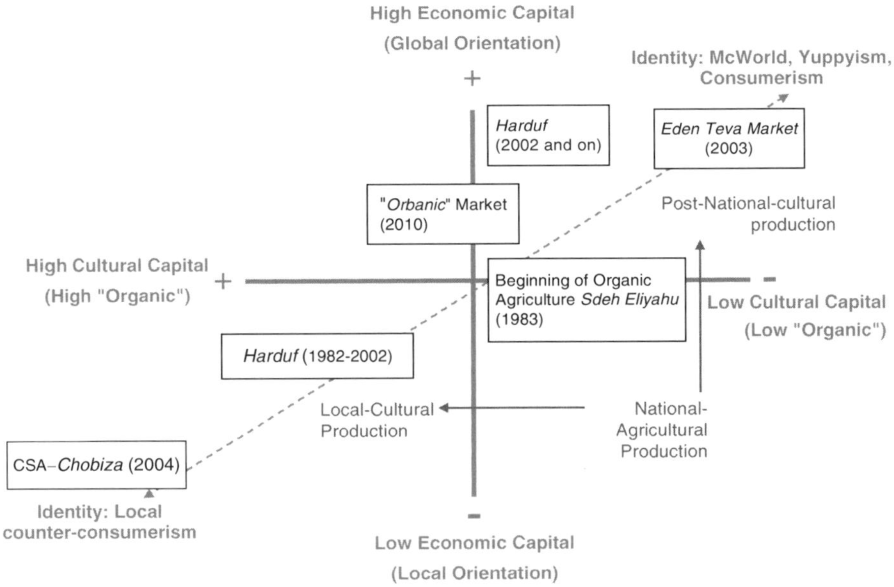

Fig. 8.2 Organic food in Israel, 1983–2011

a high commitment to "organic" and aesthetic-culinary complexity embodied in the product. On the other side, the negative, are actors (producers) who adopt clear marketing-capitalist practices, perceived as motivated by "extra-organic" reasons: capital accumulation and expansion of consumption extent. The second axis (vertical in the Fig. 8.2) refers to the degree of globalism attributed to the actions of the actors, that is, their aim to integrate in the global commodity circulation or loading the organic food produced by them with significant global implications.

Over the years the field of organic food in Israel went through processes of change and differentiation, whose outcome is the polarization of identities represented in the field. This polarization is appropriate to the characteristic dialectic of the "glocalization" era: on the one hand – the strengthening of popular culture, motivated by the force of commerce striving to expand (Barber 2010, p. 56 [1995]), a culture seeking to be tied to the neo-liberal market forces and promoting global consumer ideas (MacWorld to use Benjamin Barber's term). This movement reveals, in the case in front of us, the tendency of actors such as *Eden teva Market*, *Orbanic* Market and *Harduf* (2002 onward) to move toward commodification of "the organic" and toward global orientation and low organic cultural capital.

On the other hand – is the emergence of a local, community, or ethno-national culture, which opposes industrial modernization and economic globalization (Barber 2010, p. 46). This could be seen in the operations of the Israeli CSA, the anthroposophist approach toward organic agriculture in *Harduf* (1983–2002) and in neo-national projects such as the establishment of organic farms in Jewish settlements in the Palestinian territories.

In any case, the global socio-economic processes are the basis for the formation of the new categories that comprise the field. Moreover – those factors are the ones that convert the essence of the field from a field of agricultural production (as it was in the genesis of the Israeli organic field) to a field of consumer-culture production.

Acknowledgements The author wishes to thank Uri Ram for his valuable comments and support. Special thanks to Yehudit Shoham. This article greatly benefited from her insights and assistance.

References

Arce, A., and T.K. Marsden. 1993. The social construction of international food: A new research agenda. *Economic Geography* 69(3): 293–311.

Barber, B.R. 2010. *Jihad vs McWorld*. Tel-Aviv: Babel [in Hebrew].

Bourdieu, P. 1993. *The field of cultural production*. Cambridge: Polity.

Bourdieu, P. 2005. *Question in sociology*. Tel-Aviv: Resling [in Hebrew].

Buck, D., C. Getz, and J. Guthman. 1997. From farm to table: The organic vegetable commodity chain of Northern California. *Sociologia Ruralis* 37(1): 3–20.

Castells, M. 2003. *The power of identity: The information age: Economy, society and culture*. Malden: Blackwell.

Cone, C.A., and A. Myhre. 2000. Community-supported agriculture: A sustainable alternative to industrial agriculture? *Human Organization* 59(2): 187–197.

Coombes, B., and H. Campbell. 1998. Dependent reproduction of alternative modes of agriculture: Organic farming in New Zealand. *Sociologia Ruralis* 38(2): 127–145.
Engler, M. 2012. Hijacked organic, limited local, faulty fair trade: What's a radical to eat? *Dissent* 59(2): 20–25.
Eytan, K. 1993. Organic produce export summary *Renewed Agriculture*. Israeli Bio-Organic Agriculture Organization. leaflet. Tel-Aviv [in Hebrew].
FAO/ITC/CTA. 2001. World markets for organic fruit and vegetables – Opportunities for developing countries in the production and export of organic horticultural products. *FAO Corporate Document Repository*. Retrieved from http://www.fao.org/docrep/004/Y1669E/y1669e00.htm#Contents. Accessed 7 Nov 2012.
Guthman, J. 1998. Regulating meaning, appropriating nature: The codification of California organic agriculture. *Antipode* 30(2): 135–154.
Guthman, J. 2003. Fast food/organic food: Reflexive tastes and the making of 'yuppie chow'. *Social & Cultural Geography* 4(1): 45–58.
Guthman, J. 2004a. The trouble with 'organic lite' in California: A rejoinder to the 'conventionalisation' debate. *Sociologia Ruralis* 44(3): 301–316.
Guthman, J. 2004b. *Agrarian dreams: The paradox of organic farming in California*. Berkeley/Los Angeles/London: University of California Press.
Knight, A.J. 2010. Supersizing farms: The McDonaldization of agriculture. In *McDonaldization – The reader*, 3rd ed, ed. G. Ritzer, 183–195. Thousand Oaks: Pine Forge Press.
Kristal, M. 2009. Sharp increase in exports organic fruits and vegetables. Retrieved from "http://www.ynet.co.il" www.ynet.co.il/Ext/Comp/ArticleLayout/CdaArticlePrintPreview/1,2506,L-3757629,00.html. Accessed 12 Sept 2013 [in Hebrew].
Lash, S., and J. Urry. 1987. *The end of organized capitalism*. Madison: University of Wisconsin.
Lavi, A. 2008. Not only organic. Retrieved from http://www.nrg.co.il/online/1/ART1/824/656.html. Accessed 12 Sept 2013.
Maroz, G. (2010). The organic experiment. Retrieved from http://www.nrg.co.il/online/1/ART2/154/585.html. Accessed 3 Nov 2013 [in Hebrew].
Marsden, T.K., and A. Arce. 1995. Constructing quality: Emerging food networks in the rural transition. *Environment and Planning A* 27(8): 1261–1279.
Mazori, D. 2005. Organic food consumption in Israel has grown by 40% over the past two years. Retrieved from http://www.nrg.co.il/online/16/ART/886/920.html. Accessed 12 Sept 2013 [in Hebrew].
Murdoch, J., T. Marsden, and J. Banks. 2000. Quality, nature, and embeddedness: Some theoretical considerations in the context of the food sector*. *Economic Geography* 76(2): 107–125.
Pollan, M. 2006. *The omnivore's dilemma: A natural history of four meals*. New York: Penguin.
Ram, U. 2005. *The globalization of Israel: McWorld in Tel Aviv, Jihad in Jerusalem*. Tel-Aviv: Resling [in Hebrew].
Raynolds, L.T. 2004. The globalization of organic agro-food networks. *World Development* 32(5): 725–743.
Robertson, R. 1995. Glocalization: Time-space and homogeneity-heterogeneity. In *Global modernities*, ed. M. Featherston, S. Lash, and R. Robertson, 25–44. London: Sage Publications.
Ruah-Midbar, M. 2006. *The New Age culture in Israel: A methodological introduction and the conceptual network*. Dissertation, Bar Ilan University, Ramat Gan, Israel [in Hebrew].
Sikoller N. 2009. Does not grow: Organic export rose by only 12% in 2008. Retrieved from http://www.calcalist.co.il/local/articles/1,7340,L-3208586,00.html. Accessed 12 Sept 2013 [in Hebrew].
Stanford, L. 2006. The role of ideology in New Mexico's CSA (community supported agriculture) organizations: Conflicting visions between growers and members. In *Fast food/slow food: The cultural economy of the global food system*, ed. R. Wilk, 181–200. Lanham: Altamira Press.
Svirsky, S. 2004. Israel in the global space. In *The Israeli society in the global era*, ed. D. Filk and U. Ram, 34–56. Jerusalem: Hakibbutz Hameuhad [in Hebrew].
Tal, A. 2006. *The environment in Israel: Natural resources, crises, campaigns and policy from the advent of zionism until twenty-first century*. Tel-Aviv: HaKibbutz HaMeuhad Press [in Hebrew].

Weitz, R. 1969. Change of values in our agriculture. In *The social structure of Israel – Collection of papers and research*, ed. S.N. Eisenstadt, H. Adler, R. Bar-Yosef, and R. Kahana, 162–172. Jerusalem: The Hebrew University [in Hebrew].

Willer, H., and L. Kilcher. 2009. *The world of organic agriculture. Statistics and emerging trends 2009*. Bonn/Frick: IFOAM/FIBL.

Yustman, M. 2001. Changes in the Israeli economic market. In *The Israeli economy, 1985–1998: From government intervention to market economics*, ed. A. Ben-Basat, 558–601. Tel-Aviv: Am-Oved Press [in Hebrew].

Zali, R. 1993. From the secretariat *renewed agriculture*. Israeli bio-organic agriculture organization leaflet. Tel-Aviv [in Hebrew].

Zimmerman, C.C. 1932. Ernst Engel's law of expenditures for food. *The Quarterly Journal of Economics* 47(1): 78–101.

Zvi, L. 2010. Two environmental recommendations. Retrieved from http://saloona.co.il. Accessed 21 Nov 2010 [in Hebrew].

Part IV
The Interplay of Conventional and Organic

Overview and Observations

The chapters in this part discuss the interplay between conventional and organic characteristics from rather different perspectives, and highlight different issues. The first two chapters, 'Engaging the Organic Conventionalization Debate', by Douglas Constance et al., and 'Organic Farmers: Contributing to the Resilience of the Food System?' by Lesley Hunt et al. examine the organic movement and its multiple relationships with conventional agriculture. The third chapter, 'From The Ground Up? The Principles of Australian Organic Agriculture', by Rebecca Jones discusses the Australian history of organic from the perspective of the original principles and values of the Australian organic movement and how they have changed over time. In the fourth chapter, Brock and Barham offer an example from a religious movement in which both organic and non-organic coexist and are justified within the Amish movement and values.

Constance et al. start with an overview of conventionalization and bifurcation in organic referring to examples from different continents. The authors distinguish two models characterizing agriculture. The agrarian perspective is built on social engagement, is community oriented (civic agriculture) and entails diverse types of cooperative relationships between farmers and consumers. These relationships are based on the idea that food is from "somewhere", and that as such, has a high transformative potential to bring farmers and citizens together. They conclude that in the US, government organic regulations have led to the exclusion of the mainly small and socially oriented farmers from the retail market. The commodity-oriented model is characterizing industrially organized agriculture designed to produce at the lowest cost for large retailer markets. This kind of diversification orients the discussion in the conventionalization debate. However, they point out that the characteristics of conventionalization differ between country, region and continent and therefore do not confirm the bifurcation thesis.

In their survey of "pragmatic conventional" organic farmers in Texas, the lack of financial support for the conversion period was found to be the main barrier to convert to organic. Furthermore, uncertainties concerning the viability of organic production, marketing, information, and certification were also seen as constraints on moving to organic. Interestingly, these larger farms were unsatisfied with conventional farming and sympathized with the organic philosophy. The authors conclude that better governmental incentives for organic conversion would help these farmers to convert. However, it is not clear if these farmers would be attracted positively to the original idea of organic as a social movement that links farming and community.

Hunt et al. draw upon research conducted by a transdisciplinary program to compare the sustainability of organic, integrated and conventional farming systems in dairy, sheep and beef, and kiwifruit sectors of New Zealand. They illustrate that organic farmers can contribute to the resilience of the organic sector. The authors compare three management systems with a view to understand the ability of organic practices to bring resilience through diversity in production methods. In particular, they note that resilience is related to the social and cultural acceptance of 'good farmers' from other farmers and actors. They further discuss the future of the organic movement and ask under what conditions could there be a more resilient organic production. They observe that the more practices between organic, integrated and conventional agriculture overlap, the more organic becomes socially accepted. Hunt et al. argue that shared knowledge with farmers from other management systems increases the social acceptance of organic farmers in their rural environment. This openness allows also the diffusion of the organic model to conventional farmers. The closeness between different management systems leads to the social acceptance of organic, e.g. in kiwifruit production, and allows non-organic farmers to learn the organics environmental approach. This is less the case with dairy production because of lower organic production, and it is excluded in sheep/beef production because of significant differences in the management systems. Both systems are deeply embedded in traditions without much flexibility, and modifications in the management practices. Thus, there are few opportunities for moving toward organic. Furthermore, they note the significance of the social relationships among all kiwi-farmers that facilitates communication between organic and non-organic growers that do exit in the dairy and sheep/beef.

Hunt et al. also address the conventionalization debate. Large organic export oriented farms are often criticized as conventionalized organic farms. But in this case, they remind us that New Zealand's agriculture is export oriented. Organic and conventional farmers are confronted with similar market conditions. Furthermore, lacking subsidies for environmentally friendly production, international markets are of high relevance for the organic farmers.

Jones examines the founding principles of organic in Australia. These include: humus rich, fertile soil; chemical free; and biodiversity and ecological wellbeing. These continue to be important for organic farmers. She acknowledges that while the modern organic movement in many countries has had to adapt standards and certification that encourage the conventionalization of 'the organic industry,' this process has not had a profound impact on the underlying beliefs of Australia's organic farmers or the principles upon which their organic practices are based.

Jones concludes that the values of the Australian movement today are similar to those followed in the past, and continue to focus on ecology and health. But she also observes that organic is moving toward input based agriculture. Compared to the IFOAM Principles, the Australian movement has not specifically address the principles of fairness and care, either in their early years or today. This is consistent with many other observations in this volume.

Brock and Barham discuss the diversity among Amish farmers with respect to adopting organic practices and their understanding of a range of issues in the agrofood chain. Amish farmers justify their ethical participation in either conventional or organic agriculture in largely anthropocentric and altruistic terms based on Christian values. They describe organic practices largely in terms of stewardship and traditional techniques. But they do not see the IFOAM Principles as especially relevant for their organic practices. In contrast, many social and spiritual ideas and values, and especially in Christian doctrine, were understood to support the organic movement (Massingham 1942; Balfour 1943; Conford 1988). Christian and spiritual values continue to influence the ideas that many farmers have about their occupation (Stock 2007).

However, only a few organic farms exist within the Amish communities, and Amish values tend to be concerned more with modern, visible and mechanized agricultural practices (e.g., the tractor), the use of chemical fertilizers and hybrid seeds. Amish farmers raise animals and grow crops similar to their non-Amish neighbors who follow conventional farming practices. Amish farmers keep their interactions with public and government representatives to a minimum. In this regard, organic certification or registration requirements can be problematic. Moreover, the public differentiation created by the organic price premium represents a publicly recognized differentiation that jeopardizes the values of the community. The authors conclude that Amish religious beliefs allow individual farmers to follow different practices based on different understandings of the relationship between humans and nature.

To summarize the four contributions: organic is in a process of differentiation. The intensification of production has become dominant, but is context specific in order to consider the ecological dimension. Further, social and cultural values often risk playing a secondary role in organic practices. Constance et al. and Hunt et al. confront issues on the border of organic and conventional. In examining the organic movement and its multiple relationships with conventional agriculture, they bring in a new perspective. Instead of focusing solely on the conventionalization of organic agriculture, they consider the similarities or connections of organic with conventional practices, and how government policy has shaped these connections.

These two chapters offer different perspectives on the co-existence of conventional and organic in which farmers could work together, share information, or explore new opportunities for promoting organic as part of broader rural development strategies. In both contributions, the limited government subsidies and specific support for conversion hampers to the willingness to convert to organic as well as the relationships between organic and conventional farmers. Hunt et al. show

how the social and cultural influences of a farming style are imperative in accepting 'good practices.' Specifically those farming systems most similar to organic are able to accept organic as a model of "good farming".

Jones' chapter on the history of organic in Australia underlines health and ecological issues as dominant elements of the Australian foundation for organic. Interestingly, the social dimension appears to be less important and might explain the tendency toward conventionalization. Brock and Barham highlight social and cultural barriers in a religious movement that lead to different attitudes concerning how to farm with respect to environmental impacts. Interestingly, the Amish religious foundation allows both organic and conventional approaches. Viewed from outside this movement, it is of interest to note how, from a social, cultural and religious perspective, it bridges both management systems.

All contributions show how important it may be for understanding organic diversification, to discuss multiple farming styles, time horizons, regional, national, continental or religious influences. Social and cultural commonalities could bridge conventional and organic farming as well as within the organic movement. The IFOAM Principles could serve as an ethical platform to bring the social and cultural characteristics into the debate. But the neo-liberal dominance and governmental rules that create market conditions, influence the industrialization of the organic sector, and marginalize social and cultural dimensions, therefore should be more considered.

References

Balfour, E.B. 1943. *The living soil*. London: Faber & Faber.

Conford, P. 1988. *The organic tradition: An anthology of writings on organic farming, 1900–1950*. Bideford: Green Books.

Massingham, H.J. 1942. Remembrance. In *A mirror of England*, ed. E. Abelson. Bideford: Green Books.

Stock, P.V. 2007. 'Good farmers' as reflexive producers: An examination of family organic farmers in the US Midwest. *Sociologia Ruralis* 47(2): 83–102.

Chapter 9
Engaging the Organic Conventionalization Debate

Douglas H. Constance, Jin Young Choi, and Damian Lara

> *Organics without a social vision is dangerously incomplete (DeLind 2000, p. 24).*

9.1 Introduction

Organic agriculture in the US emerged in California in the 1960s as part of the environmental social movement response to the negative externalities of industrialism (Belasco 1989). The first organic standard developed in California in 1990 is the model for the US standard (Guthman 2004a). Although opposed by conventional agriculture, organics is now part of the mainstream, available in the majority of supermarkets. The success of organics is a great victory for the environmental movement and other critics of conventional agriculture. Sociologically, the success is problematic due to conventionalization, or the process whereby organics takes on many of the characteristics of mainstream agriculture regarding scale and structure. The scholarly discussion regarding the extent and implications of conventionalization has generated a substantial literature in agrifood studies.

In this chapter we engage the conventionalization debate, informed by empirical evidence from Texas and a political economy of agrifood studies framework. The remainder of this introduction provides a brief overview of organic agriculture, ending with a discussion of the role of government policies in organic adoption. Then we present our review of the literature on the conventionalization of organics, which critically investigates the transformative potential of organics. Next, we present the research from Texas, focusing on "pragmatic conventional" producers, with analysis informed by the conventionalization debate. Finally, we discuss the debate within the context of a sociology of agrifood interpretative framework.

D.H. Constance (✉) • J.Y. Choi • D. Lara
Department of Sociology, Sam Houston State University, Campus Box 2446, Huntsville, TX 77341, USA
e-mail: Soc_dhc@shsu.edu; Jyc002@shsu.edu; Damian.G.Lara@gmail.com

B. Freyer, J. Bingen (eds.), *Re-Thinking Organic Food and Farming in a Changing World*, The International Library of Environmental, Agricultural and Food Ethics 22, DOI 10.1007/978-94-017-9190-8_9

Organic agriculture was seen as a hopeful trend with positive consequences: better for the environment, farmers, farmworkers, and consumers. It limits synthetic chemical usage and environmental contamination; it supports prosperous communities through smaller-scale operations, price premiums, and direct markets. Organics was seen as a transformative alternative to the negative impacts of the industrial food system; it would improve our quality of life in numerous ways (see Lampkin 1990; Clunies-Ross and Cox 1994; Goodman 1999, 2000; Allen and Kovach 2000; DeLind 2000; Vos 2000; Michelsen 2001; Pollan 2001; Lyson and Guptill 2004).

9.2 Background

In the 1980s, research began to document the feasibility of organic production as an alternative to chemical-based agriculture (National Research Council 1980; United States Department of Agriculture 1980; United States Government Accounting Office 1990). The USDA LISA/SARE programs were grounded in organic philosophy, but employed the term sustainable agriculture to be more politically palatable (Allen 2004; Constance 2010). After a long battle and resulting compromise, in 2002 the USDA National Organic Program (NOP) created the certified organic label (Guthman 2004a). Supporters of conventional agriculture systematically opposed official government support for organics (Guthman 1998; Madden 1998). Their attempt to insert GMOs into the "official definition" was a failed attempt to coopt the label, and it galvanized the organic movement in opposition. While the NOP Final Rule provided regulatory underpinning for organics, the resulting certified organic label was a market label with no claims to superiority to conventional systems (Lohr and Salomonsson 2000; Greene et al. 2009).

The US organic market grew at double-digit rates through the 1990s and 2000s, increasing from \$3.6B to \$31.5B in sales from 1997 to 2011 (Dimitri and Oberholtzer 2009; Organic Trade Association 2012). Globally, the market reached \$39B in 2008, increasing at a rate of about \$5B per year (Willer and Kilcher 2009). Some estimates report \$59B in global sales for 2010 (Willer and Kilcher 2012). The growth attracted the entry of conventional farmers and mainstream retail grocery chains (Guthman 2004a; Greene et al. 2009; Howard 2009b). As of 2010, mass-market grocery stores such as WalMart and Kroger accounted for 54 % of organic food sales, followed by natural retailers at 39 % (Heller 2007; Organic Trade Association 2011). In 2011 the wholesale market accounted for 81 % of organic sales, followed by 13 % in the direct-to-retail market and 6 % for the direct-to-consumer market (National Agricultural Statistics Service 2012). The organic distribution system was transformed from one characterized largely by direct sales combined with natural foods retailers such as Whole Foods to one fully incorporated into the conventional system, including mass retailers with their own private-label brands[1] and the rapid growth of middlemen "handlers" that coordinate the organic

[1]New organic private-label products increased from 35 in 2003 to 540 in 2007 (Driftmier 2009).

Table 9.1 U.S. organic certified farm operations: 1992–2007; Certified organic farmland: 1992–2005 (in thousands of acres) and Certified livestock: 1992–2005 (in thousands)

						% change			
Item	1992	1997	2002	2007	2008	92–97	97–02	02–07	07–08
Operations[a]	3,587	5,021	7,323	11,352	12,941	40	46	55	14
Total farmland	935.5	1,346.5	1,925.5	4,290.0	4,816.0	45	43	123	12
Pasture/ rangeland	532.1	496.4	625.9	2,005.0	2,005.0	−7	26	220	8
Cropland	403.4	850.1	1,299.6	2,284.9	2,655.4	111	53	76	16
Animals									
Livestock	11.6	18.5	108.4	363.3	475.8	59	485	235	31
Poultry	61.4	798.3	6,270.2	12,184.6	15,518.1	1,201	685	94	27

Source: Economic Research Service (2012)
[a]Does not include subcontracted organic farm operations

supply chains (Dimitri and Greene 2000; Organic Trade Association 2008; Dimitri and Oberholtzer 2009; Driftmier 2009). Significant entry into the organics market is expected to continue because it remains one of the fastest growing sectors of the agrifood system (Organic Trade Association 2008).

Although certified organic land in the US doubled between 1992 and 2002 and then doubled again by 2007 (see Table 9.1), domestic supply still lags substantially behind domestic demand, especially in the area of organic inputs for agrifood manufacturing (Greene et al. 2009). Data reported in Table 9.1 reveal that the number of certified-organic operations and total farmland acres increased more rapidly in the post-NOP years than before, providing some indication that the NOP increased adoption and production. The rate of increase was higher for cropland than for operations, indicating that the size of certified-organic operations increased after the NOP. The average size of the certified-organic farm increased from 268 acres in 1997 to 477 acres in 2005, supported by the rapid increase in certified-organic pastureland. In 2007, 866 of the 11,352 operations had 500 or more acres and accounted for 60 % of the certified-organic farmland (Dimitri and Oberholtzer 2009; Economic Research Service 2012).

Comparable data is not available for the years after 2008 but the National Agricultural Statistics Service (2012) "2011 Certified Organic Production Survey" provides some useful information. They conducted a survey of all USDA-certified organic producers, but did not include in this research the floriculture, Christmas trees, and mushroom producers that were included in the 2008 and previous surveys. Based on a 76 % response rate, the 9,140 USDA-certified organic farms reported \$3.53B in sales: \$2.22B from crops and \$1.31 from livestock. The average sales per farm was \$414,725, compared to \$217,675 in 2008. California was by far the largest producer with 39.3 % of sales, followed by Washington (8.4 %), Oregon (6.6 %), Texas (4.7 %), and Wisconsin (3.8 %). Crops made up 62.9 % of sales, followed by livestock and poultry products (29.5 %) and livestock and poultry (7.6 %) (National Agricultural Statistics Service 2012). By the end of 2011, there were 17,281 USDA-certified organic farms and processing facilities in the US, a 240 % increase from 2002 (Agricultural Marketing Service 2012).

The critical factor limiting organic growth changed from a lack of consumer demand during the 1990s to a lack of sufficient inputs and organic products in the 2000s (Organic Farming Research Foundation 1993, 1996; Dimitri and Richman 2000; Dimitri and Oberholtzer 2009; Greene et al. 2009). The lack of other forms of government support for organics beyond the market label – such as economic support for transition or research – hindered the entry of new organic farmers and conversion by conventional farmers, resulting in an increase in organic imports to match rising domestic consumer demand (Food and Agriculture Service 2005; Thilmany 2006; Greene 2007). The mainstream food companies entering the market were constrained by the lack of reliable supplies of organic raw materials (Organic Trade Association 2006; Heller 2007; Greene et al. 2009).

The NOP certified-organic standard allows organic farmers and handlers anywhere in the world to export to the US, as long as the products meet the NOP standards. Of the 27,000 producers and handlers certified in 2007 by USDA-accredited certifiers, 11,000 were from over 100 foreign countries, mostly from Canada, Italy, Turkey, China and Mexico (Greene et al. 2009). In 2011 there were 28,386 certified producers and handlers from 133 countries (Agricultural Marketing Service 2012). Organic foods, and especially the organic components for processed organic foods, sold in the US are increasing sourced globally (Dimitri and Oberholtzer 2009). At the national level, (Howard 2009a, b) found rapid consolidation through mergers, acquisitions and internal development and growing economic concentration in the organics industry.

Whereas Europe and other countries provided subsidies for organic conversion, the US focused instead on market-mechanisms to support the growth of the organic sector. The organic standards and USDA label were designed to facilitate market signals between producers and consumers, but there was no direct subsidy for the 3-year conversion (Lohr and Salomonsson 2000; Greene et al. 2009). Again, unlike Europe, the NOP included no official government position stating that organic foods were superior to conventional foods. This lack of direct support for conversion represents a major barrier to organic adoption in the US (Lipson 1997; Padel 2001; Duram 2006; Greene et al. 2009).

The hostile political environment and bias against organics dampened adoption (Padel 2001; Klintman and Boström 2004; Duram 2006; Volpe III 2006; Constance and Choi 2010). The first national study that demonstrated organic feasibility and profitability of organics included several recommendations regarding research, education, and public policy support for existing organic farmers and for conventional farmers interesting in conversion (United States Department of Agriculture 1980). However, the incoming Reagan Administration rejected the report and abolished the Organic Resources Coordinator position in USDA. The Land Grant University system openly criticized organics. Conventional agricultural interests opposed the organics program, specifically lobbying to ensure that the Final Rule focused only on market-based incentives and no claims for organics as a preferred or superior approach to agriculture (Guthman 2004a).

Because of these barriers, the U.S. lags behind Europe in organics research, education and production (Padel and Lampkin 1994; Lohr and Salomonsson 2000; Thilmany 2006). In response to the growing concern over the demand/supply gap and to growing criticism about the lack of government support for organics, the 2008 Farm Bill included $78 M in research, education, and extension for organics subsidized conversion. The Farm Bill also overtly acknowledges the environmental benefits of organics farming (Thilmany 2006; Organic Farming Research Foundation 2008; Greene et al. 2009). The organic movement applauded the increased government support for organics. The failure of the 2012 Farm Bill to pass threatens the continued government support for organics due to a lack of funding (National Sustainable Agriculture Coalition 2012).

During the 1990s, the optimistic view of the transformative potential of organics was called into question. Research suggested that the institutionalization of organics via the creation of organic certification standards and other government policies to support adoption was diluting the social movement components and replacing them with an industrial approach (Clunies-Ross 1990; Buck et al. 1997; Tovey 1997; Guthman 1998). Researchers also predicted that the adoption of the NOP standards in the US would accelerate conventionalization as many large firms were waiting for a system of national standards before moving into the organic market (Buck et al. 1997; Guthman 1998; DeLind 2000; Klonsky 2000; Vos 2000).

The research from California (Buck et al. 1997; Guthman 1998) introduced the concepts of conventionalization and bifurcation to interpret these trends. Conventionalization refers to the process by which organic agriculture increasingly takes on the characteristics of mainstream industrial agriculture. Bifurcation refers to the process by which the organic agriculture adopts a dual-structure of smaller, lifestyle-oriented producers and larger, industrial-scale producers. In response to the Guthman thesis, researchers criticized the conventionalization thesis as being overly deterministic and not well supported empirically (Coombes and Campbell 1998; Campbell and Coombes 1999; Campbell and Liepins 2001; Lynggaard 2001; Michelsen 2001; Hall and Mogyorody 2002; Kaltoft 2002). Some early research supported conventionalization (Lyons 1999; Lockie et al. 2000), but later research produced mixed results (Lockie and Halpin 2005; Best 2008; Constance et al. 2008; Guptill 2009). Guthman's later research and responses to the critiques of conventionalization (1998, 2000, 2004a, b, c) has extended the discourse and literature that critically evaluates the conventionalization thesis.

9.3 Conventionalization and Bifurcation

Buck et al. (1997) introduced the concept of conventionalization to analyze the changes in organic food production in California. They operationalized conventionalization through the concepts appropriation and substitutionism (Goodman et al. 1987; Goodman 2000). Firms practice appropriation when they reduce the risks of investing in agriculture by relocating processes and practices off the farm such as chemical fertilizers and pesticides. Similarly, substitutionism refers to the practice whereby agribusiness firms expand post-production activities to capture a higher proportion of the total value of the commodity. Through appropriation and substitutionism, agribusiness penetrates organic agriculture through the production of inputs and the processing of outputs. As a result, organic agriculture becomes more conventional.

Bifurcation is an outcome of conventionalization. As agribusiness enters organics, a bi-polar production system emerges made up of larger conventional operations that mix input substitution strategies with monoculture production of high value crops targeted to indirect markets while smaller farms employ artisanal practices to grow a variety of crops using more sustainable agronomic practices targeted to direct markets. The categories describing the bifurcation of organics have been referred to as "pragmatic" versus "pure" (Clunies-Ross 1990; Clunies-Ross and Cox 1994); "conventional" versus "artisanal" (Buck et al. 1997); "agribusiness" versus "lifestyle" (Guthman 1998); "lifestyle" and "conventional"; ("interdependent lifestyle/domestic/small-scale" versus "export/commercialized") (Coombes and Campbell 1998); "chemical-lite" versus "movement" (Goodman 2000); "philosophical" versus "pragmatic/instrumental" (Lockie et al. 2000); "productivist/reductionist" versus "holistic" (Vos 2000); "lifestyle" and "conventional" (Campbell and Liepins 2001); "lifestyle/domestic/small-scale" versus "export/commercialized" (Coombes and Campbell 1998); "organic lite/shallow" versus "deep organic" (Guthman 2004a); "lifestyle/domestic/small-scale" versus "export/commercialized" (Coombes and Campbell 1998); and "old guard" versus "new entrants" (Guthman 2004b).

Although Buck et al. (1997) were not the first to question the transformative ability of organics(see Clunies-Ross 1990; Lampkin 1990; Friedmann 1993; MacRae et al. 1993; Clunies-Ross and Cox 1994; Rosset and Altieri 1997; Tovey 1997), their research in California was the first to systematically document the structural trends taking place in organics. As organics moved beyond its niche status in California, agribusiness entered the market to capture the monopoly rents associated with the price premium (Buck et al. 1997). The formal organic standards that emphasized inputs over processes allowed agribusiness to employ input substitution practices that met the minimum organic standards but avoided the costly agronomic practices associated with ecological sustainability. By focusing on allowable inputs, organic regulation preempted broader agronomic processes and encouraged entry by institutions with "questionable commitment" to sustainable agriculture (Guthman 1998, p. 147). The end result was a form of agriculture that differs from conventional systems only by the use of organic inputs (Buck et al. 1997; Guthman 1998). As part of conventionalization, the organic label was coopted by large firms, thereby blunting its transformative potential as it was appropriated and subsumed (see Goodman et al. 1987; Goodman 1999) by corporate actors (Buck et al. 1997; Guthman 1998).

Buck et al. (1997) also found a "bifurcation" of organic producers in California characterized by large operations specializing in the mass production of a few high profit crops and smaller farms that employ artisanal methods to grow a variety of marketable crops. In this system the smaller operations tend to occupy the more marginal lands while the larger ones secure the expanses of certified organic land. The agribusiness ventures were more likely to employ larger numbers of migrant labor, although due to the greater mixture of crops, the smaller operations were more likely to provide year-round work.

Buck et al. (1997) note that it was in the marketing and distribution of organics that conventionalization was most dramatic. Large organic food retailers' preference for certified organic products limited their access by non-certified producers. As a result, the growing bifurcation relegated the smaller operations to the marginal markets, i.e. farmers markets and subscription farming, as the larger farms serviced the retailers and indirect markets. They argued that the localized, direct-marketing arrangements that "illustrate the promise of local networks of direct grower-to-consumer links, are effectively default choices for growers with few resources" (Buck et al. 1997, p. 14).

The politics of organic regulation influenced what kinds of producers (agribusiness or lifestyle) benefit. Guthman argued that the technical approach and resulting conventionalization contributed little to "sustainability – either socially or ecologically" (1998, p. 143). While admitting the California focus of the research, they predicted that national organic standards would accelerate conventionalization as agribusiness re-shaped "organic agriculture to its own advantage" (Buck et al. 1997, pp. 16, 17). Guthman (1998) concluded that California is the model of a broader process whereby agribusiness appropriates nature through the regulation and cooption of the organic label; California is the future of organics.

Research from Australia and New Zealand provided support for the conventionalization thesis (Lyons 1997, 1999; Lockie et al. 2000). Lyons (1999) noted recruited to convert to organics tended to have a "pragmatic/instrumental approach" whereby organic farming meant compliance with minimum certification requirements rather than a "philosophical" approach (Lyons 1997). The Heinz Wattie "corporate greening" system typified conventionalization (Lockie et al. 2000). The opportunistic corporate greening (see Buttel 1996) in Australia and New Zealand was incorporating the organic industry within conventional agricultural networks (Lyons 1999).

While Campbell and his associates (Coombes and Campbell 1998; Campbell and Coombes 1999; Campbell and Liepins 2001) agreed that organics was experiencing conventionalization and bifurcation, they disagreed that the impacts were necessarily negative and inevitable. They found a relatively stable bifurcation of the organic industry in New Zealand characterized by an interdependent lifestyle/domestic/small-scale sector of perishable goods and an export/commercialized/conventional sector of green durable goods (Coombes and Campbell 1998). They did note that the impetus for national certification standards was the export industry focused on "green products" to Northern markets (Campbell and Coombes 1999). Although some smaller producers opted out of certification because of this shift, the export industry expansion benefited the smaller growers because it enhanced the legitimacy of organics. They saw this as a durable arrangement with no signs of marginalization of the smaller growers.

In their criticism of the linearity of the conventionalization thesis, they noted that the meaning of organics changed over time as different actors entered the policy arena. They maintained that these shifts show "clearly that the local industry is not engaged in a linear trajectory towards 'conventionalization' or the uncontested assumption of industrial agricultural forms" (Campbell and Liepins 2001, p. 36). They concluded that contrary to Guthman's view, organics in New Zealand reveals a "peculiar quality" about organics that enabled it to continue as a counter-point to a globalizing food system. They called for more empirical studies before the construction of prescriptive theories.

In 2001 research expanded on the early work from California and Ireland (Tovey 1997) that argued that organics was losing its alternative characteristic. (Michelsen 2001) used the term "institutionalization" to describe the quantitative changes in the social organization of organic production. He also criticized the conclusions of the early studies for generalizing from too limited data. Research in Denmark and Belgium (Lynggaard 2001) reported that variations in institutional factors produced very different institutional arrangements, which casts doubts on universalistic interpretations of the trajectory of organics and highlights the importance of national/regional contexts. Further research from Denmark revealed that the process of the organic institutionalization through government adoption of certification standards and incentives for organic conversion reduced the broader, value-laden, and ideological formulations to technical and quantitative definitions and rules (Kaltoft 2002). With institutionalization, secondary production, processing, distribution and retailing through conventional venues developed rapidly. Kaltoft concluded that organics stopped being a social movement once it became institutionalized and integrated into the global food system. While certain organic producers might have strong ideological orientation and would resist corporate penetration of organics, for the government and industry organic farming becomes a technical solution to environmental problems (see Tovey 1997).

Research from Ontario, Canada (Hall and Mogyorody 2002) reported mixed support for the conventionalization thesis. While new entrants to organics tended to be larger and have a more economic philosophy, those operations did not fit the pattern of specialized monoculture for indirect markets. Migrant labor patterns did not support conventionalization. They noted different patterns of destination markets by commodity, but no evidence of a bifurcation between large and small growers targeted to different markets. They attributed these results to the particular institutional and bio-physical arrangement of organics in Ontario and Canada, but noted that the situation could change quickly.

In addressing the critics of conventionalization, Guthman (2004c) argued that the situation in California provided further support for her thesis (Guthman 2004a). Through mergers, acquisitions, and contracting agribusiness had rapidly increased their organic operations in California, as well as New Zealand and Australia, with most of the growth from converted conventional operations. These highly capitalized operations out-competed smaller producers via economies of scale. Industry entry produced increased price competition, a drop in price premiums, a lowering of farm-gate premiums, and a weeding out of some lifestyle producers (Guthman 2004c; see also Smith and Marsden 2004). The farm-gate price squeeze creates further pressure to intensify.

Agro-industrialization affected all organic growers due to the incorporation of organic premium values in land prices that forced growers to farm more intensively to pay for the land. (Guthman 2004c). In California, state supports for irrigation, cheap labor, and agro-technologies enhanced intensification, i.e., more crops per year in less time, which is then capitalized into land values. The industrial "organic lite" model constrained the continuance of the "deep organic" lifestyle model. For Guthman (2004c, p. 525) this paradox "is hardly the recipe for the spread of sustainable agriculture". She noted that conventionalization was not necessarily inevitable, but rather it would take creative state policies to blunt the trend of agro-industrialization in organics. For Guthman, it were these wider processes of agro-industrialization that casts doubts on the long-term viability of the "multiple paths to sustainability" put forth by Campbell and associates.

Smith and Marsden (2004) provide some support for Guthman's point by documenting the emerging negative trend in organics in the UK whereby the "farm-gate price squeeze" restricts the positive contribution of organic agriculture as a means to rural development. They link the squeeze to the growing oligopsonistic position of major supermarkets in organic retailing, a phenomenon associated with conventional food supply chains whereby the supermarkets increasingly "drive the chain" and producers have to adopt more intensive production strategies to compete with imports and stay in business. Price wars to gain market share generated lower prices paid for organics resulting "farm-gate price squeeze" that drove the smaller/indigenous producers out of business. Left to the free market, the "value capture" of organics had shifted from producers to retailers. They predicted that without supportive government policies, organics would lose its contributive role regarding rural development.

Lockie and Halpin (2005) conducted an empirical assessment of the Australian organic sector to evaluate to what degree conventionalization was inevitable or was there room for social movement resistance and/or strong state-support to avoid Guthman's prognosis. Their research problematized the bifurcation between small-scale/artisanal/lifestyle/deep organic producers and large-scale ex-conventional/industrial/shallow organic producers as part of conventionalization. Although noting differences across commodities, they found little support for bifurcation. Most operations sold a small amount of production direct to consumers and the rest in indirect markets. Motivations and attitudes about organic farming were different across groups, but it was more related to intensity of support for organics rather than direction. They found no evidence of increasing polarization into expanding large operations and marginal small operations, but again with notable differences across commodities. Lockie and Halpin (2005) conclude that while the expansion in Australia fits the "agro-industrialization thesis" of conventionalization (Guthman 2004a), there is no evidence regarding bifurcation that the smaller farms are being marginalized. These findings are "sufficient to throw doubt on the bifurcation elements (both ideological and structural) of the conventionalization thesis" (Lockie and Halpin 2005, p. 304).

Research from Texas provided mixed support for the conventionalization thesis. Constance and associates (Constance et al. 2008) found that certified and non-certified organic producers did often align with the predicted bifurcation types. While both groups exhibited similar and strong ideological support for organics, the certified producers tended to be larger, have a more economic orientation, and focus on indirect markets. The non-certified producers tended to be smaller operations that sold in direct markets. Certified organic farmers tended to have farmed longer overall, farm full time, and use more hired labor. Constance et al. (2008) did not find the expected differences in length of time farming organically, the path to organic farming (start organic or transition), tenure pattern (own/rent), or plans for expansion. They did find that the two groups expressed opposing opinions of the value and necessity of organic certification.

Best's (2008) research in Germany found some support for conventionalization. The newer organic farms tended to be larger and more specialized, with a growing proportion of the organic farmers who do not share pro-environmental attitudes. He found a trend toward less diversified organic farms and away from traditional organic marketing channels. Since 2000 several large and highly specialized livestock and poultry farms had entered the organic market. Like other authors, Best argues that the California case may be unique and care should be taken in trying to generalize the California model to other regions or countries.

Guptill's (2009) research on the dairy industry in New York revealed mixed support for the conventionalization thesis. She found that government regulations supported the commodification and conventionalization of organic milk and the cost-price squeeze has accelerated in recent years as major firms expanded into to market. She also found that in response to conventionalization, many organic producers embraced the movement aspects of organic through deeper commitment to local sourcing and a value-driven lifestyle.

The conventionalization thesis has generated a lively and valuable discussion. The evidence indicates substantial evidence of conventionalization, but less so for bifurcation. The evidence on bifurcation reveals competing manifestations and interpretations of the structure. Whereas the California model argues it marginalizes deep-organic producers, research from Australia and New Zealand reports a stable and relatively virtuous arrangement. Additionally, there are significant regional, national, and commodity differences that in the short term call into question the inevitability of Guthman's California model. Finally, research indicates that the role of the state in creating an institutional environment that supports "deep organics" or "organic lite" is crucial regarding the process of conventionalization. Whether organic social movement advocates can influence the state to hold conventionalization at bay is a topic ripe for future research. The recent change in the Farm Bill designed to enhance adoption is salient evidence of the role of the state in facilitating conventionalization.

9.4 Pragmatic Conventional Producer

The historical trend of increased government support for organic production intersects the theoretical concerns regarding the negative aspects of conventionalization and bifurcation. As governments mobilize to increase production, social scientists warn about the negative impacts of the industrialization of organics. As noted above (see Dimitri and Oberholtzer 2009; Greene et al. 2009), in 2008 the US Farm Bill amended organic policies and programs to increase domestic organic production. USDA provided research monies and subsidies to offset the lack of information and costs associated with conversion by conventional producers. The characteristics of conventional growers, who might be interested in organics, including the barriers to adoption, were a central focus of the research funding. This section of the chapter is based on research in Texas funded by those USDA programs to investigate farmers' attitudes regarding organics (York et al. 2007).

Conventional producers interested in organics have been referred to as "pragmatic conventionals" (Fairweather 1999; Darnhofer et al. 2005; Constance and Choi 2010). These producers tend to exhibit neutral or positive ideological attitudes towards organics, but are concerned about the uncertainty and risks of organic production. They are a potential pool of converters because they don't rule out organic farming. More research is needed on the characteristics of pragmatic conventional producers (Padel 2001).

While numerous technical, ideological, and financial barriers to organic adoption have been identified globally (Freyer et al. 1994; Padel and Lampkin 1994; Fairweather 1999; Lohr and Salomonsson 2000; Rigby et al. 2001; Schneeberger et al. 2002; Costa et al. 2005; Darnhofer et al. 2005), in the US the lack of subsidies for the 3-year conversion period was identified as the key constraint (Lohr and Salomonsson 2000; Padel 2001; Duram 2006; Greene et al. 2009). The 2008 Farm Bill removed that constraint in hopes that pragmatic conventional farmers in the US would convert to organics.

9.5 The Research from Texas

In 2008, Texas ranked fourth in total cropland acres (155,957 acres) and second in pasture acres (294,749 acres) (USDA/ERS 2012). Data reported in Table 9.2 reveal that the big increase in total certified organic acres in crops and pastureland and number of operations in Texas occurred in the 1997–2002 period, prior to the establishment of the NOP. Following the national trend outlined earlier in the paper (see Table 9.1), the number of certified organic operations in Texas has increased at about the same rate as the US overall (51.3 % and 55.0 %, respectively). Total certified pasture and cropland increased slower in Texas than the US (49.8 % versus 123 %, respectively), with comparable increases for livestock and poultry.

Table 9.2 Texas organic certified farm operations: 1997–2008; Certified organic farmland: 1997–2008 (in acres) and Certified livestock/poultry: 1997–2008

					% change		
Item	1997	2002	2007	2008	97–02	02–07	07–08
Operations[a]	2	150	227	279	7,400	51.3	22.9
Pasture & Crops	30,880	279,506	418,652	450,706	805	49.8	7.7
Livestock	n/d	6,065	23,099	39,535	n/d	280.9	71.2
Poultry	n/d	21,000	51,000	329,378	n/d	142.9	545.8

Source: Economic Research Service (2012), Table 9.2: based on information from USDA-accredited State and private organic certifiers
[a]Does not include subcontracted organic farm operations

In 2008 a mail survey was conducted with a representative sample of producers in Texas, stratified by commodity (York et al. 2007). Based on the orientation toward organics, the respondents were classified into three groups: Conventionals (no interest in organics); Pragmatic Conventionals (at least some interest in organics); and Organic (self-identified as organic producers). Of the 897 respondents used in this analysis, 100 producers were organic, 464 were conventional producers, and 334 were pragmatic conventional producers. Tables 9.3, 9.4, and 9.5 report significant differences among the three groups regarding the structural and attitudinal aspects of organics.

Data reported in Table 9.3 indicate that Pragmatic Conventionals fall between the Conventionals and Organics regarding years farming, but in general they are more similar to the Conventionals. Regarding annual sales, notice that Pragmatic Conventionals (44.2 %) are most likely to be larger operations than Conventionals (39.8 %) and the Organics (18.8 %). Pragmatics are also more likely to be expanding, least satisfied with their operation, and over 80 % indicate that revenue support increases the likelihood of adoption.

Table 9.4 presents attitudinal information regarding philosophical and production aspects of organics. It reveals that while the Pragmatics Conventionals are more similar to Organics regarding their positive philosophical support for organic farming, still about one-third are "not sure" or "disagree." While Pragmatic Conventionals indicate broad philosophical support for organics, they are much less sure about its feasibility in organic production. The high levels of "not sure" across the four questions indicate major barriers to adoption, as well as point to the need for government intervention to reduce the uncertainty. In particular, about one-third of Pragmatic Conventionals disagree organic compatibility with high production farming.

Table 9.5 reports similar information regarding the marketing, informational, and certification aspects of organics. It is noteworthy that majority of Pragmatic Conventionals have concern about the financial viability of organics and the economic risks of transition. They are similarly unsure and/or disagree that organic

Table 9.3 Texas producer characteristics by orientation toward organics (percent)

	Conventional (n = 463)	Pragmatic conventional (n = 334)	Organic (n = 100)
Years of farming*			
Less than 5 years	2.8	4.8	9.3
5–10 years	8.5	11.3	16.5
11–20 years	19.2	20.2	18.6
More than 20 years	69.5	63.7	55.6
Annual gross sales**			
Less than $50,000	46.0	42.4	68.7
$50,000 to $99,999	14.2	13.4	12.5
$100,000 to $499,999	25.5	31.8	11.5
$500,000 or more	14.3	12.4	7.3
Expected operation change within next 3 years**			
Expanding	18.0	33.2	26.7
Decreasing or closing	18.9	12.8	20.9
No changes expected	63.1	54.0	52.3
Satisfaction with current farming system**			
Satisfied	87.5	54.7	63.8
Neutral	5.9	21.4	23.4
Not satisfied	6.6	23.9	12.8
Increase in revenue facilitate adoption**			
Facilitate	26.5	80.6	46.2
Not facilitate	70.3	9.3	29.7
Not necessary	3.2	10.2	24.2

Note: *P* values report significance levels for Chi-square test of producers' characteristics by their orientation toward organics: *$p < 0.05$; **$p < 0.01$

markets are reliable, that the necessary information is available, and that lenders support organics. Finally, it is interesting to note that all three groups reports very low levels of understanding of organic certification.

In summary, the research on Pragmatic Conventionals in Texas reveals that they tend to be larger producers who are unhappy with their current farming situation, interested in expanding, and that an increase in revenue would support organic adoption. While the majority of Pragmatic Conventionals present philosophical support for organics, their reported levels of uncertainty and concerns about organic production, marketing, information, and certification highlights significant barriers to adoption. Overall, Pragmatic Conventionals in Texas report that an increase in revenue would support adoption, especially if their concerns about financial viability and economic risks of transition were addressed. The changes in US organic policies included in the 2008 Farm Bill appear to be designed to ameliorate these uncertainties.

Table 9.4 Attitudes regarding philosophy and production (percent)

	Conventional (n = 463)	Pragmatic vonventional (n = 334)	Organic (n = 100)
I support the philosophy of organic farming**			
Agree	26.8	66.3	73.6
Disagree	43.5	5.4	6.6
Not sure	29.7	28.2	19.8
I can successfully farm without the use of synthetic chemicals**			
Agree	14.3	18.2	55.8
Disagree	52.7	22.4	16.3
Not sure	33.0	59.4	27.9
Organic farming is technically viable**			
Agree	6.4	25.2	55.6
Disagree	62.1	15.4	15.6
Not sure	31.5	59.4	28.9
Organic farming is a feasible long-term production method**			
Agree	3.6	17.9	48.2
Disagree	68.5	14.6	18.8
Not Sure	27.9	67.5	32.9
Organic production is compatible with my high production system of farming**			
Agree	3.6	11.6	33.3
Disagree	65.0	31.5	32.1
Not sure	31.4	57.0	34.5

Note: *P* values report significance levels for Chi-square test of producers' attitudes toward organics by their orientation toward organics: $*p < 0.05$; $**p < 0.01$ Tables 9.4, 9.5 based on information from USDA-accredited State and private organic certifiers

9.6 Sociology of Agrifood

For rural social scientists, the alternative agrifood system in general and organics in particular is a central topic of discussion (Buttel 1987, 1996; Tovey 1997; Allen and Kovach 2000; DuPuis 2000; Campbell and Liepins 2001; Allen 2004; Guthman 2004c; Lyson 2004; Lockie and Halpin 2005; Hinrichs and Lyson 2008; Jordan and Constance 2008). Organic production and markets have grown rapidly and continue to do so. The certified-organic label is well-known, but the success is problematic sociologically. While the origins of organics included a transformative dimension that addressed the environmental, economic, and social externalities of modern agriculture, some authors maintain that conventionalization calls into question the original social movement aspects as organics is institutionalized (DeLind 2000; Klonsky 2000; Pugliese 2001; Allen 2004; Guthman 2004b; Klintman and Boström 2004; Constance and Choi 2010; National Research Council 2010).

Evidence presented in this chapter speaks to these concerns about conventionalization. First, it appears that due to opposition from conventional agriculture in the US, the original NOP regulations that focused only on the certified-organic label as a market signal with no conversion support was too conservative regarding

Table 9.5 Attitudes regarding marketing, information, and certification (percent)

	Conventional (n = 463)	Pragmatic conventional (n = 334)	Organic (n = 100)
Organic farming is financially viable**			
Agree	4.4	14.9	40.7
Disagree	62.9	20.1	20.9
Not sure	32.7	64.9	38.4
I am concerned about the economic risks of transitioning to organic methods**			
Agree	45.7	59.9	28.0
Disagree	25.3	9.8	37.8
Not sure	28.9	30.3	34.1
Organic markets are reliable**			
Agree	9.4	22.3	36.7
Disagree	50.9	18.7	14.4
Not sure	39.6	59.0	48.9
My lenders support the idea of organic production**			
Agree	1.1	3.1	15.6
Disagree	42.7	23.2	24.7
Not sure	56.2	73.7	59.7
I feel the necessary informational support for organic farming is available**			
Agree	17.6	17.5	33.3
Disagree	34.9	26.3	27.4
Not sure	47.5	56.2	39.3
I understand the process of organic certification			
Agree	16.1	10.9	19.1
Disagree	33.5	38.1	31.5
Not sure	50.4	51.0	49.4

Note: *P* values report significance levels for Chi-square test of producers' attitudes toward organics by their orientation toward organics: $*p < 0.05$; $**p < 0.01$

enhancing domestic organic adoption to compete in global markets, resulting in a growing import/export imbalance. To overcome this barrier, organics companies in the US and elsewhere sourced globally to service their supply chains to meeting rising consumer demand. The 2008 Farm Bill attempted to address this shortcoming with direct subsidies, research support, and overt language on the environmental benefits of organic agriculture. Recall that Guthman argues that conventionalization is inevitable without state policies to stop it. Similarly, other researchers report that the particular institutionally trajectory of organics varies by nation and commodity. Research for this chapter indicates that the institutional trajectory in the US is characterized by government policies that are designed to support conventionalization.

Second, and similarly, the USDA/ERS data do show that after the NOP in 2002, the rate of adoption increased and the size of the firms increased, supporting the conventionalization thesis. Additionally, data reported in Table 9.1 on the increases from 2007 to 2008 indicate that the rate of adoption continues to increase. Third, Howard's and other research reveal that conventionalization is progressing in the manufacturing and retail sectors as major retailers offer organic products, develop their own store brands, and account for increasing percentages of organic sales. These data support Guthman's concerns over agro-industrialization. The organic premium has attracted corporate entry, which will compete on economics of scale, which will lead to industry integration, consolidation, and concentration and an increasing cost-price squeeze, which shrinks the organic premium, thereby forcing smaller producers out of business or relegating them to marginal indirect markets. Fourth, the research from Texas indicates that indeed it is the larger conventional operations that are interested in organics, and that economic concerns are the major barrier to entry. The 2008 Farm Bill policies are designed to address the concerns of the Pragmatic Conventionals in Texas.

A useful framework to interpret the conventionalization debate is the "Four Questions in Agrifood Studies" (Constance 2008).[2] The first question is the "Environmental Question" which asks: "What is the relationship between modern agriculture and the quality of the environment?" (see Carson 1962; Buttel 1987, 1996; Klonsky 2000). The environmental dimension of the crisis of modern agriculture was the first to reach critical mass, generate a social movement critical of reductionist science and chemical monoculture, and produce legislation (SCS, EPA, SARE, NOP) to address the negative externalities (Constance and Choi 2010). Organics is the most far reaching of these programs as the certified organic label stretches beyond national borders and restricts the huge US market to those producers that meet the USDA/NOP standards. Organics is a good example of what German philosopher Ulrich Beck (1992) calls reflexive modernization, which would argue that upon reflection we realized that chemical intensive monoculture generates significant negative externalities. It was a mistake to blindly adopt it but we can fix this error through the reflexive use of science and appropriate technologies, such as growing food using organic methods. Most all agree that organics improves environmental quality as it drastically reduces chemical contamination. Many observers also agree that the compromise over the USDA/NOP standards created a structure that favored eco-input substitution over agro-ecological farming. Guthman and associates argue that with conventionalization organics loses its broader social and agro-ecological potential to transform food production and instead becomes a system of eco-input substitution and farming to the list.

The second question is the "Agrarian Question" which asks: "What is the relationship between the structure of modern agriculture and the quality of life in rural communities" (see Buttel and Newby 1980; Lobao 1990; Heffernan 2000; Lyson 2004). The quality of life for farmers and rural peoples is influenced by the structure of agriculture and the type of commodity chains they are linked to (Lyson 2004; Gibbon and Ponte 2005; Constance 2008). Research indicates that family-based, middle-class farm operations tend to support a higher quality of life in rural

[2]The Four Questions are: Environmental, Agrarian, Food, and Emancipatory (see Constance 2008). While all four questions do apply to organics, in this analysis we employ the Environmental and Agrarian Questions.

communities than do industrial forms of agricultural production based on large-scale units and hired labor (Goldschmidt 1947; Lobao 1990; Buttel 1996; Lyson 2004; Lyson et al. 2008).

Initially organics was based on local and regional systems dominated by direct sales and an anti-industrial philosophy. The value chains were local and regional as opposed to national and global. Because of the power of conventional agriculture, the NOP rule focused on farming practices and a market label, rather than scale or quality of life issues. Organic bifurcation creates a system whereby larger certified operations supply the major retailers via indirect commodity-like markets and smaller, non-certified operations service localized, direct markets. One group converts to organics and farms to the list to capture the organic premium, while the other farms agro-ecologically and sells on trust, resulting in "organic lite" and "deep organic" archetypes of production (Guthman 2004a, b; Constance et al. 2008). The result is organic global supply chains based on comparative advantage and global sourcing – not that dissimilar from conventional global supply chains (Gereffi and Korzeniewicz 1994; Gibbon and Ponte 2005; McMichael 2005; Bonanno and Constance 2008). Research (Howard 2009a, b) documents the growing consolidation in the organic industry over the past 10 years as the dominant conventional food companies enter the market through acquisition and expansion.

The sociology of agrifood literature often employs a commodity systems or commodity chain methodology to analyze trends in the food system and the relationships among actors along the supply chain (Friedland 1984; Gereffi and Korzeniewicz 1994; Gibbon and Ponte 2005). There are several kinds of commodity chains, also called supply chains and/or value chains. Global commodity chains are long chains that deal in undifferentiated commodities in indirect markets. These chains tend to be characterized by unequal power where corporations drive the chain and capture more of the profits (Heffernan 2000; Fishman 2006; Hinrichs and Lyson 2008; Burch and Lawrence 2009). McMichael (2005) calls this "Food from Nowhere." Global value chains are problematic sociologically in that they tend to externalize social, economic, and ecological costs.

Alternatives such as Fair Trade value chains (Raynolds et al. 2007) tend to be built on a cooperative philosophy that encourages transparency along the chain and reduces the middleman functions in an effort to transfer wealth from the corporations to the producers. Local value chains are based on direct sales, smaller scale, and community embeddedness. These types of value chains are more likely to support the ecological, economic, and social dimensions of sustainability (Kloppenberg et al. 1996; Lyson 2004; Hinrichs and Lyson 2008). CSAs, farmers' markets, and farm to institution value chains are some examples. Agriculture in the Middle (Lyson et al. 2008) value chains focus on operations that are too large for direct sales and too small to compete in global commodity markets. They propose the development of regional fair trade value chains as a mechanism to support sustainable rural development by repopulating rural areas with moderate-sized operations.

The conventionalization debate can also be interpreted within the macro discussion of the globalization of the agrifood system from a regimes perspective (see Friedmann and McMichael 1989; McMichael 2009). Organics fits well with Friedmann's (2005) "corporate-environmental regime." At various levels, countries and companies are mobilizing to enter the growing organics market and capture the green premium. The corporate greening referred to by Lyons in Australia fits this perspective, as does the acquisition and development of green labels and companies reported by Howard. From the perspective of McMichael's (2005) "corporate food regime," conventionalization is replacing "Food from Somewhere" (read "deep organic") grounded in indigenous and local relations of production with "Food From Nowhere" (read "organic lite") organized around global value chains and regulated by the WTO and other supra-national forms of the state. In general, these global value chains organized by TNCs link producers in the global South to consumers in the global North.

From Burch and Lawrence's "financialization regime" (2009) approach, conventionalization is indicated by the growth of the major agrifood retailers that drive the global value chains and develop their own organic store brands. Evidence by Smith and Marsden in the UK supports this position as retailers compete for the organic market by rationalizing the supply chain and pushing the cost-price squeeze down to the farm gate. Eventually, the global comparative advantage will obtain at the production and retailing levels as organics is incorporated into the global agrifood system. Following the bifurcation thesis, inefficient producers and regions will leave the market and/or be relegated to marginal direct markets. From Pechlaner and Otero's "neo-liberal regime" (2010) view, conventionalization points to the role of the state in facilitating a neo-regulation process that favors the interests of capital over subordinate groups as part of globalization, including a focus on supporting the use of genetically-modified organisms. Several organic scholars have invoked the crucial role the state plays in the structure and development of organics. In the US, while the state's support for organics was initially limited due to the power of agribusiness, in 2008 the Farm Bill corrected this mistake. Additionally, conventional agriculture's unsuccessful attempt, with the help of the USDA, to include GMOs in the national organic standard and the social movement resistance that thwarted this effort, fit well with this perspective.

9.7 The Future of Organics

Agricultural philosopher Paul Thompson (2010) describes the two contrasting viewpoints of agriculture as the industrial and agrarian perspectives. The industrial perspective views agriculture as just another part of industrial society where commodities are produced at the lowest cost possible. The agrarian perspective, sometimes called alternative, views agriculture as having important social functions beyond its efficient production of commodities. From this view, a major departure from the conventional agriculture model is needed because it is not sustainable. Similarly, the National Research Council (2010) calls for moving beyond an incremental approach to improving agricultural sustainability to a transformative approach that goes beyond environmental fixes to conventional agriculture to incorporate the economic and social dimensions of sustainability.

Both the theory and evidence presented in this chapter indicate that the trend in organics is problematic regarding both the quality of life for rural communities and the overall sustainability of agriculture. Organics contributes positively to environmental quality, including food quality but the compromise that became the NOP purged much of the agro-ecological dimensions, as well as any scale or structure preferences towards "deep organics." The NOP does address the ecological leg of the sustainability stool, but does little to improve the negative economic and social dimensions of industrial agriculture. As conventionalization advances, organics loses more of its transformative ability and becomes less sustainable (Allen 2004; Constance 2010; National Research Council 2010). If conventionalization follows the California model as predicted, and not all researchers agree, then the vast majority of the organic foods sold in the major supermarkets will be sourced globally. This organic agrifood system based on global commodity chains becomes increasingly similar to the conventional system, criticized as a race to bottom facing a growing legitimation crisis (McMichael 2005). The early hopes for organics as a source of transformative change grounded in agrarian values gives way to incremental improvements in the ecological externalities of conventional agriculture grounded in industrial values (National Research Council 2010; Thompson 2010).

It is not that organic conventionalization and the California model are inevitable. In fact, we should expect the rationalized organic model to be perfected in California, the center of industrial agriculture in the US. While not inevitable, conventionalization is supported by the current global political economic regime dominated by neoliberal restructuring and state support for accumulation instead of protections for small-scale producers from global forces (Harvey 2005; McMichael 2005). The 2002 NOP set the organic bar to sell in the US and countries in the South mobilized to service the lucrative US market. The problem was that the US organic policies did not provide direct supports for conversion for domestic producers, which hindered production increases in the US and supported the growth of imports. The 2008 Farm Bill is an attempt to correct the mistake. The recent failure to pass the 2012 Farm Bill calls into question the continued political support for this course correction regarding organics in the US. The Cali model is not inevitable, but it is the preferred model of the agrifood TNCs constructing the global organic value chains, as well as the global regulatory organizations such as the WTO. Fred Buttel (2006) warned us of the uncanny ability of conventional agricultural to sustain the unsustainable. This chapter provides evidence in support of Fred's warning.

Deeper reflection regarding the externalities of the conventional agrifood systems combined with visioning a truly sustainable alternative agrifood system is needed to bring about transformative change grounded in agrarian values (Beck 1992; Thompson 2010). The alternative agrifood system made up of CSAs, farmers' markets, farm to institution, food sheds, food circles, Ag in the Middle, organics, food policy councils, urban agriculture, community gardens, and cooperatives are notable examples of "beyond organics" initiatives with transformative potential. Many of these alternative agrifood movements are more likely to create horizontal linkages that build social capital, because they are embedded in community (Lyson and Guptill 2004; Morgan et al. 2006; Hinrichs and Lyson 2008; Lyson et al. 2008). This system grounded in agrarian values is more likely to be based on participatory research methods and holistic conceptual frameworks with a transformative agenda (National Research Council 2010). This holistic agrifood system would not only enhance the agro-ecological dimensions of agriculture, but add the economic and social legs to the stool of sustainable agriculture (Jordan and Constance 2008).

From a rural social sciences perspective, Lyson's (2004) "Civic Agriculture" is a good model to start to relink agriculture and community: "Food from Somewhere" (McMichael 2005) with transformative potential (National Research Council 2010) grounded in an agrarian perspective (Thompson 2010). Lyson argues Civic Agriculture is the logical next step towards sustainable agriculture. The Ag in the Middle regional fair trade value chains also do a good job of incorporating some of the economic and social dimensions of sustainability (Lyson et al. 2008). Fair Trade incorporates an ethical dimension to wealth creation and extraction along the value chain, while regional addresses scale and structure issues. The region is probably the appropriate unit of analysis for the development of sustainable agrifood systems (Clancy and Ruhf 2010).

The organic conventionalization debate provides an interesting case for applying sociology of agrifood conceptual frames to the historical trends in the structure of organic production, marketing, and consumption. As organics is institutionalized into the mainstream agrifood system, new labels, standards, and metrics are exploding as part of a "beyond organics" push by the alternative agrifood movement (Constance 2010). There is a new hope that these initiatives will reinsert the transformative agenda and thereby address DeLind's (2000) concerns about the lack of social dimensions of conventionalized organics. The state will continue to play an important function in regulation, but especially at the global level, private governance is replacing public government as the regulatory venue (Busch 2011). Like organics, these organizational ventures will have to survive within a global neoliberal political economic regime that favors capital over subordinate groups.

There is much interesting work for academics and activists interested in the implications and predictive value of the Cali model. How can the transformative agenda aspects of alternative agrifood initiatives be preserved in the face of conventionalization? How can conventionalization and bifurcation be structured to be more virtuous than vicious? These questions are the focus of the "Emancipatory Question." How can we vision and create value chains that protect the civil rights of the actors who participate in the chain?

The contours of the new agrifood regime are in play, being contested by social movements, corporate interests, nation-states, and supra-national organizations. Organics provides a valuable case study of the dimensions of the contest. Agrifood social scientists can make two important contributions to the conventionalization debate. Through research we can provide a better understanding of the regulatory regime for organics and alternative agrifoods. Through public sociology we can volunteer to work with alternative agrifood organizations to create the holistic, sustainable agrifood system.

References

Agricultural Marketing Service. 2012. *2011 list of certified operations*. Retrieved from http://www.ams.usda.gov/AMSv1.0/ams.fetchTemplateData.do?template=TemplateF&navID=FactSheetslinkNOPNationalOrganicProgramHome&rightNav1=FactSheetslinkNOPNationalOrganicProgramHome&topNav=&leftNav=&page=NOPFactSheets&resultType=&acct=noppub. Accessed 23 Sept 2012.

Allen, P. 2004. *Together at the table: Sustainability and sustenance in the American agrifood system*. University Park: Pennsylvania State University Press.

Allen, P., and M. Kovach. 2000. The capitalist composition of organic: The potential of markets in fulfilling the promise of organic agriculture. *Agriculture and Human Values* 17(3): 221–232.

Beck, U. 1992. *Risk society: Towards a new modernity*, vol. 17. Thousand Oaks: Sage Publications.

Belasco, W.J. 1989. *Appetite for change*. New York: Pantheon Books.

Best, H. 2008. Organic agriculture and the conventionalization hypothesis: A case study from West Germany. *Agriculture and Human Values* 25(1): 95–106.

Bonanno, A., and D.H. Constance. 2008. *Stories of globalization: Transnational corporations, resistance, and the state*. University Park: Pennsylvania State University Press.

Buck, D., C. Getz, and J. Guthman. 1997. From farm to table: The organic vegetable commodity chain of Northern California. *Sociologia Ruralis* 37(1): 3–20.

Burch, D., and G. Lawrence. 2009. Towards a third food regime: Behind the transformation. *Agriculture and Human Values* 26(4): 267–279.

Busch, L. 2011. *Standards: Recipes for reality*. Cambridge, MA: MIT Press.

Buttel, F.H. 1987. New directions in environmental sociology. *Annual Review of Sociology* 13: 465–488.

Buttel, F.H. 1996. Environmental and resource sociology: Theoretical issues and opportunities for synthesis. *Rural Sociology* 61: 56–76.

Buttel, F.H. 2006. Sustaining the unsustainable: Agro-food systems and the environment in the modern world. In *Handbook of rural studies*, ed. P. Cloke, T. Marsden, and P. Mooney. London/Thousand Oaks: Sage Publications.

Buttel, F.H., and H. Newby. 1980. *The rural sociology of the advanced societies: Critical perspectives*. Montclair: Allanheld, Osmun & Co. Publishers Inc.

Campbell, H., and B. Coombes. 1999. Green protectionism and organic food exporting from New Zealand: Crisis experiments in the breakdown of Fordist trade and agricultural policies. *Rural Sociology* 64(2): 302–319.

Campbell, H., and R. Liepins. 2001. Naming organics: Understanding organic standards in New Zealand as a discursive field. *Sociologia Ruralis* 41(1): 22–39.

Carson, R. 1962. *Silent spring*. Greenwich: Fawcett Publications.

Clancy, K., and K. Ruhf. 2010. Is local enough? Some arguments for regional food systems. *Choices* 25(1): 123–135.

Clunies-Ross, T. 1990. Organic food: Swimming against the tide. In *Political, social and economic perspectives on the international food system*, ed. T. Marsden and J. Little. Aldershot: Avebury Press.

Clunies-Ross, T., and G. Cox. 1994. Regulating agriculture. In *Regulating agriculture*, ed. P. Lowe, T. Marsden, and S. Whatmore. London: David Fulton Publishers Ltd.

Constance, D.H. 2008. The emancipatory question: The next step in the sociology of agrifood systems? *Agriculture and Human Values* 25(2): 151–155.

Constance, D.H. 2010. Sustainable agriculture in the United States: A critical examination of a contested process. *Sustainability* 2(1): 48–72.

Constance, D.H., and J.Y. Choi. 2010. Overcoming the barriers to organic adoption in the United States: A look at pragmatic conventional producers in Texas. *Sustainability* 2(1): 163–188.

Constance, D.H., J.Y. Choi, and H. Lyke-Ho-Gland. 2008. Conventionalization, bifurcation, and quality of life: Certified and non-certified organic farmers in Texas. *Southern Rural Sociology* 23(1): 208–234.

Coombes, B., and H. Campbell. 1998. Dependent reproduction of alternative modes of agriculture: Organic farming in New Zealand. *Sociologia Ruralis* 38: 127–145.
Costa, L., M. Sottomayor, and R. Ribeiro. 2005. *Conversion to organic farming in mainland Portugal*, Economic working paper, vol. 10. Porto: Portuguese Catholic University.
Darnhofer, I., W. Schneeberger, and B. Freyer. 2005. Converting or not converting to organic farming in Austria: Farmer types and their rationale. *Agriculture and Human Values* 22(1): 39–52.
DeLind, L.B. 2000. Transforming organic agriculture into industrial organic products: Reconsidering national organic standards. *Human Organization* 59(2): 198–208.
Dimitri, C., and C. Greene. 2000. *Recent growth patterns in the US organic foods market*, Agriculture information bulletin, vol. 777. Washington, DC: US Department of Agriculture, Economic Research Service.
Dimitri, C., and L. Oberholtzer. 2009. *Marketing US organic foods*, Economic information bulletin, vol. 58. Washington, DC: US Department of Agriculture, Economic Research Service.
Dimitri, C., and N.J. Richman. 2000. *Organic food markets in transition*, Wallace center policy studies report, vol. 14. Arlington: Winrock Corporation.
Driftmier, C.F. 2009. *10 ways to thrive in turbulent times*. 6/1. Retrieved from http://www.organicprocessing.com/opjanfeb09/opjf09CoverStory.htm. Accessed 13 Jan 2012.
DuPuis, M. 2000. Not in my body: rBGH and the rise of organic milk. *Agriculture and Human Values* 17(3): 285–295.
Duram, L.A. 2006. Organic farmers in the U.S.: Opportunities, realities, and barriers. *Crop Management*. doi:10-1094/CM-2006-0921-03-RV. Available online: http://www.plantmanagementnetwork.org/pub/cm/symposium/organics/Duram/. Accessed 26 June 2009.
Economic Research Service. 2012. *Organic production: Data sets overview*. Retrieved from http://www.ers.usda.gov/Data/Organic. Accessed 1 Mar 2012.
Fairweather, J.R. 1999. Understanding how farmers choose between organic and conventional production: Results from New Zealand and policy implications. *Agriculture and Human Values* 16(1): 51–63.
Fishman, C. 2006. *The Wal-Mart effect: How the world's most powerful company really works–and how it's transforming the American economy*. New York: Penguin.
Food and Agriculture Service. 2005. *U.S. Market Profile for Organic Food Products*. Retrieved from http://www.fas.usda.gov/agx/organics/USMarketProfileOrganicFoodFeb2005.pdf. Accessed 14 Mar 2009.
Freyer, B., R. Rantzau, and H. Vogtmann. 1994. Case studies of farms converting to organic agriculture in Germany. In *The economics of organic farming: An international perspective*, ed. N.H. Lampkin and S. Padel, 243–263. Wallingford: CAB International.
Friedland, W.H. 1984. Commodity systems analysis: An approach to the sociology of agriculture. In *Research in rural sociology and development*, ed. H. Schwarzweller, 221–235. Greenwich: JAI Press.
Friedmann, H. 1993. After Midas's feast: Alternative food regimes for the future. In *Food for the future: Conditions and contradictions of sustainability*, ed. P. Allen, 213–233. New York: Wiley.
Friedmann, H. 2005. From colonialism to green capitalism: Social movements and emergence of food regimes. In *Research in rural sociology and development*, vol. 11, ed. F.H. Buttel and A.J. McMichael, 227–264. Oxford: Elsevier.
Friedmann, H., and P. McMichael. 1989. Agriculture and the state system: The rise and decline of national agriculture, 1879 to present. *Sociologia Ruralis* 29(2): 93–117.
Gereffi, G., and M. Korzeniewicz (eds.). 1994. *Commodity chains and global capitalism*. Westport: Praeger.
Gibbon, P., and S. Ponte. 2005. *Trading down: Africa, value chains, and the global economy*. Philadelphia: Temple University Press.
Goldschmidt, W.R. 1947. *As you sow*. Glencoe: The Free Press.
Goodman, D. 1999. Agro-food studies in the 'age of ecology': Nature, corporeality, bio-politics. *Sociologia Ruralis* 39(1): 17–38.

Goodman, D. 2000. Organic and conventional agriculture: Materializing discourse and agro-ecological managerialism. *Agriculture and Human Values* 17(3): 215–219.

Goodman, D., B. Sorj, and J. Wilkinson. 1987. *From farming to biotechnology: A theory of agro-industrial development*. Oxford: Blackwell.

Greene, C. 2007. *US organic farm sector continues to expand*. Retrieved from http://www.ers.usda.gov/AmberWaves/April076/Findings/Organic.htm. Accessed 14 Mar 2009.

Greene, C., C. Dimitri, B.-H. Lin, W.D. McBride, L. Oberholtzer, and T.A. Smith. 2009. *Emerging issues in the US organic industry*, Economic information bulletin, vol. 55. Washington, DC: US Department of Agriculture, Economic Research Service.

Guptill, A. 2009. Exploring the conventionalization of organic dairy: Trends and counter-trends in upstate New York. *Agriculture and Human Values* 26(1): 29–42.

Guthman, J. 1998. Regulating meaning, appropriating nature: The codification of California organic agriculture. *Antipode* 30(2): 135–154.

Guthman, J. 2000. Raising organic: An agro-ecological assessment of grower practices in California. *Agriculture and Human Values* 17(3): 257–266.

Guthman, J. 2004a. Back to the land: The paradox of organic food standards. *Environment and Planning A* 36(3): 511–528.

Guthman, J. 2004b. The trouble with 'organic lite' in California: A rejoinder to the 'conventionalisation' debate. *Sociologia Ruralis* 44(3): 301–316.

Guthman, J. 2004c. *Agrarian dreams: The paradox of organic farming in California*, vol. 11. Berkeley/Los Angeles/London: University of California Press.

Hall, A., and V. Mogyorody. 2002. Organic farmers in Ontario: An examination of the conventionalization argument. *Sociologia Ruralis* 41(4): 399–322.

Harvey, D. 2005. *A brief history of neoliberalism*. Oxford: Oxford University Press.

Heffernan, W.D. 2000. Concentration of ownership in agriculture. In *Hungry for profit: The agribusiness threat to farmers, food, and the environment*, ed. F. Magdoff, J.B. Foster, and F.H. Buttel, 61–76. New York: NYU Press.

Heller, L. 2007. *Organic undersupply stunts market growth, OTA*. Retrieved from http://ofrf.org/pressroom/organic_news_clips/071114_foodnav_organicundersupply.pdf. Accessed 14 Mar 2009.

Hinrichs, C.C., and T.A. Lyson. 2008. *Remaking the North American food system: Strategies for sustainability*. Lincoln/London: University of Nebraska Press.

Howard, P. 2009a. Organic industry structure. *Journal of the New Media Caucus* 5(3).

Howard, P. 2009b. Consolidation in the North American organic food processing sector, 1997 to 2007. *International Journal of Sociology of Agriculture and Food* 16(1): 13–30.

Jordan, J.L., and D.H. Constance. 2008. Sustainable agriculture and the social sciences: Getting beyond best management practices and into food systems. *Southern Rural Sociology* 23(1): 1–22.

Kaltoft, P. 2002. Organic farming in late modernity: At the frontier of modernity or opposing modernity? *Sociologia Ruralis* 41(1): 146–158.

Klintman, M., and M. Boström. 2004. Framings of science and ideology: Organic food labelling in the US and Sweden. *Environmental Politics* 13(3): 612–634.

Klonsky, K. 2000. Forces impacting the production of organic foods. *Agriculture and Human Values* 17(3): 233–243.

Kloppenberg, J., J. Hendrickson, and G.W. Stevenson. 1996. Coming in to the foodshed. In *Rooted in the land: Essays on community and place*, ed. W. Vitek and W. Jackson, 113–123. New Haven: Yale University Press.

Lampkin, N. 1990. *Organic farming*. Ipswich: Farming Press.

Lipson, M. 1997. *Search for the O – Word*. Santa Cruz: Organic Farming Research Foundation.

Lobao, L.M. 1990. *Locality and inequality: Farm and industry structure and socioeconomic conditions*. Albany: The State University of New York Press.

Lockie, S., and D. Halpin. 2005. The 'conventionalisation' thesis reconsidered: Structural and ideological transformation of Australian organic agriculture. *Sociologia Ruralis* 45(4): 284–307.

Lockie, S., K. Lyons, and G. Lawrence. 2000. Constructing "green" foods: Corporate capital, risk, and organic farming in Australia and New Zealand. *Agriculture and Human Values* 17(4): 315–322.

Lohr, L., and L. Salomonsson. 2000. Conversion subsidies for organic production: Results from Sweden and lessons for the United States. *Agricultural Economics* 22(2): 133–146.

Lynggaard, K.S. 2001. The farmer within an institutional environment: Comparing Danish and Belgian organic farming. *Sociologia Ruralis* 41(1): 85–111.

Lyons, K. 1997. *What shade the greenwash? Corporate environmental commitment to organic agriculture or constructing a green fascade*. Environmental justice: Global ethics for the twenty first century conference. Melbourne University, Victoria, Australia.

Lyons, K. 1999. Corporate environmentalism and organic agriculture in Australia: The case of Uncle Tobys. *Rural Sociology* 64(2): 251–265.

Lyson, T.A. 2004. *Civic agriculture: Reconnecting farm, food, and community*. Lebanon: Tufts University.

Lyson, T.A., and A. Guptill. 2004. Commodity agriculture, civic agriculture and the future of US farming*. *Rural Sociology* 69(3): 370–385.

Lyson, T.A., G. Stevenson, and R. Welsh (eds.). 2008. *Food and the mid-level farm: Renewing an agriculture of the middle*. Cambridge, MA: MIT Press.

MacRae, R., J. Henning, and S. Hill. 1993. Strategies to overcome barriers to the development of sustainable agriculture in Canada: The role of agribusiness. *Journal of Agricultural and Environmental Ethics* 6(1): 21–51.

Madden, J.P. 1998. *The early years of LISA, SARE, and ACE programs: Phase I*. Logan: Utah State University.

McMichael, P. 2005. Global development and the corporate food regime. In *Research in rural sociology and development*, vol. 11, ed. F.H. Buttel and P. McMichael, 265–299. Oxford: Elsevier Press.

McMichael, P. 2009. A food regime genealogy. *The Journal of Peasant Studies* 36(1): 139–169.

Michelsen, J. 2001. Recent development and political acceptance of organic farming in Europe. *Sociologia Ruralis* 41(1): 3–20.

Morgan, K., T. Marsden, and J. Murdoch. 2006. *Worlds of food: Place, power, and provenance in the food chain: Place, power, and provenance in the food chain*. Oxford/New York: Oxford University Press.

National Agricultural Statistics Service. 2012. *2011 Certified Organic Production Survey*. Retrieved from http://www.nass.usda.gov/Surveys/Organic_Production_Survey/. Accessed 23 Sept 2012.

National Research Council. 1980. *Alternative agriculture*. Washington, DC: National Academic Press.

National Research Council. 2010. *Toward sustainable agricultural systems in the 21st century*. Washington, DC: National Academic Press.

National Sustainable Agriculture Coalition. 2012. *Path to 2012 farm bill: Farm bill expires on Monday – What does it mean and what happens now?* Retrieved from http://sustainableagriculture.net/blog/faq-on-farm-bill-expiration/. Accessed 3 Nov 2012.

Organic Farming Research Foundation. 1993. *National organic farmers' survey results*. Santa Cruz: Organic Farming Research Foundation.

Organic Farming Research Foundation. 1996. *Final report of the 1995 national organic farmers' survey*. Santa Cruz: Organic Farming Research Foundation.

Organic Farming Research Foundation. 2008. *Organic Farming Research Foundation applauds bill victories for organic famers and ranchers*. Retrieved from http://ofrf.org/pressroom/releases/080527_farm-bill-victories.html. Accessed 14 Mar 2009.

Organic Trade Association. 2006. *Manufacturer survey*. Retrieved from http://www.ota.com/pics/documents/short%20overview%20MMS.pdf. Accessed 4 June 2007.

Organic Trade Association. 2008. *Industry statistics and projected growth*. Retrieved from http://www.ota/com/organic/business.html. Accessed 14 Mar 2009.

Organic Trade Association. 2011. *Organic Trade Association 2010 organic. industry survey*. Retrieved from http://www.ota.com/organic/mt/business.html. Accessed 13 Nov 2011.

Organic Trade Association. 2012. *Organic Trade Association 2011 organic. Industry survey*. Retrieved from http://www.ota.com/pics/documents/2010OrganicIndustrySurveySummary.pdf. Accessed 12 Oct 2012.

Padel, S. 2001. Conversion to organic farming: A typical example of the diffusion of an innovation? *Sociologia Ruralis* 41(1): 40–61.

Padel, S., and N.H. Lampkin. 1994. Conversion to organic farming: An overview. In *The economics of organic farming: An international perspective*, ed. S. Padel and N.H. Lampkin, 295–313. Wallingford: CAB International.

Pechlaner, G., and G. Otero. 2010. The neoliberal food regime: Neoregulation and the new division of labor in North America. *Rural Sociology* 75(2): 179–208.

Pollan, M. 2001. *Behind the organic-industrial complex*. Retrieved from http://www.wesjones.com/pollan2.htm. Accessed 4 July 2007.

Pugliese, P. 2001. Organic farming and sustainable rural development: A multifaceted and promising convergence. *Sociologia Ruralis* 41(1): 112–130.

Raynolds, L.T., D.L. Murray, and J. Wilkinson (eds.). 2007. *Fair trade: The challenges of transforming globalization*. New York: Routledge.

Rigby, D., T. Young, and M. Burton. 2001. The development of and prospects for organic farming in the UK. *Food Policy* 26(6): 599–613.

Rosset, P.M., and M.A. Altieri. 1997. Agroecology versus input substitution: A fundamental contradiction of sustainable agriculture. *Society & Natural Resources* 10(3): 283–295.

Schneeberger, W., I. Darnhofer, and M. Eder. 2002. Barriers to the adoption of organic farming by cash-crop producers in Austria. *American Journal of Alternative Agriculture* 17(01): 24–31.

Smith, E., and T. Marsden. 2004. Exploring the 'limits to growth' in UK organics: Beyond the statistical image. *Journal of Rural Studies* 20(3): 345–357.

Thilmany, D.D. 2006. *The US organic industry: Important trends and emerging issues for the USDA*. Fort Collins: Colorado State University, Department of Agricultural and Resource Economics.

Thompson, P. 2010. *The agrarian vision: Sustainability and environmental ethics*. Lexington: University Press of Kentucky.

Tovey, H. 1997. Food, environmentalism and rural sociology: On the organic farming movement in Ireland. *Sociologia Ruralis* 37(1): 21–37.

United States Department of Agriculture. 1980. *Report on recommendations on organic farming*. Washington, DC: United States Department of Agriculture/National Agricultural Statistics Service.

United States Government Accounting Office. 1990. *Alternative agriculture*. Washington, DC: U.S. Government Accounting Office.

Volpe III, R.J. 2006. *Exploring the potential effects of organic production on contracting in American agribusiness*. 2006 annual meeting, July 23–26, Long Beach, CA. No. 21086. American Agricultural Economics Association.

Vos, T. 2000. Visions of the middle landscape: Organic farming and the politics of nature. *Agriculture and Human Values* 17(3): 245–256.

Willer, H., and L. Kilcher. 2009. *The world of organic agriculture: Statistics and emerging trends 2009*. Bonn/Frick: IFOAM/FIBL.

Willer, H., and L. Kilcher. 2012. *The world of organic agriculture: Statistics and emerging trends 2012*. Bonn/Frick: IFOAM/FIBL.

York, M., M.H. Lau, R. Hanagriff, and D.H. Constance. 2007. Identifying barriers to entry into the organic market and possible strategies to increase the likelihood of success for potential organic producers. *Final report to the Texas Department of Agriculture for the 2007 Federal State Marketing Improvement Project*: Texas Department of Agriculture.

Chapter 10
Organic Farmers: Contributing to the Resilience of the Food System?

Lesley Hunt, Chris Rosin, Hugh Campell, and John Fairweather

10.1 Introduction

Advocates of organic practices claim that organics should play a greater role in growing our food.[1] If this is so, we need to ask how such practices contribute to our food system (Campbell 1997) in a way that will enable it to better feed the people on our planet with safe food produced in a sustainable and resilient way. It is widely acknowledged that the context in which food is produced is changing rapidly and food producers are facing enormous challenges in very uncertain times (Urry 2005; McIntyre et al. 2009; Pretty et al. 2010; National Academy of Sciences 2011). According to Darnhofer et al. (2010a, p. 546) present and future uncertainty "may increasingly require farmers to keep their farms flexible to be able to respond to new challenges as they arise." If the practice of organics lives up to the rhetoric associated with it from its beginnings as a social movement, then it will have a lot to

[1]For example, Organics New Zealand claims "... organic production is the foundation of better crops, better livestock and better futures" (www.oanz.org.nz).

L. Hunt (✉) • J. Fairweather
Agribusiness and Economics Research Unit, Lincoln University, P.O. Box 84, Lincoln 7647, New Zealand
e-mail: lesley.margaret.hunt@gmail.com; johnrobyn@netsurf.co.nz

C. Rosin
Centre for Sustainability: Agriculture, Food, Energy, Environment (CSAFE), University of Otago, PO Box 56, Dunedin 9054, New Zealand
e-mail: Chris.Rosin@otago.ac.nz

H. Campell
Department of Sociology, Gender and Social Work, University of Otago, Dunedin, New Zealand
e-mail: Hugh.Campbell@otago.ac.nz

B. Freyer, J. Bingen (eds.), *Re-Thinking Organic Food and Farming in a Changing World*, The International Library of Environmental, Agricultural and Food Ethics 22, DOI 10.1007/978-94-017-9190-8_10

offer in the present and future in terms of its contribution to the possible pathways to adaptation and flexibility it offers to agricultural practices in general.

In this chapter, New Zealand agriculture is used to illustrate three theses: the contribution of organic farming is more apparent when it is studied in a farming sector context rather than in isolation; organic farmers and their farms add to the resilience of the food system; and, when organic farmers are seen as "good farmers" they have more influence on others. New Zealand is a small country, highly dependent on exporting agricultural and horticultural products, but geographically positioned some distance from its markets. Hence, the government encourages the production of quality niche products for which it can obtain a market premium. However, many of its products presently are commodities that can be manufactured in some way within New Zealand and shipped in large volumes. Since the mid-1980s, the New Zealand government has pursued neo-liberal market-led, user-pay policies. Primary production is not subsidized nor is environmental welfare encouraged by subsidies. Instead, environmental welfare is regulated by industry, local or national government, or as part of a market-led audit system associated with the production of particular products such as kiwifruit and lamb for specific buyers such as supermarkets. Commercial large-scale organic farming is relatively recent and it is export oriented. The government has invested very few resources in organics – either through regulation (until the recent Ministry of Agriculture and Forestry (MAF) arrangement for AsureQuality[2] to audit organic farms) or through scientific research. This context makes New Zealand an interesting and useful exemplar for the comparison of agricultural resilience and sustainability across some of the different agriculture sectors and the role of organics within them.

We seek to demonstrate that when organic practitioners are considered as part of a community, an agricultural sector, a supply chain and a nation, we can better recognize the contribution they and their practices make to the resilience of the production system. In addition, the impact of their performance on the sector is likely to be greater when their organic practices are seen to provide a model of farming (and farmers) when organic farmers are seen as "good farmers" in their local communities and in a particular agricultural sector.

As uncertainty and change now and in the future affects the world's food production, the provision of a sustainable and resilient supply of safe, good quality food in sufficient quantity may be best achieved by maximizing the contribution of organic practices (Milestad and Darnhofer 2003). Others also see a role for organics in differing pathways to sustainability. De Schutter (2010), the Special Rapporteur on the Right to Food to the United Nations, and supporter of agro-ecology presents the use of organic fertilizer as the ultimate goal, but one that may not be reached without the use of subsidized inorganic fertilizers in poor countries. Pretty et al. (2010) support the use of both GMOs and organics if agriculture is to be intensified enough to feed the world while being sustainable.

---

[2]AsureQuality is commercial company 100 % owned by the New Zealand government to provide food safety and biosecurity services to the food and primary production sectors, http://www.organiccertification.co.nz/

10.2 Organics in New Zealand

Along with many others throughout the western world, in the late 1920s some New Zealand farmers began using organic principles and became part of the organic movement concerned about “the direction of modernity and the growth of materialism; the relationship between food quality and health; environment; sustainable use of resources; and spirituality” (Stuart and Campbell 2004, pp. 228, 229). The first organic society, the Humic Compost Club was launched in 1941, and as its name suggests, promoted the relationship between healthy food and the use of compost to improve soil health. Early supporters saw organics as a sustainable agriculture practice, a spiritual and responsible connection between people and the land, “promoting better health, improving and sustaining soil fertility in ways that conserved resources and lessened environmental damage and to significantly reduce both household and national dependence on foreign trade” (Stuart and Campbell 2004, p. 230). The latter political emphasis was at odds with New Zealand farmers’ avowed role of “feeding the world” (Fairweather et al. 2001, 2007; Saunders 2009).

In the 1970s, interest in organic agriculture re-emerged. Ritchie and Campbell (1997, pp. 10, 11) propose that this was a combination of four historical processes: the 1930s response to the “development of “scientific” agriculture”; the emergence of the American environmentalist movement in the 1960s; the influence of migrants from Europe between 1950 and 1970; and, the emergence of alternative land use patterns around urban centers such as lifestyle blocks or hobby farms. Most organic produce was sold locally on a trust system where the buyer knew the grower. In 1983, three organizations formed BioGro to provide a certification system for organic production (Campbell and Liepins 2001, p. 28). Certification was considered appropriate for a variety of reasons including, the large variation in grower definitions of organic; a need for a guarantee of quality; a related concern arising out of food scares in Europe and the U.S.A.; the relationship and support of the bio-dynamic movement; international links with the bio-dynamic organization, the Soil Association and IFOAM; and finally, the absence of government interest (Campbell and Liepins 2001, p. 29).

The scene changed in 1990 with the involvement of two corporate actors who wished to use the BioGro label for exported products, and “by 1999 BioGro was certifying over 90 % of organic producers in New Zealand” (Campbell and Liepins 2001, p. 30). Over this period, organic production increased and some conventional growers converted to organics, though overall, organic growers remained a small minority. However, some members of the certification board were concerned the involvement of corporates would lower standards. In response BioGro moved to “a more professional and formal inspection process [. . .]” (Campbell and Liepins 2001, p. 31) and aligned local standards with accreditation to IFOAM. Since the inception of BioGro, the Ministry of Agriculture and Forestry has also become involved (at an arm’s length) in organic certification (along with other auditing and assessment services) through AsureQuality. BioGro and AsureQuality now also conform to the USDA organic standards, important for the acceptability of

New Zealand's organic exports.[3] Organic standards in the kiwifruit industry are recognized as meeting GLOBALGAP requirements.

One of the unintended consequences of the development of inspection and compliance processes has been that one discourse of organic became institutionalized.[4] Despite the hegemony of the corporate actors through their alliance with BioGro, the companies involved did not attempt to water down the standards partly because they did not want to be involved in any hint of green protectionism. However, the effect was to reduce the meaning of organics to a piece of text and the flexibility of the past was lost and any contestation that occurred was over the meaning of that text (Campbell and Liepins 2001).

10.3 The Meaning of What It Is to Be 'Organic'

Campbell and Liepins (2001) dispute that there ever has been a clear and fixed understanding of what it means to be organic or to practice organics. They argue that the flow of development from the social movement organics to industry, i.e., from alternative agriculture to conventionalized agriculture, is not linear. It fails "to allow enough space for the circular, contested and complex development and ongoing negotiations around organics" (Campbell and Liepins 2001, p. 24) The meaning of organic and the development of organic standards are not necessarily fixed objects but "derived from a plethora of social, scientific, economic and political contexts" (Campbell and Liepins 2001, p. 26). Seppänen and Helenius (2004) suggest that in Finland compliance was negotiated rather than fixed. This negotiation around text is illustrated internationally. IFOAM has produced a definition of organic agriculture (2009) that has been developed with extensive research and consultation since 2005:

> Organic agriculture is a production system that sustains the health of soils, ecosystems and people. It relies on ecological processes, biodiversity and cycles adapted to local conditions, rather than the use of inputs with adverse effects. Organic agriculture combines tradition, innovation and science to benefit the shared environment and promote fair relationships and a good quality of life for all involved (p. 1).

This development continues. IFOAM is now going through a period of consultation over a draft of the standards and norms of organic agriculture so it is a very pertinent time to be considering the issue of organic agriculture and how it fits in a changing world. Campbell and Liepins (2001, p. 32) suggest, "there is an ongoing

[3]In 1998 MAF, as part of the government's user pays policy, spun off two companies Asure New Zealand and AgriQuality, with AgriQuality having a specific company CERTENZ responsible for organic certification. CERTENZ achieved ISO 65 certification in 2001 that gave entry into the EU (www.organic-register.com). In 2007 Asure New Zealand and AgriQuality merged to form AsureQuality (www.stuff.co.nz/business/38232, "Food safety merger plan" will cost few jobs').

[4]For examples from other countries see Seppänen and Helenius (2004) (Finland) and Guthman (2000) (California, U.S.A.).

interplay between the meanings/standards of organics and the contexts in which they are practiced". Rosin and Campbell (2009) suggest that the changes in organics are more complex and involve relationships with markets, economics and politics. That is, they are positioned within a wider agri-food system and have a strategic function.

10.4 Description of ARGOS as Empirical Basis for Arguments

The findings reported in this chapter are taken from the Agriculture Research Group on Sustainability (ARGOS)[5] research project (2003 to 2012). The objective of ARGOS is to advance understanding of sustainable agriculture through the trans-disciplinary comparison of different management systems (conventional, integrated and organic) in three main sectors of New Zealand agriculture (sheep/beef, dairy and horticulture).[6] Farms or orchards were selected at 12 different geographical locations in each sector, with each location having a cluster of farms representing different management systems – organic, integrated and conventional management, as a way of accounting for ecological variation in analyses of the data. Each farmer or orchardist was interviewed at least twice (Hunt et al. 2005, 2006; Rosin et al. 2007a, b) and provided values for many economic, social and farm/orchard management variables. Farms and orchards were also assessed ecologically.

ARGOS sheep/beef and dairy farms tend to be generational family farms, whereas the orchard businesses are mainly individually owned but managed by different combinations of owners, managers and pack house employees. ARGOS farms and orchards are all operated as commercial-scale businesses, supplying international markets with their products and so the organic farmers within ARGOS could be said to form part of Pollan's "Industrial Organic" (Campbell and Coombes 1999; Pollan 2001, p. 32).[7]

10.5 Studying Organic Farming in Isolation Limits the Understanding of the Breadth of Its Contribution

We are not going to argue for or against different views of what ways of farming are or are not organic. Instead we will assert that organic farming is encompassed within a much larger farming field/arena/domain of many different farm management

[5] See www.argos.org.nz for more information.

[6] Hence the ARGOS study complies with the assertion that trans-disciplinary approaches are needed to study sustainable agriculture (e.g., Cousins et al. 2007; McIntyre et al. 2009; Hunt et al. 2010).

[7] Pollan also talks of "Big Organic" and "Little Organic" (Pollan 2001, p. 34).

systems which all interact (Campbell and Rosin 2011; Lamine 2011), and that it is better studied within that field where it can be compared and contrasted with other management systems in the same field. All these systems fit within the food supply chain (Friedmann and McMichael 1989; Friedland 2004). "Yet if the word organic means anything, it means that all these things are ultimately connected: that the way we grow food is inseparable from the way we distribute food, which is inseparable from the way we eat food" (Pollan 2001, p. 63).

The following quote is taken from the Royal Society (2009) report, *Reaping the Benefits: Science and the Sustainable Intensification of Global Agriculture*, which was produced to promote the contribution the biological sciences could make to food crop production in the face of global food insecurity. It was intended to influence the U.K.'s research priorities in the future. It is reproduced here to illustrate the role organics is seen to play in that future.

> Past debates about the use of new technologies for agriculture have tended to adopt an either/or approach, emphasizing the merits of particular agricultural systems or technological approaches and the downsides of others. This has been seen most obviously with respect to genetically modified (GM) crops, the use of pesticides and the arguments for and against organic modes of production. These debates have failed to acknowledge that there is no technological panacea for the global challenge of sustainable and secure global food production. There will always be trade-offs and local complexities [. . .] No techniques or technologies should be ruled out. Global agriculture demands a diversity of approaches, specific to crops, localities, cultures and other circumstances. Such diversity demands that the breadth of relevant scientific enquiry is equally diverse, and that science needs to be combined with social, economic and political perspectives (The Royal Society 2009, p. IX).

Three things are notable about this quote. First, organic production is not seen as playing a dominant role in future food production, but as part of the diversity of approaches available. Second, the word organic used with respect to the use of an organic management system in agriculture, only appears once in the whole report (in this summary) and is never referred to again. (The other uses of it are to do with soil organic matter). Finally, it suggests that social, economic and political perspectives have a role to play in determining the future sustainability of the world's food system, that is it is not just about science.

According to Campbell et al. (2009) there are three perspectives/groups of proponents to the anti- and pro-organic debate: those who support mainstream agriculture and the role of science and are against organics; those who support commercially oriented organics – big organics – and also use science; and those within the organic movement who feel that organic agriculture has sold out to commercial interests. In promoting organics as a panacea for all ills the third grouping has presented themselves as disciples in a social movement, saying this is the only way to save the world and that all agriculture should use organic practices. Others have questioned the role of organics as an exemplar of sustainable agricultural practices as commercially oriented organic businesses moved into the global arena (e.g., Pollan 2001; Trewavas 2001).

As indicated above, this is, of course, contested. There is debate about what is real organics, whether an organic practitioner has to adhere to a certain philosophical approach and whether organic practices have been supplanted by corporate and institutional interests that are primarily motivated by profit (Guthman 2000; Pollan 2001; Campbell et al. 2009). Pollan (2001, p. 63) writes how he buys local food which is not necessarily organically grown: "[. . .] I'm resolved to do it anyway. Because organic is not the last word [. . .]". Similarly, it is assumed that any research on agriculture tends to support the status quo and the dominance of answers provided by science (especially the use of GM technology), which makes organic agricultural practices invisible within mainstream agricultural science,[8] as is apparent in the Royal Society report referred to earlier.

One response to this is to question the either/or approach to organics as the only pathway to sustainability (Campbell et al. 2009). The synthesis report of the International Assessment of Agricultural Knowledge, Science and Technology for Development (IAASTD) supports the use of policies that facilitate and provide "incentives for alternative markets such as . . . certification for sustainable forests and fisheries practices and organic agriculture" (McIntyre et al. 2009, p. 6) which suggests that there is fourth group to those mentioned above – the position that we are advocating – that a diversity of approaches and practices is important for resilience.

Commercial organic farming is part of a primary industry field (a complex system) that has developed its own norms and rules about what it means to be a farmer. The organic farming field intersects with this more dominant field (e.g., Carolan 2005) sharing a common body of knowledge with all producers of products of the same type as their own, about the basic process used to produce their product (Hunt et al. 2005, 2006).[9] We represent this in Fig. 10.1 where the dotted lines indicate that the boundaries are not distinct and ideas and practices are able to

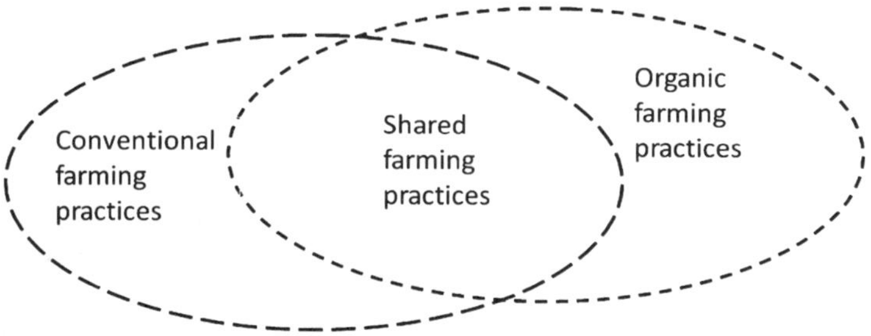

Fig. 10.1 Shared knowledge and practices between practitioners of different farming systems (Source: Own development)

[8]See a collection of responses to GM in farming provided by organizations such as the Soil Association.

[9]We have also demonstrated that practices and attitudes within a particular management system are not uniform in our paper, "Are conventional farmers conventional?" (Fairweather et al. 2009b).

filter between different practices and are open to influences external to both. For example, all kiwifruit growers – organic or non-organic – have to prune their vines and manage their canopies, all dairy farmers milk their cows and along with sheep/beef farmers manage the grass supply to feed their stock. Organic farms and orchards are not found in only one geographical area and they are very much in the minority; they co-exist with farms and orchards using other management systems.

Farmers look over the fences of the farms they pass in their cars, and orchardists over their boundaries at their neighbor's properties. They talk to each other about each other and may modify their practices accordingly (Hunt et al. 2005, 2006), as illustrated in the following two quotes taken from interviews with ARGOS farmers and orchardists.

> Interviewer: So how do you know how your farm is doing financially?
> Farmer (male): The end result, I guess.
> Interviewer: And what's that? What measure can detect all that?
> Farmer (male): Whether it was profitable, isn't it?
> Farmer (female): But that's not really how you tell.
> Farmer (male): Yeah, I know. Mmmm (pause).
> Farmer (female): So, um, your stock would be a measure wouldn't it? Whether the neighbors have poorer stock than you - you can compare your stock with his. You know if you're doing something better. A lot of its watching neighbors over the fence, isn't it? It's seeing whether you're doing something that's good or whether they're doing something that's bad. (conventional sheep/beef farmers).
>
> Ah well, the fact that it's organic, there's an awful lot of people that are sort of looking over the fence, watching what's going on. (male, organic sheep/beef farmer).

In ARGOS we are developing a concept we have named "breadth of view" (Hunt et al. 2009, 2011; Rosin et al. 2009) to describe the awareness a farmer/orchardist may have of the extent of the impact of their farming practices on social and environmental wellbeing. This developed partly out of noticing in interviews that many organic kiwifruit orchardists claimed that growing organically meant the world became a better place because it was being supplied with a healthy, "good for you" product and improved the environment: "It [growing organic kiwifruit] is for the good of the planet as well as the environment as well as myself – it's not just for the good of my ego" (male, organic orchardist). In a survey, organic farmers (Fairweather et al. 2009a; Hunt et al. 2011, pp. 173–178) assessed themselves as having a broader social and environmental breadth of view than their conventional or integrated counterparts. This attribute may contribute to farmers being more open to change (Rosin et al. 2009). Often, because of their physical distance from other organic practitioners, their social and business/supply networks were more geographically distant and often international indicating that they were more likely to come across new ideas and different ways of doing things. However, new practices were often constrained by organic certification bodies struggling to keep up with the products and technologies available, particularly in areas of farming fairly new to organic certification in New Zealand such as sheep/beef and dairy farming.

It does not mean, however, that they were more sympathetic to playing a part in alleviating climate change or that they felt in any way responsible for it (Fairweather et al. 2009a, pp. 201–205). In fact, some of the organic farmers denied anthropogenic climate change. They also did not seem to be any more sympathetic than other farmers to the use of native plants and trees compared with exotic ones, or the presence and useful functions native birds may perform in the landscape compared with introduced species (except for organic farmers in the sheep/beef sector) (Fairweather et al. 2009a, pp. 68, 78).

In 1990 and 1991 the industry was in crisis when Italian trade officials expressed concern about the level of chemical residues in some New Zealand kiwifruit (Campbell et al. 1997). Then, in 1992 the price crashed. An integrated pest (IPM) management program, KiwiGreen, was born out of this and undoubtedly generated benefits for organics (see Campbell and Fairweather 1997, p. 21). KiwiGreen is now integrated into GLOBALGAP certification, a requirement for all exported kiwifruit, and part of what enables New Zealand kiwifruit to claim a large premium in world markets as organic and IPM fruit. It is a guaranteed way of differentiating high quality New Zealand fruit from that of other countries and of negotiating Maximum Residue Level (MRL) barriers (Campbell and Fairweather 1997, pp. 18, 19).[10]

Through comparing different management systems within and across sectors (sheep/beef, dairy, kiwifruit) we are able to make claims about the contributions organics can make to the primary sector as a whole. Thus we suggest, that organic farming practices need to be viewed as part of a much wider system that encompasses many different practices and that there is some degree of synergy or interaction between these practices. In order to make claims about the sustainability and resilience of organic farming, we need to be able to compare it with other forms of farming, while at the same time assuming that like other farming systems, organic practices are not static in time or necessarily completely distinctive. Whatever way it is considered it seems that the organic industry cannot be "disentangled from the organic social movement" and "will continue to act as a counterpoint, moment of contestation, or site of dialogue with the globalizing food system" (Campbell and Liepins 2001, p. 36; also see Rosin and Campbell 2009).

10.6 Organic Farmers and Their Farms Contribute to the Resilience of the Food System

Resilience implies adaptability and redundancy at many levels – farm/farmer, community, national and global. We propose that organic farms and farmers perform a useful function as part of a resilient farming system because they serve the purpose

[10]Growing kiwifruit organically was very difficult until the development of ultra-fine mineral oil sprays for the control of scale (1993) and the approval of their use in organics (Tomkins et al. 1996). These examples illustrate the cross-fertilization and transfer of technology that can occur from IPM practices to organic.

of introducing alternative practices and providing alternative possibilities to non-organic farmers, and expand the diversity of an agriculture sector. "Resilience is the capacity of a system to absorb disturbance and reorganize while undergoing change so as to still retain essentially the same function, structure, identity and feedbacks" (Walker et al. 2004, p. 2). The resilience approach conceptualizes the world as a complex, constantly evolving, dynamic, adaptive system (Rammel et al. 2007; Darnhofer et al. 2010b).

According to Darnhofer et al. (2010a) the literature indicates that agro-ecosystems also need to be thought of in this way, and that their adaptability can be strengthened by farmers "learning through experimenting and monitoring" (Darnhofer et al. 2010a, p. 549); enhancing both the short and long-term flexibility by enrolling on-farm and off-farm resources; and increasing the diversity of activities of those in the farm household, the products produced and the methods or practices by which they are produced. If the farm system is expanded to include the supply chain or the farming field, then these qualities can be seen as affecting farming's resilience in general. A good quality primary product on the farm is only the first stage in a supply chain that also needs to have the capacity to adapt to changing circumstances and demands of the market.

The resilience of organic farming has been demonstrated in the ARGOS program by the findings that the soils of organic kiwifruit orchards are adequately maintained under organic management and they had several important qualities that were considered to be better than their non-organic counterparts – a larger microbial biomass and organic matter content, and improved soil condition (Carey et al. 2009). They were also a source of greater invertebrate biodiversity (Todd et al. 2011), which it is hoped indicates a more resilient provision of ecosystem services. As these results become more widely known they will influence non- organic orchardists, and demonstrate that organic growers do not "mine" the soil as is popularly claimed (Campbell et al. 2009).

Organic kiwifruit orchardists provide knowledge that is available to non-organic growers by developing and experimenting with non-chemical fertilizers, such as compost teas and seaweed and fish mixes, and growing kiwifruit without the use of bud break sprays. When chemical fertilizers become too expensive because of rising oil prices or disadvantageous exchange rates, or a chemical spray is banned because it is not approved of by consumers, organic orchardists can provide practices which make the industry as a whole more resilient. In our most recent ARGOS interviews with dairy farmers it was apparent that many so-called conventional dairy farmers were starting to use homeopathic treatments acceptable in organic certification for animal health treatment and prevention of lameness and mastitis (somatic cell counts) and introducing more variety into their pastures. Organic sheep farmers are farming without chemical fertilizers, without drenches and trying to breed worm resistant stock, thus helping conventional farmers in the future who face the risk of drench resistance.

Mainly we have described examples of how organic practices can influence conventional practices directly, or indirectly, as in this quote: "I've got no desire to go organic, um, but I don't believe in using pesticides or things where they aren't necessary, or I think they're unnecessary . . . " (conventional male sheep/beef farmer). But the exchange can happen in the opposite direction: organic farming can also be influenced by more common agricultural practices. For example, on one ARGOS sheep/beef farm, which has been biodynamic, the son, who has recently taken over management from his father, is changing from his father's practices. He is introducing sheep breeds and grass cultivars from outside the farm (and seen in the conventional and integrated management practices of his neighbors and friends). To his father the farm was a closed system.

However, it is not just the change of practices that is of interest but how the differing management systems and types of farmers or growers serve different purposes within the farming system. Hunt (2010) has shown how conservative IPM and organic kiwifruit orchardists tend to stay with kiwifruit production through good times and bad, while those who like to take risks and seek high returns will exit the industry when times are not so good, which also serves the industry well by reducing supply. The latter will also be the ones who introduce costly innovations. Organic orchardists model practices that show a greater concern for environmental wellbeing. All of these work together to provide a greater number of acceptable ways of "being" a kiwifruit orchardist, and support a more resilient industry even if it is not sustainable for some individuals in certain difficult times.

In the kiwifruit industry,[11] it has not been traditional for diversification of kiwifruit production to occur on the orchard unless the orchard also grows other non-kiwifruit products. Kiwifruit orchardists usually grow only green or only gold kiwifruit, not both and it is rare for an orchard to produce both organic and non-organic fruit though some businesses do include a suite of differently managed individual orchards. Thus diversification, until recently with the introduction of new varieties of kiwifruit, has tended to happen at the supply chain level where ZESPRI[12] is able to market organic and non-organic green and gold fruit.

Diversity of products and the enhanced resilience it can bring to producers can occur at different points in the supply chain. By increasing the number of varieties of kiwifruit available, and by having the options of organic or IPM fruit, ZESPRI hopes to be more resilient to market demands. The dairy sector has a similar dynamic. Fonterra[13] is continually experimenting with and introducing new milk products to their customers which are often ingredients that enhance the products of other manufacturers. Organic milk could extend their options in the market place. The sheep/beef sector is quite different. It has a competitive and diverse supply chain. Sheep/beef producers in New Zealand supply 15 meat processors (in 2009–2010

---

[11] Darnhofer et al. (2010b) have produced a study of resilience based on the New Zealand kiwifruit industry.

[12] ZEPSRI is the single desk-monitoring organization for the export of kiwifruit from New Zealand. It is regarded as a monopsony.

[13] Fonterra is New Zealand's largest dairy company, a cooperative which is one of the world's largest exporters of dairy products (http://business.newzealand.com/common/files/Dairy-industry-in-New-Zealand.pdf).

(Davison 2010)). Two hundred and thirteen (213) registered meat export companies (current September 2010 (New Zealand Meat Board 2010, p. 34)) then compete with each other in international markets to sell meat. Within these companies there are probably only several that sell organic meat. Non-organic farmers often do not commit to contracts to a particular company but shop around for the best price, with the result that a processing company may not have a consistent market supply and can end up being at the mercy of the market demand at the time. The competition may benefit the buyers at the top end of the supply chain but the meat producer, the farmer, who is at the bottom of the supply chain has to take what is left over after all others in the chain have had their cut.

In a sense, some organic farmers in New Zealand cannot exist economically without the other farmers. Though we have shown that organic farmers and orchardists do not appear to make any more or any less profit than others[14] (Greer and Hunt 2011), we also know that in the dairy and the kiwifruit sectors the premiums paid for organic are decided by the industry not by the premium obtained in the market. This rather negates the usefulness of this comparison. The data from ARGOS have helped industry to decide what premium maintains an equivalence between farmers whether organic or not. However, it indicates the dependence of organic farmers and orchardists on the much larger, less alternative majority in these two sectors. In other ways organics is also dependent. Fonterra was growing its organic dairy suppliers in order to provide a wider array of products for the market, adding to its resilience. ZESPRI similarly, uses organic kiwifruit to expand its product range and to extend its marketing season because the organic fruit has had longer keeping qualities.

A sector without organic practices would miss out on possible and alternative resilient practices to conventional or normalized practices that organics make visible and thinkable (Bourdieu 1990, p. 59; Shucksmith 1993, p. 468). Organic practices have the potential to increase the biodiversity of birds and plants as well as soil fauna and other soil characteristics. We have shown how farmers and orchardists who practice different management systems such as organics, work to make a supply chain or field of agriculture more resilient. Different parts of the supply chain can enhance the resilience of producers who may produce less diverse products such as in the kiwifruit and dairy industries, whereas the large number of competing companies marketing meat products may reduce the resilience of the sheep/beef farmers by keeping product prices low. Just as importantly, the practices of organic and non-organic farmers can be exchanged and the diversity, availability and redundancy of these practices add to the resilience of the supply chain at the farm level. The extent to which organic farmers are likely to influence the resilience of a given supply network is a factor of their relative level of acceptance as 'good farmers'.

The good farming literature suggests that the status of good farmer is awarded by farmers to other farmers who follow cultural rules established over time by the farming community and government policies/regulations. Because ARGOS studies

---

[14]However, all results are so variable that the low statistical power of these comparisons could be masking underlying differences.

three agricultural sectors we can demonstrate that in the kiwifruit sector, organic management is one of many models of good orcharding whereas in the other sectors organic practices challenge the precepts of good farming.

10.7 The Good Farming Model

A recent trend in social analysis has included a more culturally-oriented perspective (Holloway 2002) focused on "language, meaning, representation, identity and difference" (Burton 2004b, p. 361), using qualitative research to study farmers as legitimate actors and seeking to more fully understand why farmers do what they do. One strand of this approach focuses on the concept of the "good" farmer and raises the question of what it means to farm well. According to Burton (2004a, p. 195) the "production- oriented roles came to symbolize, both to farmers and the country, the notion of good farming practice that enabled farmers to claim a high social position as caretakers of the nation's food supply". This status is now being contested by concerns about the environmental impact of intensive, production-oriented farming (Silvasti 2003; Setten 2004). Similarly, in New Zealand farmer identities are being challenged in a society placing an increasing emphasis on a knowledge economy, not on the production of agricultural commodities (Jay 2007; Rosin 2008; Hunt 2009b). The link between the "good farm", the "good farmer" and their community context is described by Burton (2004a):

> [...] for many farmers it [the landscape] represents a picture of good farming practice, displayed in a manner that enables the farmer to obtain social status and recognition within the community as a "good farmer" and to judge the credentials of others. The farm is not simply an object, it is consubstantial with the farmer and, importantly, it is the very part of the farmer that is used to express his/her and his/her family's identities, both to other members of the farming community and to the world in general (p. 207).

Good farming cannot be practiced in isolation. Farmers live in a social context in which they strive[15] to be seen as good farmers and practice in ways that reinforce and maintain their identities in particular ways (Silvasti 2003; Burton 2004a; Setten 2004; Stock 2007; Hunt 2010). Hence the good farmer approach is one way of explaining why farmers do or do not change their practices. In this respect, the significance of experimentation as an attribute of resilience can be perceived as social as well as scientific. It may be more about "understanding the "rules of the game" of farming and how these rules change" (Darnhofer et al. 2010a, p. 549) as well as watching for feedback loops through monitoring on-farm processes. Experimentation may also be a community or collegial activity – as farmers meet in discussion groups or on the side of the road, or when someone or something outside of farming sparks their curiosity.

[15] We use the term farmer in the generic sense to include all types of primary producers, such as orchardists.

We have also demonstrated that it is important to have different ways of producing a product and different models for the ways of being a farmer/orchardist because choice and redundancy make for greater resilience. A greater range of types of people with a variety of knowledge and backgrounds will enter farming/orcharding if there are different models or ways of being a good farmer/orchardist. In the kiwifruit sector this is more apparent because people becoming kiwifruit orchardists come from many different backgrounds and experiences. They bring these experiences with them (Hunt 2009a) and seek acceptable ways of using this knowledge to do something different. For example, in the first quote below an organic orchardist describes the practices of his neighbor, while the following two give their reasons for having "tidy" orchards.

> I might mow here three times a year. I give the neighbor this side [a hard time]. He's just got a new mower and it's like a bowling green, you know [. . .] got an hour to spare and he's out killing the place. But to me, the longer grass – there's creatures in it as well, you know – bugs and birds and bits and pieces running round out there [. . .] (male, organic orchardist).
>
> I try to run a pretty good hygiene program on my orchard [. . .] keeping the canopy very tidy, and keeping the floor of the canopy tidy. No high weeds that will form a bridge for insects to multiply. Mowing my orchard often, but not too often [. . .] (male, IPM orchardist).
>
> [. . .] [the] shelter's nice and trimmed and it's even [. . .] being tidy is important to me . . . that's part of the health of the place, I believe (male, IPM orchardist).

Organic farming doesn't happen in isolation – organic orchardists have neighbors. The orchardists here contribute to the resilience of orcharding by being sources of diversity as the first orchardist provides an alternative way of looking after his orchard from the other two. The quotes also show redundancy – orchards do not exist just for the purpose of growing kiwifruit. In the first quote, other things are allowed to exist alongside the kiwifruit vines! In the other quotes the orchardists get satisfaction from the tidiness of their orchards. In terms of good farming, the quotes also demonstrate that all models are socially acceptable – in the first the two neighbors appear to have a relationship such that they can tease each other and they presume that each is a "good" orchardist (Hunt 2010). In addition, an owner of an orchard has access to many different ways of getting the work done on the orchard, all supported by the kiwifruit industry structure. For example, the owner may do much of the work just getting in contract labor for the big pruning jobs, or the owner may only mow the orchard and have a manager who organizes the rest of the work. An owner may live on the orchard or may live anywhere in the world!

The protocols surrounding kiwifruit growing are fairly narrow so it is important that people are still able to see themselves as autonomous, having choices. In dairying, a focus on giving status to those who produce the most milk solids (and therefore earn the most) is dominant (Jay 2007) and as organic production is likely to be considerably less productive (Greer and Hunt 2011), this can result an "outsider" status being assigned to organic dairy farmers, as the following, contrasting quotes indicate, even though in practice, the premium structure may mean that they earn just as much as a conventional farmer.

So I like the little bit of the competitive thing about it. Yeah and I just love, you know, trying to grow as much grass and I can. I love having healthy cows. I love being able to feed them well. Yeah and, you know, and you get a milk docket[16] every day which tells you, hey, you're doing a good job or otherwise (male, conventional dairy farmer).

The funniest thing that makes me [laugh]. Everybody asks you, are you up for the year? And we're down for the year but we're probably $30,000 better off [than them], but I should say I'm up, but I love saying I've gone backwards, you know. But that's how they gauge you. Because the dairy company's got you brainwashed. Every day you're looking at the ticket to see whether you're plus on last year and plus on last month ... And it's just this big game to get more production. But it does cost - a lot of things - mostly the environment (male, organic dairy farmer).

Taking on the risk of lower production could be a difficult hurdle for dairy farmers who would like to practice organically. The model of profitable dairy farming while producing less is a very important alternative in the dairy industry. As in the kiwifruit sector, there are many different relationships between ownership and performed labor on dairy farms in New Zealand, where there is a strong culture of starting the path to farm ownership by being a share milker. In this arrangement a person or a group may own a proportion of the herd and has an agreement with the owner about who pays for what and how the income is assigned. Increasingly too there are so-called corporate farms, which have very large herds and employ staff to do particular tasks, and even manage the enterprise.

In sheep/beef farming the culture of the good farmer is even more restrictive and farmers who appear different can become quite isolated, as the following quotes illustrate.

I mean my father and some people predicted doom [when I went organic] [...] but you know it's just carried on. I think it's just as sustainable a system (male, organic mixed farmer).

Ohhh sometimes, sometimes I'll have a smile to myself, you know, if you drive past a paddock and there's shitty ewes or something there 'cos you sort of think oh well, it's not just organic people that have shit on their sheep, mm [...] people tend think that's what's gonna' happen with [...] organic so, mmm (female, organic sheep/beef farmer).

Ah, and my plan was just go and copy off somebody ... But, being like the first year, that anybody had ever paid a premium for [organic] livestock, um, all the people that were in it prior to me were doing it for deep, philosophical reasons and, you could hardly get in the door for leather sandals and, um, kaftans, but there was not many on the case farmers doing it because there was no reason to do it (male, organic sheep/beef farmer).

You know um, it's a pretty conservative area here so when the farms [are run] through families and um land gets [passed on], Presbyterian farming stock you know, yeah, lot of "Mc" names and things, you know, but um and ah some farmers don't wear it, don't ever talk to us about organics. They pretend we don't do it (male, organic sheep/beef).

We also grow red clover here for seed and that's basically pollinated by honey bees and the bumbles. Well I know a lot of farmers can have a struggle to do that now, um that they struggle with the bumble bees – they haven't got them and that's partly to do with the old tree lines and things they've taken out and [...] not having wilderness areas on their farm

[16]Each day in the milking season a farmer gets a docket when the milk is collected giving the quantity of milk solids produced that day and comparing total production up to that time with the last year and so on.

> ... you know these modern farmers like to clean things up all the time and um take out old trees and unproductive land and put a fence up and spray out the blackberry and stuff. Whereas we've left areas [...] it's quite steep, three or four hectares [...]. So now it's a block of gorse, um, with flowers and stuff. It upsets the neighbors but I don't care. Bee keepers should be pleased, because they're moaning about a lack of pollen sources in the winter time (male, organic mixed farmer).

There was a strong feeling that if you were organic you not only had dirty looking sheep, you were also alternative in other ways – had long hair, wore sandals and probably smoked marijuana – demonstrating both the older age on average of sheep/beef farmers and the more traditional values. The views of the latter majority group demonstrate how they think it is important to make the land productive and for it to look tidy, as the flowing quotes indicate.

> But you know, generally our farming policies are all pretty kosher with what you are allowed to do, if you had to pull us up on some things I would say, they might say we're working too much steep ground but you know it's very minimal, the damage that's done there. We're actually developing ground that was in gorse and manuka and stuff that's now becoming productive land (male, conventional sheep/beef).
>
> We had this goal of a tidy farm, well run and efficient [...]. This farm – we're fortunate considering the shape with the layout but it wasn't really the tidy farm we wanted and probably, because of our financial restrictions, some areas aren't yet. You know [...] I envy the people that buy an ugly block and have got the finances - they can just put a bulldozer through the whole thing and start again (male conventional sheep/beef farmer).

These examples illustrate what are deemed to be acceptable farming practices in the good farming model and how they are presented visually on a farm can determine the acceptability of certain ways of farming or orcharding. In the kiwifruit industry organic is a socially acceptable way to be a kiwifruit orchardist, but in the dairy and sheep/beef sectors there are preconceived negative ideas about what organic farmers and their farms are going to be like, which acts to discourage many farmers from using organic practices.

10.8 Some Statistics: Comparing Organic and Non-organic Farms Across Three Different Sectors

The context in which farming takes place needs to be taken into account as it shapes the diversity of the options available. Tables 10.1 and 10.2 demonstrate how it is likely that organic farms are more visible in the kiwifruit sector than in the pastoral sectors of dairying and sheep/beef,[17] whichever way it is considered – by number of farms, or by area. This is accentuated by the fact that a kiwifruit orchard is much smaller than a pastoral farm, being more intensive, so there are many more orchards

[17] Note the figures in the following tables are derived from so many different sources that when making comparisons between values from different sources, the values given must be regarded as approximate.

Table 10.1 Number of farms/orchards

Sector	No. organic farms/orchards	Total no. farms/orchards
Sheep/beef	30–90[a] (0.1–0.3 %)	29,241[b]
Dairy	80[c] (0.7 %)	11,400[b]
Kiwifruit[d]	160–190 (5.2–6.2 %)	3,077

Source:
[a]It was estimated in 2012 that there were 170 organic livestock properties over 50 ha (which would include dairy). Estimated by looking at BioGro and AsureQuality figures (pers. comm. Jon Manhire, 18/04/11)
[b]2007 Meat and Wool New Zealand (2010)
[c]2010, Business New Zealand (2010)
[d]2010/11 pers. comm., via Jayson Benge from person in ZESPRI Group Limited (2010)

Table 10.2 Area in production (sown pasture/crop) in 2007

Sector	Area in organic certification (ha)[a]	Total area (1000 ha)[b]
Horticulture and cropping/ cropping and other[c]	5,045 (0.2 %)	2,517
(Kiwifruit as subset of above)[d]	505 (4.0 %)	13
Livestock[e]	52,070 (0.4 %)	12,173
Mixed and other[c]	6,768	–

Source:
[a]Cooper et al. (2010)
[b]Meat and Wool New Zealand (2010, pp. 19, 20)
[c]Cooper et al. (2010) use "horticulture and cropping" and "mixed and other" as separate categories Meat and Wool New Zealand (2010) use "cropping" and "other". It is hard to know whether "other" includes horticulture or not. It does include non-commercial small holdings so may well be inflated in value for our purposes
[d]ZESPRI Group Limited (2010, p. 26)
[e]Cooper et al. (2010) use "livestock farming". Meat and Wool New Zealand (2010) use dairying, sheep and beef farming, deer farming, pig farming, mixed livestock farming and poultry

in a given area which means that it is more likely to come across an organic orchard in a kiwifruit growing area than it is to come across an organic farm in dairy or sheep/beef farming areas. Though there are more organic kiwifruit orchards than is apparent in the orchard area statistic, this will be because on average they are probably smaller enterprises (ZESPRI Group Limited 2010). It may be that the greater visibility of organics in the kiwifruit sector helps contribute to their social acceptability.

10.9 Discussion

Dedieu et al. (2009, p. 108) make clear in their introduction to a special issue of *Outlook on Agriculture* titled "Innovations in farm system approaches", farmers "[. . .] are expected to produce high-quality products at competitive prices using environmentally friendly production methods that maintain cultural landscapes". As the context in which farmers are to achieve this feat is increasingly complex and uncertain, it is an understatement to say that this is going to be difficult. In this paper, we have shown how the presence of organic practices in farming supply chains can bring greater diversity and choice and with the constant interplay between farmers, a farming sector can change and become more robust and resilient.

In New Zealand the presence of organic farming has challenged mainstream agriculture (Stuart and Campbell 2004, p. 234) and its visibility was boosted in New Zealand's history at crisis times – the recognition of declining soil fertility at the beginning of the twentieth century and during the depression of the 1920s and 1930s. Declining soil fertility was linked to many social ills including "declining nutrition, colonial dependency, and reductionist technical solutions to environmental problems" (Stuart and Campbell 2004, p. 235) and, as far as its adherents were concerned, practicing organics was going to protect the growing population. However, in some parts of the world when organic products started moving into the national and global market place, questions arose about whether the suppliers of these markets could really be organic, so certification was pursued by the organic sector to support its own legitimacy. Alongside that we have shown that in fact the definition of what it means to be organic has been negotiated and changed over time (Rosin and Campbell 2009). It is not static and the practices involved become influenced by and influence non-organic farming practices.

We suggest that organics is part of a farming supply chain and as such its presence enables and nurtures diversity (Folke et al. 2003; Darnhofer 2010; Darnhofer et al. 2010b). The existence of diversity makes visible and possible new options to act (Ison et al. 2000; Ondersteijn et al. 2006). If farmers are going to learn from each other their practices have to be seen as thinkable, as in the realm of possibility, so it is better if their practitioners and their farms appear as socially acceptable, at least to some. Whichever way the sectors are compared it is apparent that organic practices will be far more visible in the kiwifruit industry than in the sheep/beef and dairy sectors – not only because of the higher participation of organic practitioners but also because kiwifruit orchards use much less land (and so there are more of them within a fixed area) and are more concentrated in a particular region of New Zealand (the Bay of Plenty). We have also illustrated how organics is more socially acceptable, even taken-for-granted (Bourdieu and Wacquant 1992, p. 127) in the kiwifruit industry and organic orchards can be considered good, and seen as one example of the right thing to do (Bourdieu 1998, p. 8). In the dairy industry it has obviously been difficult to get farmers to convert to organics. Fonterra have been stating how quickly it wished to grow the number of organic dairy farms but have not been meeting their targets. The numbers they wished to achieve were specifically mentioned in annual reports two years in a row (2005, 2006) and as the targets have not been met organic dairying is no longer being encouraged.

Organics keeps visible the possibility of how things can be done in a different way. (Some organic dairy and sheep/beef farmers told us that they were just doing the same thing as their father or grandfather, thus keeping active that memory. This may be seen as redundancy as retaining institutional memory of how things used to be done could be seen as unnecessary in a practical sense). In kiwifruit growing, fertilizers used can be synthetic or provided by using different manures, composts and compost tea. In sheep/beef farms fertilizer can be synthetic or derived from rock phosphate. Because of a scarcity of organic inputs, many organic farmers have to provide their own feed stocks such as ensilage, baleage and maize, practices to produce feed which for conventional dairy farmers can be sourced off-farm, some of whom are becoming dependent on imported palm kernel meal. Organic dairy farmers do not use antibiotics but replace them with other "natural" remedies and pasture mixtures containing herbal leys. In this way a "function" on the farm is not dependent on one method (see Elmqvist et al. 2003; Walker et al. 2006). This is important at a time of peak oil and rising costs, particularly rising costs of imports.

Organics introduce greater flexibility into the agriculture sector. According to Tarondeau (1999 cited in: Darnhofer et al. 2010b) there are three sources of flexibility – the products, the processes and the input specificity. We have shown that often in an agricultural sector diversity of products is not happening on the farm such as in dairying and kiwifruit, but can be provided further down the supply chain. ZESPRI, the export marketer for kiwifruit, delivers a mix of products and markets for the producer. In the dairy sector, the manufacturers are continually introducing new milk products to other manufacturers further down the supply chain.

The situation in the sheep/beef sector is different with farmers having to make their own product choices, but their supply chain sells a limited meat and wool product range through a myriad of sellers. In terms of processes, kiwifruit orchardists have developed many different models of being a kiwifruit orchardist, whereas in sheep and beef and dairy farming there is a dominant model of what it means to be a "good farmer" and this typically excludes organic farmers. However, the presence of even a fledgling organic dairy industry is showing how practices can cross the organic/non-organic divide. Throughout all sectors, organic practices demonstrate alternative inputs to chemical fertilizers and sprays, and other ways of "doing" farming. As long as these alternatives are kept alive, and although replacing practices used by conventional farmers and therefore in a way redundant, they can be there ready for use when a crisis strikes, such as when there is a shortage of chemical fertilizers, or costs become too expensive, or a chemical used for pest control or increasing growth is banned. Diversity and flexibility are linked. As Darnhofer et al. (2010a, p. 551) state: "managing complex systems and uncertain future developments implies spreading risks and creating buffers". Magne and Cerf (2009) also point out that this diversity is just not to do with practices, but can be associated with information – the different ways in which farmers perceive and interpret information can increase the capacity to act.

Production of a sole product by a producer makes their business enterprise more brittle and less robust than that of a producer who is able to spread risk by producing a range of products. When organic products are part of a larger supply chain organic farmers have greater security and the whole system is able to produce a more diverse product range for the market, and is therefore more resilient through spreading risk. If one product does not go well one year, or its production is decreased because of an extreme weather event for example, the industry as a whole can still survive.

Both the kiwifruit and dairy industries follow this premise of diversification at the market end of the supply chain (as does one meat processing company which is able to provide products of differing provenance), because at the farm/orchard end only one thing is produced – a single kiwifruit variety or milk, under organic or non-organic management. Sheep/beef farmers, on the other hand, have a range of choices available to them about what products to produce, from cropping through to livestock production, to servicing the dairy industry – all dependent on their geographical locations.

It was quite a shock to us at a stakeholder meeting to have our assumptions about organic practitioners challenged. We had assumed that they would be of the philosophical persuasion that would want all food production to be organic. When one of our ARGOS researchers suggested that it would be better for the resilience of the kiwifruit industry environmentally if a greater proportion of orchards became organic, there was a murmur of disagreement within the organic stalwarts. They were very concerned that such an action would reduce the organic premium they receive for their fruit. For them, they walk the balance of producing a regular and large enough supply for a constant market but not producing so much that they lose their market scarcity. However, rather than wishing the whole world's food production to become organically grown, to be the method regarded as best practice, perhaps having organic production as just one of the ways of being a "good farmer" keeps the global food system dynamic and alive to ever-changing alternative possibilities, more able to respond to the changes and challenges we face now and in the future.

10.10 Conclusion

In this chapter, under the premise that organic farming needs to be studied in the context in which it is practiced within the global agricultural supply chain, we have discussed how the practices and presence of organic farming can contribute to the resilience and sustainability of the food production system. To do this we examined three interlinked theses using the results from the ARGOS research program as the empirical example. First, we provided support for the argument that studying organic farming in isolation limits the understanding of the breadth of its contribution. Obviously it can be studied in isolation, but the study is richer if organics is studied in the context of the agricultural sector as a whole, where links and contrasts can be made to other farm management systems, across different sectors within agriculture and within a supply chain. Even organic practices and compliance are in a constant state of negotiation as certifiers face questions and challenges from within and outside their jurisdiction.

Second, we considered how organic practices enhance the resilience of the sectors in which organic products are produced. The ability to supply organic kiwifruit to the global market has strengthened the resilience of the kiwifruit supply chain. Organic practitioners can be seen as repositories of redundant knowledge not needed by their conventional counterparts (and vice versa, of course). However, as change and crises will continue to challenge agricultural producers in the future, this knowledge may come into its own. We have also emphasized that all farmers share a common body of knowledge and the more permeable the boundaries between the distinctions made between different management systems over and above this commonality, the better for our future. By adding to diversity, organic practitioners and their farms add to the resilience of global agriculture.

Finally, we examined the cultural context of "good farming" in which organics is practiced. The second and third theses are closely related and fit within the first. If mutual learning is to take place between farmers, their practices have to be seen as possible or "thinkable", which will be more likely if organic farmers and their farms are socially acceptable, at least to some. By comparison within and across agricultural sectors in the ARGOS program, we showed that organic practices are most visible and more socially acceptable, and organic orchards can be considered "good". In dairying, even though the "increase production at all costs" ethos rules, there has been some crossover of practices. While there is potential for this to happen in the sheep/beef sector, it is most difficult for organic sheep/beef farmers to be seen as good farmers.

Therefore, we conclude that the presence of organic farming does add to the resilience of the farming sector by providing diversity and making visible and possible different and more sustainable ways of farming. If organic farming becomes more socially acceptable then the potential for this contribution can only increase so that all ways of farming are able to learn from each other to enable the production of safe food for the world's population in our changing and challenging times.

References

Bourdieu, P. 1990. *The logic of practice*. Redwood: Stanford University Press.

Bourdieu, P. 1998. *Practical reason: On the theory of action*. Redwood: Stanford University Press.

Bourdieu, P., and L.J. Wacquant. 1992. *An invitation to reflexive sociology*. Chicago: University of Chicago Press.

Burton, R.J.F. 2004a. Seeing through the 'good farmer's' eyes: Towards developing an understanding of the social symbolic value of 'productivist' behaviour. *Sociologia Ruralis* 44(2): 195–215.

Burton, R.J.F. 2004b. Reconceptualising the 'behavioural approach' in agricultural studies: A socio-psychological perspective. *Journal of Rural Studies* 20(3): 359–371.

Business New Zealand. 2010. *New Zealand dairy industry: June 2010*. Wellington: Business New Zealand.

Campbell, H. 1997. Organic food exporting in New Zealand: Sustainable agriculture, corporate agribusiness and globalizing food networks. In *Sustainable rural development*, ed. H. De Haan, B. Kasimis, and M. Redclift, 51–72. Aldershot: Ashgate Publishing.

Campbell, H., and B. Coombes. 1999. Green protectionism and organic food exporting from New Zealand: Crisis experiments in the breakdown of Fordist trade and agricultural policies. *Rural Sociology* 64(2): 302–319.

Campbell, H., and J. Fairweather. 1997. *The development of organic horticultural exports in Nez Zealand*, AERU research report, vol. 238. Lincoln: Lincoln University.

Campbell, H., and R. Liepins. 2001. Naming organics: Understanding organic standards in New Zealand as a discursive field. *Sociologia Ruralis* 41(1): 22–39.

Campbell, H., and C. Rosin. 2011. After the 'organic industrial complex': An ontological expedition through commercial organic agriculture in New Zealand. *Journal of Rural Studies* 27(4): 350–361.

Campbell, H., J. Fairweather, and D. Steven. 1997. *Recent developments in organic food production in New Zealand: Part 2, Kiwifruit in the Bay of Plenty*, Studies in rural sustainability, research report, vol. 2. Dunedin: Department of Anthropology, University of Otago.

Campbell, H., C. Rosin, S. Norton, P. Carey, J. Benge, and H. Moller. 2009. Examining the mythologies of organics: Moving beyond the organic/conventional binary? In *Food security, nutrition and sustainability*, ed. G. Lawrence, K. Lyons, and T. Wallington. London: Earthscan.

Carey, P., J. Benge, and R. Haynes. 2009. Comparison of soil quality and nutrient budgets between organic and conventional kiwifruit orchards. *Agriculture, Ecosystems & Environment* 132(1): 7–15.

Carolan, M.S. 2005. Barriers to the adoption of sustainable agriculture on rented land: An examination of contesting social fields*. *Rural Sociology* 70(3): 387–413.

Cooper, M., H. Campbell, J. Manhire, H. Moller, C. Rosin, S. Norton, and L. Hunt. 2010. *New Zealand Organic Sector report 2010*. Wellington: Organic Products Exporters of New Zealand Inc. (OPENZ).

Cousins, B., M. Hoffman, N. Allsopp, and R. Rohde. 2007. A synthesis of sociological and biological perspectives on sustainable land use in Namaqualand. *Journal of Arid Environments* 70(4): 834–846.

Darnhofer, I. 2010. Strategies of family farms to strengthen their resilience. *Environmental Policy and Governance* 20(4): 212–222.

Darnhofer, I., S. Bellon, B. Dedieu, and R. Milestad. 2010a. Adaptiveness to enhance the sustainability of farming systems. A review. *Agronomy for Sustainable Development* 30(3): 545–555.

Darnhofer, I., J. Fairweather, and H. Moller. 2010b. Assessing a farm's sustainability: Insights from resilience thinking. *International Journal of Agricultural Sustainability* 8(3): 186–198.

Davison, R. 2010. *NZ meat sector domestic pastoral trends*. Paper presented at the MIA conference, Christchurch, New Zealand: Beef + Lamb New Zealand.

De Schutter, O. 2010. Report submitted by the Special Rapporteur on the right to food. In *United Nations Human Right Council*, ed. O. De Schutter. New York: United Nations Human Right Council.

Dedieu, B., I. Darnhofer, S. Bellon, K. De Greef, F. Casabianca, L. Madureira, R. Milestad, M. Paine, P. Steyaert, D.J. Stobelaar, and S. Zasser-Bedoya. 2009. Innovations in farming systems approaches: Introduction. *Outlook on Agriculture* 38: 108–110.

Elmqvist, T., C. Folke, M. Nyström, G. Peterson, J. Bengtsson, B. Walker, and J. Norberg. 2003. Response diversity, ecosystem change, and resilience. *Frontiers in Ecology and the Environment* 1(9): 488–494.

Fairweather, J.R., H.R. Campbell, C.J. Tomlinson, and A.J. Cook. 2001. *Environmental beliefs and farm practices of New Zealand organic, conventional and GE intending farmers*, AERU report, vol. 251. Lincoln: Lincoln University. Agribusiness and Economics Research Unit.

Fairweather, J.R., H. Campbell, L.M. Hunt, and A.J. Cook. 2007. *Why do some of the public reject novel scientific technologies? A synthesis of results from the Fate of biotechnology research programme*, ARGOS research report, vol. 09/02. Lincoln: Lincoln University. Agribusiness and Economics Research Unit.

Fairweather, J., L. Hunt, J. Benge, H. Campbell, G. Greer, D. Lucock, J. Manhire, S. Meadows, H. Moller, C. Rosin., C. Saunders, and Y. Fukuda. 2009a. *New Zealand farmer and orchardist attitude and opinion survey 2008: Characteristics of organic, modified conventional (integrated) and organic management, and of the sheep/beef, horticulture and dairy sectors*, vol. 09/02. Christchurch: Agricultural Research Group on Sustainability (ARGOS).

Fairweather, J.R., C.J. Rosin, L.M. Hunt, and H.R. Campbell. 2009b. Are conventional farmers conventional? Analysis of the environmental orientations of conventional New Zealand farmers*. *Rural Sociology* 74(3): 430–454.

Folke, C., J. Colding, and F. Berkes. 2003. Synthesis: Building resilience and adaptive capacity in social-ecological systems. In *Navigating social-ecological systems: Building resilience for complexity and change*, ed. F. Berkes, J. Colding, and C. Folke, 352–387. Cambridge: Cambridge University Press.

Friedland, W.H. 2004. Agrifood globalization and commodity systems. *International Journal of Sociology of Agriculture and Food* 12(1): 17–28.

Friedmann, H., and P. McMichael. 1989. Agriculture and the state system: The rise and decline of national agriculture, 1879 to present. *Sociologia Ruralis* 29(2): 93–117.

Greer, G., and L. Hunt. 2011. *Comparison of the productivity and profitability of organic and conventional farms in the New Zealand sheep/beef and kiwifruit sectors 2002/03 to 2007/08.* Paper presented at the 18th conference of the International Farm Management Congress, Christchurch.

Guthman, J. 2000. Raising organic: An agro-ecological assessment of grower practices in California. *Agriculture and Human Values* 17(3): 257–266.

Holloway, L. 2002. Smallholding, hobby-farming, and commercial farming: Ethical identities and the production of farming spaces. *Environment and Planning A* 34(11): 2055–2070.

Hunt, L. 2009a. Moving to the country for a graduated retirement: Constructing new meaningful lives. In *Tracking rural change: Community, policy and technology in Australia, New Zealand and Europe*, ed. F. Merlan and D. Raftery, 111–133. Canberra: Australian National University Press.

Hunt, L. 2009b. Doing science in a culture of accountability: Compliance through resistance to alienation and estrangement. *Kōtuitui: New Zealand Journal of Social Sciences Online* 4(1): 25–40.

Hunt, L. 2010. Interpreting orchardists' talk about their orchards: The good orchardists. *Agriculture and Human Values* 27(4): 415–426.

Hunt, L., C. Rosin, C. McLeod, J.R. Fairweather, and H. Campbell. 2005. *Understanding approaches to kiwifruit production in New Zealand: Report on first qualitative interviews of ARGOS kiwifruit participants*, ARGOS research report, vol. 06/01. Christchurch: Agricultural Research Group on Sustainability (ARGOS).

Hunt, L., C. Rosin, M. Read, J.R. Fairweather, and H. Campbell. 2006. *Understanding approaches to sheep/beef production in New Zealand: Report on first qualitative interviews of ARGOS sheep/beef participants*, ARGOS research report, vol. 06/01. Christchurch: Agricultural Research Group on Sustainability (ARGOS).

Hunt, L., C. Rosin, J. Fairweather, H. Moller, and others in ARGOS. 2009. *Doing the unthinkable: Is breadth of view important in bringing innovation to organic practices?* Paper presented at the organic sector conference, University of Waikato, Auckland, New Zealand: Organics Aotearoa New Zealand (OANZ).

Hunt, L.M., J. Benge, H. Campbell, J. Fairweather, G. Greer, D. Lucock, J. Manhire, H. Moller, J. Reid, C. Rosin, and C. Saunders. 2010. *Experiences of transdisciplinarity in research on agricultural sustainability.* Paper presented at the 9th European IFSA symposium, Vienna: International Farming Systems Association Europe Group (IFSA).

Hunt, L., J. Fairweather, C. Rosin, H. Campbell, D. Lucock, and G. Greer. 2011. *Doing the unthinkable: Linking farmer's breadth of view and adaptive propensity to the achievement of social, environmental and economic outcomes.* Proceedings of the 18th International Farm Management Association Congress Thriving in a global market: innovation, co-operation and leadership, Canterbury, U.K.: International Farm Management Association, 197–203.

IFOAM. 2009. *Definition of organic agriculture*. Retrieved from http://www.ifoam.org/en/organic-landmarks/definition-organic-agriculture. Accessed 1 Feb 2013.

Ison, R., C. High, C.P. Blackmore, and M. Cerf. 2000. Theoretical frameworks for learning-based approaches to change in industrialised-country agricultures. In *Cow up a tree: Knowing and learning for change in agriculture: Case studies from industrialised countries*, ed. M. Cerf, D.P. Gibbon, B. Hubert, R. Ison, J. Jiggins, M. Paine, J. Proost, and N. Röling, 31–53. Paris: INRA.

Jay, M. 2007. The political economy of a productivist agriculture: New Zealand dairy discourses. *Food Policy* 32(2): 266–279.

Lamine, C. 2011. Transition pathways towards a robust ecologization of agriculture and the need for system redesign. Cases from organic farming and IPM. *Journal of Rural Studies* 27(2): 209–219.

Magne, M.-A., and M. Cerf. 2009. How information becomes a resource for action in an uncertain and complex world: Sense-making and contingency in the knowing process. *Outlook on Agriculture* 38(2): 157–165.

McIntyre, B.D., H.R. Herren, J. Wakhungu, and R.T. Watson (eds.). 2009. *IAASTD international assessment of agricultural knowledge, science and technology for development: Global report*. Washington, DC: Island Press.

Meat & Wool New Zealand. 2010. *Compendium of New Zealand farm facts*, April 2010. Wellington Meat & Wool New Zealand Economic Service.

Milestad, R., and I. Darnhofer. 2003. Building farm resilience: The prospects and challenges of organic farming. *Journal of Sustainable Agriculture* 22(3): 81–97.

National Academy of Sciences. 2011. America's climate choices. *Report of National Research Council Committee*. Washington, DC: National Academy of Sciences.

New Zealand Meat Board. 2010. *Annual report 2009–10*. Wellington: Meat & Wool New Zealand Economic Service.

Ondersteijn, C., G. Giesen, and R. Huirne. 2006. Perceived environmental uncertainty in Dutch dairy farming: The effect of external farm context on strategic choice. *Agricultural Systems* 88(2): 205–226.

Pollan, M. 2001. Naturally: How organic became a marketing niche and a multibillion-dollar industry. *New York Times Magazine,* pp. 30–37, 57–58, 63–64.

Pretty, J., W.J. Sutherland, J. Ashby, J. Auburn, D. Baulcombe, M. Bell, et al.. 2010. The top 100 questions of importance to the future of global agriculture. *International Journal of Agricultural Sustainability* 8(4): 219–236.

Rammel, C., S. Stagl, and H. Wilfing. 2007. Managing complex adaptive systems—A co-evolutionary perspective on natural resource management. *Ecological Economics* 63(1): 9–21.

Ritchie, M., and H. Campbell. 1997. Historical development of the organic agriculture movement in New Zealand. In *Recent developments in organic food production in New Zealand: Part 1, organic food exporting in Canterbury*, vol. 2, ed. H. Campbell, J. Fairweather, and D. Steven, 51. Dunedin: Anthropology Department, University of Otago.

Rosin, C. 2008. The conventions of agri-environmental practice in New Zealand: Farmers, retail driven audit schemes and a new spirit of farming. *GeoJournal* 73(1): 45–54.

Rosin, C., and H. Campbell. 2009. Beyond bifurcation: Examining the conventions of organic agriculture in New Zealand. *Journal of Rural Studies* 25(1): 35–47.

Rosin, C., L.M. Hunt, H. Campbell, and J.R. Fairweather. 2007a. *Becoming the audited: Response of New Zealand sheep/beef farmers to the introduction of supermarket initiated audit schemes*, ARGOS research report, vol. 07/05. Christchurch: Agricultural Research Group on Sustainability (ARGOS).

Rosin, C., L.M. Hunt, H. Campbell, and J.R. Fairweather. 2007b. *There are audits and there are audits: Response of New Zealand kiwifruit orchardists to the implementation of supermarket initiated audit schemes*, ARGOS research report, vol. 07/06. Christchurch: Agricultural Research Group on Sustainability (ARGOS).

Rosin, C., L.M. Hunt, J.R. Fairweather, and H. Campbell. 2009. *Social objective synthesis report 2: Social differentiation and choice of management system among ARGOS Farmers/Orchardists*, ARGOS research report. Christchurch: Agricultural Research Group on Sustainability (ARGOS).

Saunders, C.M. 2009. *Feeding the world: the role of New Zealands" food exports*. Paper presented at the 44th Otago Foreign Policy School Dunedin, N.Z.: University of Otago.

Seppänen, L., and J. Helenius. 2004. Do inspection practices in organic agriculture serve organic values? A case study from Finland. *Agriculture and Human Values* 21(1): 1–13.

Setten, G. 2004. The habitus, the rule and the moral landscape. *Cultural Geographies* 11(4): 389–415.

Shucksmith, M. 1993. Farm household behaviour and the transition to post-productivism. *Journal of Agricultural Economics* 44(3): 466–478.

Silvasti, T. 2003. The cultural model of "the good farmer" and the environmental question in Finland. *Agriculture and Human Values* 20(2): 143–150.

Stock, P.V. 2007. 'Good farmers' as reflexive producers: An examination of family organic farmers in the US Midwest. *Sociologia Ruralis* 47(2): 83–102.

Stuart, A., and H. Campbell. 2004. "Business as usual": Contextualising the GM/organic conflict within the history of New Zealand agriculture. *New Zealand Sociology* 19(2): 220–239.

The Royal Society. 2009. Reaping the benefits: Science and the sustainable intensification of global agriculture. *Report of the Royal Society (UK) October 2009*. London: Royal Society.

Todd, J.H., L.A. Malone, B.H. McArdle, J. Benge, J. Poulton, S. Thorpe, and J.R. Beggs. 2011. Invertebrate community richness in New Zealand kiwifruit orchards under organic or integrated pest management. *Agriculture, Ecosystems & Environment* 141(1): 32–38.

Tomkins, A., A. Greaves, D. Wilson, and C. Thomson. 1996. *Evaluation of a mineral oil for pest control on kiwifruit*. Proceedings of the New Zealand Plant Protection Society conference. Auckland, N.Z.: New Zealand Plant Protection Society, 12–16.

Trewavas, A. 2001. Urban myths of organic farming. *Nature* 410(6827): 409–410.

Urry, J. 2005. The complexities of the global. *Theory, Culture and Society* 22(5): 235–254.

Walker, B., C.S. Holling, S.R. Carpenter, and A. Kinzig. 2004. Resilience, adaptability and transformability in social-ecological systems. *Ecology and Society* 9(2): 5.

Walker, B., L. Gunderson, A. Kinzig, C. Folke, S. Carpenter, and L. Schultz. 2006. A handful of heuristics and some propositions for understanding resilience in social-ecological systems. *Ecology and Society* 11(1): 13.

ZESPRI Group Limited. 2010. *ZESPRI annual review 2009/10: Growing innovation*. Mt Maunganui: ZESPRI Group Limited.

Chapter 11
From the Ground Up? The Principles of Australian Organic Agriculture

Rebecca Jones

11.1 Background and Introduction

Organic agriculture today is described as an "industry" and its growers as "producers", the food they grow is "an agricultural sector" and commercial organic farmers take out "certification". In Australia, like many other western nations, organic farming is part of an industrialized, capitalist, agro-economy and commercial organic production is influenced by government regulations. Despite this edifice of industrialization and institutionalization are the values upon organic growing was founded still relevant in organic production today?

In order to examine the values of organic agriculture in the past and present, this chapter will, first explore the fundamental principles upon which Australian organic agriculture was founded then, secondly, examine whether these principles still resonate with, and are relevant to Australian organic agriculture. This study analyzed the documents of Australian organic growers organizations: The Australian Organic Farming and Gardening Society of the state of New South Wales, the Victorian Compost Society and the Living Soil Association of Tasmania which were Australia's first three organic growers organizations, all founded in the 1940s. Also analyzed are the documents of the two major, current Australian organic producers' organizations: National Association for Sustainable Agriculture and Australian Certified Organic (Biological Farmers of Australia). The documents from all these organizations, which were analyzed thematically, included organizational objectives, newsletters, organic agricultural standards and other organizational publications. Documentary sources were complemented by semi-structured, in-depth

R. Jones (✉)
School of History, College of Arts and Social Sciences, Australian National University, Coombs Building, Canberra, ACT 0200, Australia
e-mail: rebecca.jones@anu.edu.au

B. Freyer, J. Bingen (eds.), *Re-Thinking Organic Food and Farming in a Changing World*, The International Library of Environmental, Agricultural and Food Ethics 22, DOI 10.1007/978-94-017-9190-8_11

interviews with organic farmers conducted by the author. This chapter is part of a larger study that examined organic farming and gardening through the lens of environmental history exploring the history of beliefs and practices in Australian organic agriculture from the early twentieth century to the present day (Jones 2010).

This chapter is divided into two sections. The first section explores the beginning of organic agriculture in Australia and identifies three fundamental principles upon which Australian organic growing was founded. The second section explores Australian organic agriculture in the 2000s. It will firstly discuss the structure and context of Australian organic agricultural certification and industrialization today then return to the founding principles of organic growing and examine whether these remain the key principles of Australian organic agriculture today.

11.2 Organic Agriculture in the 1940s and 1950s – The Beginnings of Organic Farming in Australia

Australia's first organic farming and gardening societies were founded in the 1940s: the Australian Organic Farming and Gardening Society (AOFGS) in 1944; the Victorian Compost Society in 1945 and the Living Soil Association of Tasmania in 1946. These groups were among the earliest organizations in the world devoted to the production of food using organic methods. International organizations founded at a similar time as the Australian societies and with similar aims (and to which the three Australian societies were affiliated) were the Humic Compost Club of New Zealand founded in the early 1940s and the British Soil Association founded in 1947.

The establishment of these groups has been identified as the beginning of a distinctive movement of organic growing. Although the growers of these societies drew upon many traditional agricultural practices (such as use of animal manure and compost as fertilizer), the establishment of these societies was the beginning of a conscious affiliation of growers who shared fundamental philosophies and practices of food production, were self-conscious about these principles and practices and identified themselves as a group in opposition to other farming practices (Conford 2001). All three Australian societies comprised both farmers and gardeners and drew membership from the middle class social establishment in both urban and rural areas. The Victorian and Tasmanian societies directed most attention to small farmers and gardeners while the New South Wales society made graziers their area of special interest. After only 2 years of operation, the Victorian Compost Society had over 450 financial members, increasing to 600 by the early 1950s, and for every financial member there would most likely also have been a 'submerged network' of unaffiliated supporters (Melucci 1989).

Each society produced its own publications: *Farm and Garden Digest* and *Organic Farming Digest* were published by the Australian organic farming society; *Victorian Compost News* by the Victorian Compost Society and a self-titled newsletter and occasional pamphlets by the Living Soil Association of Tasmania.

The founding of the three Australian organic growers' societies and their development during the next 10 years was based on three fundamental principles: (1) Humus-rich, fertile soil; (2) Chemical free; and, (3) Biodiversity and ecological wellbeing. Each of these three principles will be discussed, in turn, below.

11.2.1 Humus-Rich Fertile Soil

Humus-rich fertile soil, including the promotion of methods for increasing soil organic content, was the vision upon which Australia's first organic agricultural societies were founded in the 1940s. Members of these societies argued that humus-rich fertile soil was the basis of plant, animal and in turn, human health. Issues both in Australia and abroad were a catalyst for the founding of the Australian societies at this time and for their focus on soil humus and fertility. Most Australian native soils differ markedly from agricultural soils in the northern hemisphere. Phosphorus, nitrogen, copper, magnesium, iron and boron are present in different quantities in many Australian soils compared to agricultural soils in Europe and North America and most native Australian soils are characteristically low in soil organic matter (Young and Young 2001). Soils dominated by clay or sand predominate in Australia which have a very different tilth from loamy soils common in agricultural areas in other parts of the world.

European settlers, who occupied Australia in the late eighteenth and early nineteenth centuries, cleared native vegetation, plowed, sowed and reaped European staples such as potatoes, wheat and vegetables and they introduced sheep and cows as well as rabbits. On the surface of the land, settlers were attempting to domesticate and Europeanize the environment, but beneath the surface, imported farming techniques further compounded the differences between Australian and European soils. Their cultivation techniques, over time, eroded soil structure and exacerbated naturally low levels of soil organic matter and soil fertility. After only 15–20 years of cultivation, Australian soils lost half of their stored organic matter (Charman and Murphy 2000).

After 100 years of European occupation, many agricultural areas were experiencing significant erosion as stock damaged river and creek banks and wind whipped up the dry, sandy, exposed and damaged soil. Just as in the United States' Midwest where erosion rendered formerly productive agricultural areas almost uninhabitable, during the interwar years, the Mallee district of southern Australia faced its own 'dust bowl' and drought in the 1920s, 1930s and 1940s that exacerbated the problem (Keating 1992). In the final years of the Second World War, the years the first Australian organic societies were established, vast quantities of soil in semi-arid South Eastern Australia eroded. Declining soil condition became a national preoccupation (Barr and Cary 1992).

Some Australian farmers sought solutions to declining soil fertility in the increased application of agricultural fertilizers. However, British agricultural writer Albert Howard, whose books were readily available in Australia, offered an alternative solution. Albert Howard was a botanist, agricultural scientist, farmer, writer and agricultural advisor to the British government in India in the interwar years. With his botanist wife, Gabrielle, Howard was developing ways to maintain soil fertility and increase plant and animal disease resistance through large-scale production of compost. Howard, in his books *An Agricultural Testament* and *Farming and Gardening for Health and Disease* argued that humus was vital for plant growth and encouraging disease resistance. Human health, he claimed, whether plant, animal or human, was dependent on nutritious food grown in humus rich fertile soil (Howard 1943, 1945). Howard's ideas were part of a widespread renewed interest in the connection between human health and environmental influences (McMichael 2001b). The inherent challenges of Australian soils and the crisis of erosion and declining soil fertility; specific solutions offered by Albert Howard and resurgence of popular interest in the importance of nutrition to human health provided the catalysts for the founding of Australia's first organic farming and gardening societies in the 1940s.

The importance of creating humus rich, fertile soil was described by early Australian organic gardeners as the 'Rule of Return' whereby anything organic or biodegradable must be returned to soil to decay and thereby replenish the organic matter within the soil. It was promoted as one of the fundamentals of life and organic growers described the neglect of this Rule had resulted in erosion and loss of fertility in Australian soils (Living Soil Association of Tasmania 1946; Australian Organic Farming and Gardening Society 1951). Practicing the Rule of Return was the key to quality food production:

> The most amazing results that have been obtained, by obeying this law of return [...] are found in the improvement of human health. Where the food has been supplied from grains, vegetables, and fruit, milk, eggs and meat produced in obedience to this cycle, the general health, energy and power of resisting disease has been built up to an amazing extent (Australian Organic Farming and Gardening Society 1947).

Versions of the Rule of Return had been part of traditional agricultural practices worldwide (McNeill and Winiwarter 2004), but for early Australian organic farmers of the 1940s the most important methods, promoted in the magazines of the three societies were composting, use of manure and green cover crops (or 'green manure'). Composted human manure or sewage, particularly municipal sewage, was heavily promoted by the organic societies in the 1950s. They believed that the composting of human manure was an essential practice in the Rule of Return, returning nutrients and organic matter to the soil. Entire issues of the societies' magazines were devoted to the issue and they held public demonstrations of large-scale composting of municipal sewage (Victorian Compost Society 1954).

11.2.2 Chemical Free

By the late 1940s, 'chemical free' had joined humus rich, fertile soil as another key principle of organic farming. 'Chemical' was used by organic growers as a pejorative shorthand term to describe manufactured fertilizers, pesticides, herbicides and fungicides which were created in a laboratory; molecules that did not exist in a natural form except through human synthesis. To quote author Rachel Carson (1962), they were "the synthetic creations of man's inventive mind, brewed in his laboratories, and having no counterparts in nature" (Carson 1963).

It was actually fertilizers that were the first type of chemical to galvanize Australian organic farmers' attention. Chemical fertilizers based on nitrogen, phosphorus, and potassium (N, P, and K) were used in large quantities in Australia by the end of the Second World War. Australian farmers used well over a million tons of artificial fertilizers on a quarter of a million acres of crop and pasture land in the 1950s. Superphosphate, made by chemically treating phosphorus with sulfuric acid, was the chemical fertilizer used most heavily by conventional Australian farmers in the 1940s and 1950s as many Australian soils have a lower level of naturally occurring soil phosphorus than soils in Europe and North America and superphosphate, it was hoped, would rectify this 'deficiency'.

Organic growers spoke out strongly in opposition to chemical fertilizers. In 1948, the editor of the *Victorian Compost News*, stated: "The Victorian Compost Society (1948) stands 4-square in its advocacy of organic methods as a means of maintaining and increasing fertility of the soil. [...] Difficulties can be overcome without resorting to inorganic chemical fertilizers". Organic growers condemned chemical fertilizers for damaging worms and other soil life and disrupting decay and regrowth of plants (Living Soil Association of Tasmania 1946). However, their antipathy to chemical fertilizers was more fundamental. Artificial fertilizers, based on the elements nitrogen, phosphorus and potassium were derided for reducing soil fertility to three agents. In contrast, organic agriculture, they argued, was about maintaining the complex symbiotic relationship between humans, plants, animals and soil matter.

British historian Philip Conford (2001) named these contrasting perspectives on agriculture 'The Great Humus Controversy'. The chemical perspective held that minerals (particularly nitrogen, phosphorus and potassium) were the most important aspect of plant nutrition, while the biological perspective (held by organic farmers) drew on agricultural biological research and argued that soil organic matter was a crucial reservoir of nutrients and was additionally important as a promoter of bacteria and fungi necessary for plant growth. This controversy was, in essence, a philosophical battle between ecological and non-ecological perspectives. Organic growers' ecological view was that health was achieved by working with biological processes such as decay and regrowth and preventing disease by assisting biological cycles between plant and animal waste, decay and growth, animal health and human health. In contrast, supplying individual elements through the addition of manufactured chemical fertilizers provided a 'quick fix' rather than systemic prevention. Artificial fertilizers were like a drug, supplying a response to a deficiency rather than creating a soil environment that would support and provide healthy, nutrient-rich plants and animals.

By the early 1950s, the Australian organic societies were also strongly arguing against the use of chemical pesticides. The Australian Organic Farming and

Gardening Society of New South Wales enshrined opposition to all chemical pesticides in its objectives. It stated that: "the Society [. . .] condemns the use of poisonous sprays and dusts as such preparations injure the soil by killing its micro-organism, also the earthworms, bees and birds that are in the vicinity" (Australian Organic Farming and Gardening Society 1952).

Chemical pesticides, many based on arsenic or tobacco had been available in Australia at least since the late nineteenth century (Jones and Chesters 2006). Soon after the end of the Second World War arsenic based chemical pesticides were being replaced by a new generation of complex chemicals: the organochlorides, chemical compounds of carbon, hydrogen and chlorine. The most notorious member of this family of chemicals was DDT (dichlorodiphenyltrichloroethane). DDT, whose release as an agricultural chemical in Australia was roughly contemporaneous with the establishment of the organic societies was strongly opposed by organic growers. Most people associate the beginning of the campaign against DDT with the publication of Rachel Carson's *Silent Spring* in 1962, but Australian organic growers were condemning DDT over 10 years before *Silent Spring's* publication. Within 3 years of the chemical's release in Australia, organic societies launched a concerted and increasingly vocal campaign against DDT as responsible for killing beneficial insects, and for threatening human health. For example, the New South Wales society, in a cover article about pesticides in 1950, questioned the effect of DDT on human health: "If fruit trees need to be drenched with poison sprays before they can produce a crop, what is the effect of such fruit on the health and wellbeing of the people who have to consume it?" (Australian Organic Farming and Gardening Society 1950).

The use of chemical pesticides, like chemical fertilizers, was contrary to organic growers' ecological philosophy. Pesticides provided a short-term remedy to destroy the pest on an individual plant or animal rather than a systematic method of prevention of pests, diseases and deficiencies. By contrast, the organic methods of practicing the Rule of Return attempted to prevent disease by creating healthy plants, animals and people. Opposition to chemicals pesticides and fertilizers, like chemical fertilizers, concerned the dependence of human health on natural cycles of growth, decay and plant and animal resistance to disease rather than on manufactured chemical inputs. The key to health, organic growers claimed, was working with the natural environment to prevent disease rather than administering 'quick fixes'. While the main organic response to pests and diseases was to grow plants and raise animals on humus-rich soil, organic growers also promoted other systemic preventative measures to avoid pest and disease concentration. These solutions were based on the idea, expounded by ecologist Eugene Odum, among others, that nature as a whole was self-regulating and in a state of equilibrium or constantly evolving towards stable equilibrium. Individual organisms contributed to the overall balanced state of the environment. Organic growers criticized insecticides for *upsetting Nature's balance* by indiscriminately killing beneficial as well as harmful insects, destroying predator insects as well as the pests (Australian Organic Farming and Gardening Society 1951). Organic methods for avoiding accumulation of pests and diseases included raising disease resistant plants and animals through fertile soil, promoting a diversity of species of both plant and animals and rotating crops to avoid a build up of diseases.

11.2.3 Biodiversity and Ecological Wellbeing

Biodiversity was emerging as a third principle of Australian organic farming in the 1950s at the same time as soil fertility and chemical free growing were consolidating as major principles. Organic societies were increasingly advocating for the preservation of native fauna and flora in and beyond the farm environment. This approach reflected a broadening of organic growers' belief in the connection between human and environmental health to include so-called non-productive elements of the environment: The Australian Organic Farming and Gardening Society (1950) stated in its magazine in 1950s that: "The wellbeing of man is bound up with that of the animal and insect kingdoms, the trees and plants and with the living soil itself. All are inter-related and mutually dependent upon each other".

One example of the burgeoning interest in biodiversity on the farm was the promotion of the establishment and preservation of native trees on organic farms. Australian Organic Farming and Gardening Society (1952) made tree planting one of its primary objectives: "the Society advocates large-scale tree-planting operating to temper the natural aridity, conserve moisture, foster bird-life, and provide wind-breaks". The Eucalypt tree, an iconic Australian native species, became an advertisement of the sound organic farm. Trees were emblematic of the organic approach because they were a counterpoint to non-organic practice. During the 1940s and 1950s tree clearing was still strongly encouraged by government departments of agriculture throughout Australia, as it had been throughout the nineteenth and early twentieth centuries.

Organic growers' interest in native flora was not limited to trees but also included native shrubs and wildflowers. Wildflowers were recognized for their important role in attracting pollinator insects such as bees and native wasps and the 1952 objectives of the Australian Organic Farming and Gardening Society (1952) recognized the role native fauna played in maintaining a well functioning farm environment: "the Society holds that the natural function of birds is to keep insects in check. It deplores the wanton destruction of birdlife now prevalent, which gives rise to periodic plagues of insects e.g. locust etc."

During the 1950s, Australian organic farmers also extended the first tentative feelers beyond the perimeters of the farm to support the preservation of flora and fauna on non-agricultural land. They reasoned that preserving plants and animals in their natural environment enhanced the broader wellbeing of the whole environment. The Victorian Compost Society and the Australian Farming and Gardening Society of New South Wales both leant support to campaigns for the establishment of conservation reserves. In their concern about the preservation native flora and fauna in its natural environment they also formed links with nature preservation organizations dedicated to the protection of flora and fauna such as Save the Forests Campaign, the Field Naturalists Club of Victoria, and the Victorian National Parks Association.

11.2.4 Summary

All three key principles that defined Australian organic growing in its early years – humus-rich, fertile soil, chemical free production and biodiversity – were founded on the fundamental concept that human health and the biophysical environment are interdependent and intertwined. Food was a crucial site of interaction between a person and their surrounds. To be organic was to produce food in a way that growers believed promoted human health by co-operating with natural processes and enhancing environmental health. Ecological thinking inspired these ideas. During the mid twentieth century, the period when organic agriculture was becoming established, there was a burgeoning interest in ecology and humans' relationship with the natural environment (Worster 1994). Ecological thinking, as both a science and a philosophy of interrelatedness, is a way of thinking about and understanding the world that emphasizes the interdependency of all things. Humans are seen as part of, and dependent on, natural systems that are mutual, dynamic and interactive. Animate and inanimate entities and processes are interrelated and interdependent and the context in which something exists has a profound impact on the individual. Therefore changes within the natural environment can impact on human health and wellbeing and as health ecologist and epidemiologist Tony McMichael explains, human actions which damage the environment can, in turn, have detrimental effects on human health (McMichael 2001a).

Organic growers of the mid twentieth century acknowledged the inspiration of ecological thinking in their principles, making statements in their magazines which referenced ecology such as: "each life contributes its own individual part in the great symphony of nature. The science of ecology teaches us of this co-operation" (Victorian Compost Society 1952). One of the most eloquent and popular ecologists of the interwar years who inspired Australian organic farmers and gardeners was American forester and environmental philosopher Aldo Leopold. Leopold's writings were informed by his observations of the destruction of environments and over-hunting of large predators. Leopold's *A Sand County Almanac* (Leopold 1949), published posthumously, proposed a 'Land Ethic' that encouraged people to see themselves as part of a community with the non-human world. He argued that human wellbeing depended upon this circle of interrelatedness in which humans were equal members rather than conquerors. Human actions, he claimed, should therefore be limited by responsibility to the rest of the community and curtailed to those actions that promoted the integrity and stability of the whole community. Leopold described this as 'enlightened self interest' and argued that seemingly non-productive elements of this community (for example, wildflowers, birds, microscopic life) were essential to the healthy functioning and stability of the whole. Echoing Leopold's land ethic, Australian organic farmers and gardeners pleaded for the recognition of the interdependency of all life. A speaker to the Victorian Compost Society in 1957 explained: "Man is an animal. He shares this world with other humans and members of other species, both plant and animal, all drawing on its material and energy resources" (Victorian Compost Society 1957).

11.3 Organic Agriculture in the 2000s – The Regulation of Australian Organic Farming

The greatest change to transform Australian organic farming since the foundation of the first organic organizations in the 1940s has been the introduction of organic standards and certification for organic farmers. The development of standards for commercial organic farming in the late 1980s was an attempt to assure consistent and verifiable food production. Adherence to these standards would, it was hoped, guarantee the consistency of organic growing practices, the quality of produce and the quality of the agricultural land upon which it was grown. Certification of growers against these standards also allowed organic growers for the first time to profit commercially from growing food organically as they could now sell their produce as identifiably 'organic'.

There had been a series of moves towards standards for organic production in Australia since the 1960s, encouraged by the development of organic standards in Britain, Europe and California. The Australian Bio Dynamic Research Institute (founded in the 1950s) began certifying biodynamic farmers in 1967 and the institute adopted the 'Demeter' trademark, an internationally recognized logo, developed in Switzerland which signifies that produce is grown according to biodynamic traditions (Bio – Dynamic Research Institute 2005). Guidelines were also developed by organic organizations in Australia in the late 1970s and 1980s but they were not widely applied and were not monitored by the organizations but operated on goodwill (Doubleday Organic Research Association 1977; Lyons 1999).

The beginning of consistent, monitored certification began in 1986, when a coalition of approximately 30 Australian organic growers groups formed to provide a coordinated approach to developing common standards amongst organic growers and to lobby government. This group became the National Association for Sustainable Agriculture Australia (NASAA). The first farm certified according to NASAA standards was Dick and Dot McNeil's orchard, poultry and vegetable farm near Sydney.

Inspectors examined the property, production processes and animals and tested the soil for pollutants (D. McNeill 2006). This was a process which replaced the purely honor-based system previously adhered to by members of organic organizations, although this process was still not enforceable by law. As trade in organic food became more globalized there was a perceived need for guarantees of organic production standards. Organic production guidelines became more urgent when the European Union demanded that imported organic produce have a legally enforceable standard of production. NASAA, with the involvement of the Commonwealth Government Australian Quarantine and Inspection Service (AQIS) and other organic growers developed a national standard for certifying organic and biodynamic farms and produce in 1992.

During the 1990s and early 2000s various other non-government certification organizations developed their own standards based on the AQIS standards. Although these standards were originally developed to verify the quality of exported produce they also became, by default, the standard used to certify produce for domestic consumption. In 2009 the Commonwealth Government peak standards authority, Standards Australia, released new minimum standards for Australian organic and biodynamic production for the domestic market. These standards were based on and similar to the AQIS guidelines. While the Australian Government licenses organizations to certify produce and enforces the use of the certification logos, government regulation remains at arms length, with standards industry regulated.

In 2011 seven Australian organic organizations were certifying Australian organic farms. The two largest of these are NASAA and Australian Certified Organic (ACO), the certification arm of the organic growers' association Biological Farmers of Australia. These two organizations use the government standards as the basis from which they developed their own, sometimes more stringent certification requirements. NASAA and ACO, between them, now certify about 90 % of Australia's estimated two and a half thousand certified organic farms. These farms are located in all Australian states and territories and manage between 8 and 12 million hectares of land, the largest area of organically managed land per country in the world (Halpin 2004; Willer and Yussefi 2006).

Certification encouraged the development of 'the organic industry'. Whereas organic growers of the past sold their produce on the general market, certification enabled organic produce to be distinguished from non-organic produce. This allowed farmers to sell their products as identifiably and verifiably 'organic' and charge a premium for this. Certified organic cereal, livestock, fruit and vegetables sold in Australia in the 2000s commanded a price premium of between 50 and 75 % more than the equivalent conventionally grown item (Halpin 2004; Willer and Yussefi 2006). The value of organic produce grown in Australia is now estimated to be more than 100 million Australian dollars, with fruit, vegetables, cereals and milk the major domestic products and cereals and beef the most important export products (Willer and Yussefi 2006).

The development of commercially oriented organic certification schemes has caused some commentators of organic farming to question the degree to which the traditional values and practices of organic growing have remained relevant. Some academic researchers of the organic industry in North America, Western Europe and Australia (Buck et al. 1997; Tovey 1997; Guthman 1998; Lyons 1999, 2001; Jordan et al. 2006) have argued that core organic values and practices have been compromised as government bureaucracies became involved in the regulation of organic methods and large industrial corporations influenced organic production which has encouraged organic farmers to be motivated by commercial rather than ideological principles.

However, other researchers examining organic agriculture in Europe, New Zealand, Canada and Australia (Coombes and Campbell 1998; Kaltoft 1999; Campbell and Liepins 2001; Michelsen 2001b; Lund et al. 2002; Köpke 2005; Lockie and Halpin 2005) have countered that while there is *some* evidence that certification and industrialization has changed the structure of organic farming, the core organic principles have not been compromised. They argue that certification has not had a profound impact on the underlying beliefs of organic growers themselves or the principles upon which their organic methods are based. Some researchers argue that core organic principles have actually been reinforced rather than eroded by government regulated certification (DeLind 2000; Lynggaard 2001; Michelsen 2001a).

To understand whether the core principles upon which organic produce was founded in the 1940s are still applicable today, I return to each of the three key elements discussed above – soil, chemical free, ecological wellbeing– and explore their resonance with Australian organic agriculture in the 2000s.

11.3.1 Humus-Rich Fertile Soil

Fertile, humus-rich soil, the founding principle of Australian organic growing in the 1940s, continues to be the first principle of organic farming. Increasing soil humus as a means of encouraging the growth of plants and animals continues to be the most important organic practice today. Fertile, humus-rich soil features prominently in the Australian certification standards. Echoing Albert Howard's writings 70 years earlier the NASAA standards state:

> Healthy soil is the prerequisite for healthy plants, animals and products. The maintenance of soil health by ecologically sound means is at the heart of organic production systems [. . .] NASAA places great emphasis on the levels of organic matter and humus maintained in soils as an indicator of sustainability and of organic status . . . Optimum soil fertility, soil structure and biological activity are fundamental aims of organic farming (National Association for Sustainable Agriculture Australia 2004).

The production of humus-rich fertile soil also remains a core organic practice. South Australian beef farmer Brice Douglas, interviewed by the author, explained that his first action as an organic grower was to test the mineral and organic content of the soil on his farm and apply dolomite to reduce the acidity of the soil and add calcium and magnesium to the soil. He explained his belief in the connection between fertile soil and the health of plants, animals and humans:

> You can go out into the paddock anywhere [on my farm] now and dig and there are earthworms everywhere. The dung beetles have come back and here in the pond in front of our house and in our four dams there are frogs everywhere. It is just the same as the human body, which is what I did for 38 years [as a naturopath and osteopath]. You find out what is the problem, the base cause, the deficiency. You give them the minerals and vitamins

> to restore their health. That is exactly what you are doing with the soil. The end result is healthy animals that you have no problems with (Douglas 2006).

Like the organic growers of the 1940s and 1950s, Victorian dairy farmers Ron and Bev Smith describe bringing life to their soil through organic methods. When they began farming organically in 1980 "the soil smelt lifeless", they recall. After 25 years of organic practices "there was abundant soil, life, bacteria and fungi [. . .] The soil smells rich and sweet" (Smith 2008).

Compost, animal manure and green cover crops remain the stalwarts of organic farming. Australian organic standards recommend compost, animal manure and green manure as fertilizer and soil conditioner (Standards Australia 2009 2.5) just as the organic societies did in the 1940s and 1950s and these are all core practice undertaken by Australian organic farmers today. Queensland, sugar-cane farmers Anthony and Debbie Skopp, also interviewed by the author, practice sheet composting where they spread organic matter directly onto the soil to decompose and increase the humus content in the soil (Skopp and Skopp 2006). To fertilize his farm Queensland banana farmer Desmond Chappel applies composted chicken manure rather than the artificial fertilizers used by conventional banana growers and Victorian organic sheep and wheat farmer Anthony Sheldon rotates grazing animals and crops so the animal manure will fertilize the fields prior to sowing (Chappel 2006; Sheldon 2006).

Although composting, manuring and green cover crops remain essential organic methods, the use of human sewage as fertilizer, enthusiastically promoted by organic societies in the 1940s and 1950s is now prohibited by Australian organic standards on land used for the production of human or animal feed (Standards Australia 2009 4.1). Preoccupation with hygiene, concern about the heavy metal content of municipal sewage as well as fear of a negative public image of organics have all contributed to the rejection of human sewage as a fertility source by commercial organic farmers. When the United States government proposed the use of sewage sludge in its own national organic standards there was strong dissent from organic groups (Vos 2000).

The certification of aquaculture operations, such as mussel farms, in Australia has seemingly created an anomaly for organic agriculture today which continues to be defined as soil-based systems (Mansfield 2004). However, this anomaly reflects the history of organic farming in Australia as overwhelmingly land-based and the continued importance the foundation of Australian organic growing societies on belief in the importance of soil humus organic food production.

11.3.2 Chemical Free

For many people today 'chemical free' is synonymous with organic production. Raising plants and animals without the use of synthesized chemical fertilizers, pesticides, herbicides and fungicides continues to be a significant means by which today's organic farmers assert the healthfulness of their produce. Many of the pesticides and fertilizers available to farmers today are different to those used in previous decades but organic growers' aversion to chemicals has changed little since the campaigns against DDT and superphosphate in the 1950s. Organic standards prevent the use of all synthesized chemicals that have been created or changed chemically, particularly those toxic to humans and non-target species (Standards Australia 2009). Prohibition also extends to naturally occurring substances that are known to be toxic (such as nicotine) (Standards Australia 2009).

Organic farmers continue to reject artificially synthesized chemicals as unhealthy. Organic farmer Anthony Sheldon comments: "I think organic food would have to be more healthy. If something that is toxic doesn't have to be used at all it has got to be better. If you had half Chemical and Coke you'd probably get crook" (Sheldon 2006). Brice Douglas reiterates his disapproval of agricultural pesticides and fertilizers:

> It is logical. The residues left from the chemicals and the sprays, they have to affect you. If you are eating an animal that has been reared using chemicals – treating for worms, eating grass that has been grown with fertilizers and sprayed with pesticides to kill the bugs and beetles – the end result is that you are doing exactly the same thing as drinking the chemical (Douglas 2006).

The desire to farm without the use of artificial chemicals is still a significant motivation for growers to convert to organic practices. Organic turkey and macadamia farmer, Matthew Jamison grew up on a conventional beef property in Queensland and tells of his experiences of illness due, he believes, to contact with agricultural chemicals:

> All my family has died of cancers. By the time I was thirty-four I was the oldest of my descent line. I grew up at the stage when everyone was spraying 245 T and everyone in our rural community had stillborn babies, which I can only put down to all the volatile chemicals. ... We used to spray the cattle with some organophosphate. My father would be on one side and I would be on the other, spraying it on the cattle but also on each other. We put barrier cream on our hands and were wearing overalls but we'd be soaking wet (Jamieson 2006).

Surveys of organic farmers in the last 20 years confirm that organic farmers' concerns about the health of their families, consumers, animals and themselves motivated them to convert to organic farming (Wynen 1992; Halpin 2004). Stories abound among organic growers of rashes, itches, flakey skin, cancer and respiratory ailments, even death – all of which they attribute to agricultural chemicals (Chappel 2006; McNeill 2006; Smith 2008). Telling and retelling stories about chemical poisoning affirms organic growers' collective aversion to chemicals and the belief in the comparative healthfulness of organic produce.

Aversion to the unnaturalness of chemical fertilizers and pesticides also underpins current organic growers' rejection of genetically modified (or GM) plants and animals. Australian organic standards prohibit genetically modified or genetically engineered products and processes from any aspect of organic production, including the use of genetically modified seeds and plants and contamination of a crop by genetically modified organisms will result in decertification (Standards Australia 2009).

The process of genetic modification results in synthetically constructed entities achieved by joining fragments of DNA from different organisms, including organisms of different species (such as rice with daffodils or fish with tomatoes). It results in combinations of genetic material that could not occur through normal reproduction and hybridization.

Just as organic growers of the 1950s, viewed all synthesized products with suspicion and distrusted government and scientific claims about their 'safety', so do organic growers oppose genetically modified organisms on precautionary grounds as well as on principle. There is currently no conclusive evidence confirming the presence or absence of health effects of genetically modified organisms (Pretty 2001). Therefore, organic standards take a precautionary approach, reasoning that actions should be avoided if there is a possible threat of serious or irreversible environmental damage (Standards Australia 2009).

Organic growers today, unlike their counterparts in earlier decades have access to a plethora of manufactured products which are 'chemical free' and considered to be organically acceptable. During the 1990s and 2000s organic remedies began to be manufactured commercially on an industrial scale and organic farmers now have access to manufactured remedies such as concentrated seaweed fertilizer, organic fruit fly bait and garlic spray. Brice Douglas uses powdered sulfur to control pests on his cattle (Douglas 2006). Similarly, Anthony Sheldon uses a commercially manufactured, but organically certified product, to control fly-strike in his sheep (Sheldon 2006). Anthony Sheldon sees the substitution of 'unsafe' manufactured products with 'safe' manufactured products as an important part of the future of organics. He sees large chemical industries as having a role in Australian organic farming, manufacturing non-toxic agricultural products:

> If more of these products can work their way into the market to replace stuff that is toxic it can only do good. They are not putting themselves out of business because they are still making a product that does the job but is safe. Hopefully that type of thing is the way of the future. The big companies that supply chemicals now, they have the expertise and, in time, could make natural products. They could still be in business but selling stuff that is non-toxic and it has got to be a good thing (Sheldon 2006).

Relying on manufactured products to control pests and diseases, maintain fertility and contain weeds has come to be known, among organic growers, as 'input substitution'; replacing poisonous, synthesized 'conventional' chemicals with non-toxic, non- polluting 'organic' substances. Australian organic standards incorporate both an input substitution and systemic approach to organic farming. Organic standards encourage farmers to use systemic preventative measures. Organically allowable products that are not toxic to humans, animals and non-target species, not polluting and not synthesized are allowed but only 'as needed' an uncharacteristic response to unusual events (Standards Australia 2009, 4.5.2). They encourage farmers to work towards less reliance on external inputs and develop systemic management strategies to prevent pests and disease.

Some organic growers such as writer Jackie French and Hardy Vogtmann of the International Federation of Organic Agricultural Movements (IFOAM) criticize the input substitution approach to organic farming in Australia (Vogtmann 2005;

French 2006). Both insist that organic farming should be about creating systems of production that prevent pests and diseases through fertile soil, encouraging beneficial fauna and working with natural processes rather than replacing one remedy with another. This argument echoes the original soil fertility approach of organic growers in the 1940s. Although manufactured organic inputs are made of plant and rock extracts and other naturally occurring products they are manufactured using industrial processes and stretch the definition of 'natural'. The toxicity of a product such as Dipel or powdered sulfur is limited to the target species, but their use challenges the definition of 'poison'. The debate surrounding this approach to organic farming has revealed conflicting philosophies between an emphasis on systemic or ecological preventions and remedy based solutions, which sometimes sit uncomfortably within the broader ecological philosophy of organic growing.

11.3.3 Biodiversity and Ecological Wellbeing

Matthew Jamieson, NSW organic turkey and macadamia farmer interviewed by the author describes himself as 'The Farmer Giles of Ham' (Jamieson 2006). *The Farmer Giles of Ham* is a short story written by J.R.R. Tolkien about a genial farmer who battles dragons (Tolkien 1974). After studying science and entomology, Matthew worked at the Environment Centre in Darwin and became involved in campaigning against forest clearance, the destruction of rivers and the rights of West Timorese people in Indonesia. For Matthew, organic farming is an extension of environmental activism – another way for him to slay the dragons. "I feel that the world needs to change. I always felt that producing food with sustainable farming was the right thing to be doing. I wanted to produce something sound that doesn't have a negative effect on the planet" (Jamieson 2006).

The importance of ecological wellbeing and the interdependence of human wellbeing upon the wellbeing of the whole environment remain core principles of Australian organic farming today. The maintenance of ecological systems through organic agriculture is identified by IFOAM, the international peak body for organic grower organizations, as one of the key principles of organic farming: "Organic agriculture should be based on living ecological systems and cycles, work with them, emulate them and help sustain them" (IFOAM 2009). Anthony Sheldon explains the importance of working within the limits of his semi-arid environment:

> The challenge we are facing, not only in Australia but everywhere, is to live within the limits of the landscape. If we do that we'll be here for a long time and a good time. I don't believe in short term mentality. People who think like that will exhaust their resources and by the time they are forty they'll be standing there on the bones of their arse looking at another forty years. And then there won't be any good time, just a long time (Sheldon 2006).

Ecological principles have been enshrined in Australian standards for organic production. The Australian Standards state: "the aim is to enhance biodiversity on farm enabling 'eco-system services' to enhance the productivity and sustainability of organic farming operations and the surrounding environment" (Standards Australia 2009, 47.2).

Organic societies of the 1950s encouraged organic growers to protect and enhance on-farm native flora, fauna and habitats as a way of co-operating with native ecosystems. Today the national standards make it mandatory for certified organic farmers to protect primary native ecosystems on their properties such as forest, scrub, wetlands and native pastures. They are required to maintain and increase these ecosystems by setting aside at least 5 % of their property as refuges for indigenous flora and fauna and native habitat ecosystems (Standards Australia 2009). On his 40 ha property, Matthew Jamieson has reserved 3 ha of indigenous subtropical rainforest which is approximately 8 % of the total farm area (Jamieson 2006). Anthony Sheldon, in the semi-arid Mallee has fenced off and planted 250,000 native species of plants. This now covers about 250 ha which is a little less than one quarter of his farm. Although initially he was unsure about loosing this amount of productive land he recognized these areas acted as windbreaks, reduced sand drift and provided habitat for beneficial native birds, insects and other flora and fauna (Sheldon 2006).

As drought, climate change and resource conservation have become poignant issues in Australia during the 2000s, organic growers have promoted organic methods of food production as part of broader environmental solutions to ecological damage. In this context human health is defined in the broadest sense; not so much in terms of particular diseases but as a matter of human and planetary survival. During the 2000s, organic growers have begun to see themselves as having a central role in mitigating global warming. Organic organizations such the Biological Farmers of Australia, the Rodale Institute in the United States and IFOAM, as well as scientists such as Tim Flannery now speak about organic farming as a strategy for combating climate change. Organic farming is discussed as a strategy for reducing the amount of carbon dioxide, one of the most problematic greenhouse gases contributing to global warming (The Rodale Institute 2003; Kotschi and Müller-Sämann 2004; Australia 2006; Flannery 2006).

Carbon is one of the natural constituents of soil, being contained in the organic matter component of the soil. The absorption of carbon by the soil is part of the continual cycling of carbon between air, vegetation and animals in the process of growth and decay. Soil naturally contains more carbon than both vegetation and the atmosphere (Flannery 2006). Organic practices such as composting, mulching, cover crops, manuring and recycling crop wastes such as stubble all increase the organic matter in the soil, and therefore also raise carbon levels in soil, creating a long-lived secure store of carbon. By increasing soil carbon that comes largely from carbon dioxide gas, atmospheric carbon is reduced. The amount of carbon sequestered in this way varies according to soil type. Soils in arid agricultural areas, such as inland Australia are able to store up to 150 kg of carbon per hectare while soils in moist cool or humid areas such as temperate, coastal southern Australia and tropical northern Australia are able to store up to 1,000 kg per hectare. Therefore regular organic agricultural food production methods have the potential to offset global carbon dioxide emissions by an estimated 5 and 15 % (Lal 2004).

Organic organizations also argue that organic methods are a solution to high-energy consumption that also contributes to climate change. Organic growing techniques are more energy efficient methods of food production. They require less fossil fuel and contribute less to global warming and resource depletion. Large amounts of energy are required to manufacture artificial fertilizers therefore organic farming techniques, which do not use manufactured fertilizers, are a less fossil fuel intensive method of agriculture. Organic farming systems also generate fewer greenhouse gases, such as nitrous oxide, produced by soluble nitrogenous fertilizers used in conventional farming (Lal 2004).

Therefore, the importance of humus-rich soil to human health is now as much a question of ecological wellbeing as it is to the production of nutritious food. Although some organic growers remain uncertain of the nutritional superiority of food grown in humus-rich soil compared to conventional produce, the contribution of organic soils to improved human health remains secure by way of its contribution to ecological wellbeing and combating environmental damage through global warming.

11.4 Conclusion

Returning to the question posed at the beginning of this chapter: are the key principles of Australian organic growing still relevant for organic farmers today? Researchers of current organic farming who return only to the recent past of the 1970s and 1980s as the source of organic principles would inevitably conclude that the ideological basis for organic growing today has deviated from its original principles. Australian organic farmers now grapple with many issues not faced in the mid-twentieth century, such as genetic modification, climate change, a plethora of commercially manufactured organic products, certification and the existence of organic aquaculture, issues which organic growers of the 1940s and 1950s would not have recognized. Common organic practices from the 1940s, such as composting human sewerage is no longer part of organic agricultural practice and commercial organic production is now controlled by standards and, as a result, can command a premium for their produce.

However, it is not possible to understand organic agriculture in Australia today without examining the emergence of its key principles. Comparing the defining principles of organic growing in the 1940s and 1950s with Australian organic growing today reveals that the original key principles remain fundamental to organic agriculture today. Australian organic farming continues to be characterized by the production of humus-rich fertile soil, chemical free production and the maintenance of ecological wellbeing. The explanation of the term 'organic agriculture' in

NASAA's constitution illustrates the integration and endurance of these three principles into the definition of organic:

> Organic agriculture means a system of agriculture able to balance productivity with low vulnerability to problems such as pest infestation and environmental degradation while maintaining the quality of land for future generations. In practice this involves a system which avoids or largely excludes the use of synthetically compounded fertilizers, pesticides, growth regulators, livestock feed additives and other harmful or potentially harmful substances. It includes the use of technologies such as crop rotations, mechanical cultivation and biological pest control; and such material as legumes, crop residues, animal manures, green manures, other organic wastes and mineral bearing rocks. The intention is to encourage natural biological systems (National Association for Sustainable Agriculture Australia 2004).

Therefore, would an organic grower from the founding of the Victorian organic societies in the mid twentieth century recognize Australian organic farming and gardening today? Yes. They may be bemused by genetic modification, certification, industrialization and export but would note that organic agriculture continues to rest on the same belief that human health and wellbeing depend on the wellbeing of the biophysical environment, an idea which remains as relevant in the twenty-first century s as it did at the founding of Australia's first organic societies in the 1940s.

References

Australian Organic Farming and Gardening Society. 1947. Organic farming digest. *Organic Farming Digest* 1(7): 20.

Australian Organic Farming and Gardening Society. 1950. Farm and garden digest. *Farm and Garden Digest* 2(4): 3.

Australian Organic Farming and Gardening Society. 1951. Farm and garden digest. *Farm and Garden Digest* 2(10): 3.

Australian Organic Farming and Gardening Society. 1952. Farm and garden digest. *Farm and Garden Digest* 3(1): 39.

Barr, N., and J. Cary. 1992. *Greening a brown land: The Australian search for sustainable land use*. Melbourne: Macmillan.

Bio – Dynamic Research Institute. 2005. *Demeter bio – Dynamic agriculture in Australia*. Powelltown: Bio – Dynamic Research Institute.

Biological Farmers of Australia. 2006. Organics, one of our most viable solutions to climate change. *The Organic Advantage* 2–3.

Buck, D., C. Getz, and J. Guthman. 1997. From farm to table: The organic vegetable commodity chain of Northern California. *Sociologia Ruralis* 37(1): 3–20.

Campbell, H., and R. Liepins. 2001. Naming organics: Understanding organic standards in New Zealand as a discursive field. *Sociologia Ruralis* 41(1): 22–39.

Carson, R. 1963. *Silent Spring*. London: Hamish Hamilton: 6.

Chappel, D. 2006. [Interview].

Charman, P.E., and B.W. Murphy (eds.). 2000. *Soils: Their properties and management*. Oxford/Melbourne: Oxford University Press.

Conford, P. 2001. *The origins of the organic movement*, vol. 15. Edinburgh: Floris Books.

Coombes, B., and H. Campbell. 1998. Dependent reproduction of alternative modes of agriculture: Organic farming in New Zealand. *Sociologia Ruralis* 38: 127–145.

DeLind, L.B. 2000. Transforming organic agriculture into industrial organic products: Reconsidering national organic standards. *Human Organization* 59(2): 198–208.
Doubleday Organic Research Association. 1977. *Newsletter* 2.
Douglas, B. 2006. [Interview].
Flannery, T.F. 2006. *The weather makers: How man is changing the climate and what it means for life on earth*. New York: Grove.
French, J. 2006. [Interview].
Guthman, J. 1998. Regulating meaning, appropriating nature: The codification of California organic agriculture. *Antipode* 30(2): 135–154.
Halpin, D. 2004. *The Australian organic industry: A profile*. Canberra: Australian Government Department of Agriculture, Fisheries and Forestry. Canberra: Australia
Howard, A. 1943. *An agricultural testament*. Oxford: Oxford University Press.
Howard, A. 1945. *Farming and gardening for health or disease*. London: Faber and Faber.
IFOAM. 2009. *The IFOAM norms for organic production and processing – Version 2005*. Bonn: International Forum for Organic Agriculture Movements: 2.
Jamieson, M. 2006. [Interview].
Jones, R. 2010. *Green harvest: A history of organic farming and gardening in Australia*. Melbourne: CSIRO Publishing.
Jones, R., and J. Chesters. 2006. Muck, bugs and decay: The preoccupations of early Australian organic gardening. *Studies in Australian Garden History* 2: 63–79.
Jordan, S., H. Shuji, and R. Izawa. 2006. Conventionalization in the Australian organic industry: A case study of the Darling Downs Region. In *Sociological perspectives of organic agriculture: From pioneer to policy*, ed. G.C. Holt and M. Reed, 142–156. Wallingford/Cambridge, MA: CAB International.
Kaltoft, P. 1999. Values about nature in organic farming practice and knowledge. *Sociologia Ruralis* 39(1): 39–53.
Keating, J. 1992. *The drought walked through: A history of water shortage in Victoria*. Melbourne: Department of Water Resources Victoria: 123–151.
Köpke, U. 2005. *Organic agriculture: Maintaining old principles in a changing environment*. Paper presented at the Shaping Sustainable Systems; 15th IFOAM Organic World Congress, Adelaide: Australia.
Kotschi, J., and K. Müller-Sämann. 2004. *The role of organic agriculture in mitigating climate change*. Bonn: International Federation of Organic Agriculture Movements (IFOAM).
Lal, R. 2004. Soil carbon sequestration impacts on global climate change and food security. *Science* 304(5677): 1623–1627.
Leopold, A. 1949. *A sand county almanac*. New York: Oxford University Press.
Living Soil Association of Tasmania. 1946. *Compost, why and how*. Sydney: Living Soil Association of Tasmania.
Lockie, S., and D. Halpin. 2005. The 'conventionalisation' thesis reconsidered: Structural and ideological transformation of Australian organic agriculture. *Sociologia Ruralis* 45(4): 284–307.
Lund, V., S. Hemlin, and W. Lockeretz. 2002. Organic livestock production as viewed by Swedish farmers and organic initiators. *Agriculture and Human Values* 19(3): 255–268.
Lynggaard, K.S. 2001. The farmer within an institutional environment. Comparing Danish and Belgian organic farming. *Sociologia Ruralis* 41(1): 85–111.
Lyons, K. 1999. Corporate environmentalism and organic agriculture in Australia: The case of Uncle Tobys. *Rural Sociology* 64(2): 251–265.
Lyons, K. 2001. From sandals to suits: Green consumers and the institutionalisation of organics. In *Consuming Foods, Sustaining Environments*, ed. S. Lockie and B. Pritchard, 82–93. Brisbane: Australian Academic Press.
Mansfield, B. 2004. Organic views of nature: The debate over organic certification for aquatic animals. *Sociologia Ruralis* 44(2): 216–232.

McMichael, A.J. 2001a. Health and disease: An ecological perspective. In *Human frontiers, environments and disease: Past patterns, uncertain futures*, ed. A.J. McMichael, 318–340. Cambridge: Cambridge University Press.
McMichael, A.J. (ed.). 2001b. *Human frontiers, environments and disease: Past patterns, uncertain futures*. Cambridge: Cambridge University Press.
McNeill, D. 2006. [Interview].
McNeill, J.R., and V. Winiwarter. 2004. Breaking the sod: Humankind, history, and soil. *Science* 304(5677): 1627–1629.
Melucci, A. 1989. *Nomads of the present: Social movements and individual needs in contemporary society*. Philadelphia: Temple University Press.
Michelsen, J. 2001a. Organic farming in a regulatory perspective. The Danish case. *Sociologia Ruralis* 41(1): 62–84.
Michelsen, J. 2001b. Recent development and political acceptance of organic farming in Europe. *Sociologia Ruralis* 41(1): 3–20.
National Association for Sustainable Agriculture Australia. 2004. *Organic standard.* Stirling: National Association for Sustainable Agriculture Australia.
Pretty, J. 2001. The rapid emergence of genetic modification in world agriculture: Contested risks and benefits. *Environmental Conservation* 28(3): 248–262.
Sheldon, A. 2006. [Interview].
Skopp, A., and D. Skopp. 2006. [Interview].
Smith, B. 2008. The Smith family farming adventure. *Victorian Landcare and Catchment Management* 8–9.
Standards Australia. 2009. *Organic and biodynamic products,* vol. AS 6000–2009. Sydney: Standards Australia.
The Rodale Institute. 2003. *Organic agriculture yields new weapon against global warming*. Kutztown: The Rodale Institute.
Tolkien, J.R.R. 1974. *The Farmer Giles of Ham*. London: Allen & Unwin.
Tovey, H. 1997. Food, environmentalism and rural sociology: On the organic farming movement in Ireland. *Sociologia Ruralis* 37(1): 21–37.
Victorian Compost Society. 1948. Victorian compost news. *Victorian Compost News* 2(8): 70.
Victorian Compost Society. 1952. Victorian compost news. *Victorian Compost News* 6(3): 29–30.
Victorian Compost Society. 1954. Victorian compost news. *Victorian Compost News* 8(11): 1–50.
Victorian Compost Society. 1957. Victorian compost news. *Victorian Compost News* 11(5): 70.
Vogtmann, H. 2005. *Sustaining the planet: Where do we come from and where do we go?* Paper presented at the Shaping Sustainable Systems; 15th IFOAM Organic World Congress, Adelaide: Australia.
Vos, T. 2000. Visions of the middle landscape: Organic farming and the politics of nature. *Agriculture and Human Values* 17(3): 245–256.
Willer, H., and M. Yussefi. 2006. *The world of organic agriculture: Statistics and emerging trends 2009*. Bonn/Frick: IFOAM/FIBL.
Worster, D. 1994. *Nature's economy: A history of ecological ideas*. Melbourne: Cambridge University Press.
Wynen, E. 1992. *Conversion to organic agriculture in Australia: Problems and possibilities in the cereal-livestock industry*. Sydney: National Association for Sustainable Agriculture.
Young, A., and R. Young. 2001. *Soils in the Australian landscape*. Melbourne: Oxford University Press.

Chapter 12
Amish Dedication to Farming and Adoption of Organic Dairy Systems

Caroline Brock and Bradford Barham

12.1 Background on Amish Dairy Farm Adoption Decisions

Amish dairy producers have a solid and growing presence on the farm landscape, and account for over 1/8 of all U.S. dairy farms (Cross 2007). In recent decades, many Amish who have a desire to farm as a way to maintain religious and family values have emigrated from eastern urbanizing rural areas to states, such as Wisconsin, Michigan and Missouri, where non-Amish smaller-scale family farmers are leaving farming (Hostetler 1993; Cross 2004, 2007). Amish farmers are estimated to account for about 10 % of the more than 12,000 dairy farms in Wisconsin, 'America's Dairyland' (Cross 2007), and they frequently use barns that would otherwise be abandoned or torn down (Cross 2004). In fact, Wisconsin now has the second largest concentration of Amish church settlements in the U.S. (Luthy 2003). There are similar dense pockets of Amish dairy farms in other states that historically have had large numbers of small dairy herds (e.g. Iowa, Indiana, Ohio, Pennsylvania, Missouri and Kentucky) (Cross 2007).

Organic dairy farming has spread rapidly in Wisconsin, especially in the Southwestern region of the state, near to the headquarters of the nation's leading organic dairy cooperative, Organic Valley. Thirty percent of the organic dairy farmers in the state are in this region despite the fact that it only accounts for 10 % of the dairy farms, and this is also one of the denser areas of Amish farming settlements in the

C. Brock (✉)
Rural Sociology, University of Missouri, Gentry Hall #230, Columbia, MO 65211, USA
e-mail: brockcc@missouri.edu

B. Barham
Agricultural and Applied Economics, University of Wisconsin, 422 Taylor Hall, 427 Lorch Street, Madison, WI 53706, USA
e-mail: barham@mailplus.wisc.edu

B. Freyer, J. Bingen (eds.), *Re-Thinking Organic Food and Farming in a Changing World*, The International Library of Environmental, Agricultural and Food Ethics 22, DOI 10.1007/978-94-017-9190-8_12

state. Amish dairy farmers in the region may be only slightly more likely to be 'organic' as other dairy farmers in this particular region.

This might be surprising at first glance, as there is a common perception that the word "Amish" is synonymous with the word "organic" or "sustainable" due to their focus on simplicity and "natural" and "old–fashioned" style. More concretely, the Amish seem like a good fit for adopting organic dairy systems given that they use pasture as a source of feed, operate small-scale farms, and emphasize hand labor over certain kinds of machinery for reasons we explore below. However, many Amish farmers have also long since adopted "green revolution" technologies, such as chemical fertilizers and pesticides that are not permitted under organic standards.[1] Since these technologies are not addressed in their code of rules (*Ordnung*) and are not seen as a threat to their way of life, the Amish probably adopted these technologies in an earlier era for similar reasons as non-Amish farmers (i.e. labor saving, agronomic improvements, and/or economic and social pressures).

The complexity of organic dairy adoption decision-making within the Amish context is the focus of this chapter. Simple individual rational decision-making models are insufficient to understand organic dairy adoption decisions among the Amish. Instead, it is essential to explore their religious and community motivations for farming and for dairy farming in particular, and then with that background, we can examine the adoption decision of organic dairy among the Amish. To explore these questions in context, one needs to understand the particularities of their faith, church and social structure and not only how those features shape their values but also the internal and external constraints on information acquisition and processing they face when considering systematic changes in the way they run their farms.

There is also diversity in decision-making contexts across Amish communities. While they share common religious teachings and values, the establishment of local Ordinance or *Ordnung* can lead to significant differences in rules and norms and thereby create the basis for heterogeneous decisions in farming made by Amish families across communities. This diversity can be seen in comparing the adoption of organic dairy systems in two neighboring Old Order Amish communities in a region of Southwest Wisconsin known as the Driftless region.

Old Order Amish dairy family farms are defined by their Anabaptist Christian[2] religious and cultural identity. As one Wisconsin Amish farmer stated, "confess[ing]

[1]"Although the reasons are not entirely clear, by the 1970s, most Amish farmers were spreading chemical fertilizers and pesticides" (Kraybill et al. 2013, p. 286).

[2]Amish are an Anabaptist Christian religious group originating in Central and Western Europe. The Anabaptists split off from state churches during the Protestant Reformation period. One radical difference between them and other Christians was the issue of adult baptism, as Anabaptist or "to rebaptize" was originally a derogatory term that was adopted by these groups. Another generally common unifier between the Anabaptist groups is the commitment to peace through nonresistance ("turning the other cheek"). The Old Order Amish are generally considered to be the most culturally conservative of the Anabaptist groups (e.g. Mennonites, Hutterites, New Order Amish). We refer to Old Order Amish when discussing Amish throughout this paper. It is important for the Amish to remain separate from the world (Cor. 6:14) and not conform to it (Romans 12:2)

Christ with our farm" was the primary motive for farming. Many Amish farmers believe that Christian family values (e.g., humility, frugality, contentment, unity and a strong work ethic) are best fostered through a rural lifestyle, where spiritual, social, economic, and ecological values can be integrated in a framework we describe as *oikonomia*.

Oikonomia is an integrated approach to decision making that stems from the origins of the word economics *oikos* and translates as "household" (Meeks 1985; Young 1992). This household value-based dedication contributes to the Amish persistent presence in farming despite the many financial challenges. Amish values are also critical in the formation and maintenance of tight-knit communities of Amish families who support each other in daily work, family living, and religious worship. The Amish embody a connection between community, family and rural culture, and view the farm as the ideal place to raise children and live out their faith (Brock and Barham 2009; Brock 2010).

Old Order Amish settlements[3] share a commitment to support (and defend) Christian family values through locally developed rules and norms (*Ordnung*) as well as value-based sentiments. Many of these norms center on limiting the introduction of new activities, technologies or behaviors that they perceive may threaten core Christian family values. This fear of value loss illustrates one type of 'bounded rationality' behavior toward change and uncertainty that we argue below helps to explain Amish farming decisions. Thus, bounded rationality, in addition to *oikonomia*, is needed to illustrate how internal and external constraints may prevent farmers from acting in a manner consistent with neoclassical economic models of full-information, economic maximizing decisions.

No published literature to date compares farm decision-making about alternative agriculture practices across Amish settlements. This may be due in part to the logistical difficulties of working with Amish communities since they lack phone and email access and can appear closed off to outsiders. Successful fieldwork often depends upon building build trust with Amish leaders and families. The absence of literature may also be due to the complexity and subtleties involved with decisions like organic adoption. There is some research comparing sustainability outcomes of Amish farming practices (Stinner et al. 1989) with other types of

(Hostetler 1993; Dilly 1994). The Amish have chosen to respond to the tension of being Christ's example in a corrupt world with a removed lifestyle approach. This removed lifestyle explains some of their dress styles and use of technology that are so distinctive from mainstream Americans. It may also explain their distance from politics and from government assistance including farm subsidies (Dilly 1994). The Amish farmers in this report are Old Order Amish and have more restrictive ordinances (or *Ordnung*) than many of the other Amish groups. The Old Order Amish in the Kickapoo region (the focus of this study) may be different on a number of characteristics than Old Order Amish groups. Some farming practices that are distinctive of the Amish in this region include farming with horses and milking by hand.

[3]Settlements are divided up into church districts of about 25 families, and church districts are governed by a bishop and a varying number of ministers and deacons. The church district bishops govern the Ordnung of the settlement as a whole. Each settlement tends to have commonalities in the individual church's *Ordnung*.

farmers (Sommers and Napier 1993; Jackson 1998), and a recent ethnographic assessment of Amish community lifestyle contrasts their approach to sustainability with other faith groups (Vonk 2011). One study discusses the use of contour plowing patterns, a soil conservation strategy, among Amish farmers in the Kickapoo Valley in Wisconsin (Heasley 2005), but it does not consider how local norms might vary across Amish communities and affect their adoption of these practices.

Our work also diverges from farm decision-making literature. Instead of focusing solely on household level decision-making, our study explores the values and constraints operating at the community, settlement and household levels for Amish farmers. Although there are no systematic studies on the prevalence of organic agriculture among Amish farmers, discussions in Amish newsletters reveal divergent views. In some areas of the country (e.g. Ohio), organic farming among the Amish is growing (Kraybill et al. 2013), while in other areas (e.g., Missouri), it is not widespread.

This chapter explores the adoption of organic dairy systems by Amish farmers in one region of Wisconsin at a time that saw rapid growth of organic farming among smaller pasture based farms. More specifically, this chapter examines similarities and differences across two neighboring Old Order Amish settlements in order to shed light on the nuances and complexities of decision-making processes in two Amish settlements. The research in this chapter is also a unique contribution to the literature as it embeds the findings on adoption decisions of Amish farmers in a larger study that includes non-Amish farmers (Brock 2010).

The combination of *oikonomia* and bounded rationality frameworks offers an approach that describe how values and internal and external constraints (particularly around information) combine to illuminate the complexities around adopting organic dairy and the subtle differences between different church settlements. Since values and bounded rationality issues are often pertinent in other contexts, this framework is also useful for studying organic adoption and other complex system decision making in other contexts outside of the Amish settlements.

As suggested above, the Amish motivations to farm, and their management decisions, are deeply linked to their own cultural and spiritual identity, as well as their ecological and economic viability on the land. The only way one can gain insights into the particularities of the Amish church and social structure, as well as specific *oikonomia* and bounded rationality issues, is to incorporate Anabaptist theology and gain insights from conversations with church leaders and Amish farmers. We started with a survey to gain a picture of the structure of dairy farms from two Old Order Amish settlements in Southwest Wisconsin, – Cashton and Hillsboro. Although these settlements are both very similar, subtle differences between these settlements have led the Hillsboro settlement to be more open to alternative practices such as organic agriculture. One needs to look at factors that impact decisions at the community, church settlement level, and then the family level to fully understand Amish decision-making. This is a fundamentally different approach from standard decision-making studies in economics in which the individual is the center of analysis. Though our study largely focuses on the factors that shape Amish farming decisions, we also consider lessons that may carry over to other types of family farmers where social, spiritual, economic and ecological factors can all combine to shape decisions.

12.2 Conceptual Overview *Oikonomia* and Bounded Rationality

The decision to practice organic dairy among the Amish is examined here using a conceptual framework that combines an integrated value-based approach to decision making called *oikonomia* and the information heuristics of a bounded rationality framework. *Oikonomia* and bounded rationality are important frameworks for this study because the Amish faith and church structure influences household and individual values as well as their exposure to information and how information is processed. *Oikonomia* and bounded rationality prove useful as frameworks for discussing and comparing complex decision-making because they are integrated, broad and flexible.

Oikonomia[4] derives from the Greek word for economics which here consists of four key realms: economic, social, spiritual and ecological. The terms "economics," "ecology," and "ecumenism"[5] overlap in common the root word *oikos* (Meeks 1985; Young 1992), or "household". Our broader usage of the term "farm household" draws from holistic farming systems choice literature in North America (Barlett 1980; Bennett 1982; Salamon 1992; Padel 2002; Lunneryd 2003; Bell 2004). Like ours, these studies look at broader motivations for farming and for farm decision-making. The Amish offer an extreme case of how farming system choices are not separable from broader *oikonomia* preferences at the community, settlement and household levels.

Farmers and other decision makers do not always base decisions solely on *oikonomia* values, because of bounded rationality issues and constraints on information access. The high cost of information gathering and processing particularly with

[4]This use of economic argumentation resembles that of Adam Smith's *Wealth of Nations* (Smith 1776) where economics was not a separate discipline as it is considered today but a branch of philosophy which was integrated into all human activity and assessments of overall well-being (Daly and Cobb 1989; Goudzwaard 2000). t is also important to point out that the term *oikonomia* has a holistic connotation that extends beyond the farm to consider long term values for the other households and the surrounding community (Daly and Cobb 1989), which can make farmers happier in their own lives (Berry 1997). It is the integrated essence of the "family farm" experience where work, consumption, leisure, and relationship to others, the environment, and spirituality all occur largely in the same place that seem to make for a large degree of "inseparability."

[5]"Ecumenism" is the management of a household's values, morals, and spiritual resources. Economy or Oikonomia in the Christian Biblical sense is the way the Creator manages His household and the way humans are called to steward the creation is mentioned several times in the New Testament (Worster 1994; Gottfried 1995; Goudzwaard 1997, 2000).

respect to the management of organic systems makes it a particularly good case for a bounded rationality framework approach, because it too provides a context that is likely to make information, time and cultural constraints binding on fully informed decisions.[6] Farmers, as decision makers, arguably often operate using a bounded rationality framework in order to negotiate the complexities of difficult decisions under uncertainty.

Both *oikonomia* values and bounded rationality factors can influence information gathering when making adoption decisions. For example, Amish farmer decision-making emphasizes working with your hands, which can be done with the family, and de-emphasizes knowledge acquisition beyond what is needed for daily farm tasks. "The wisdom of the world is foolishness with God" is a verse in Corinthians that characterizes a view that many Amish hold with respect to acquiring knowledge (Hostetler 1993). That is, the Amish may not maximize all of the resources and information that might be expected by assuming decision-making based on an idea of rational utility. In terms of knowledge acquisition, the Amish have been portrayed as focusing on practical experiences in localized settings (Kraybill and Olshan 1994). Value-based factors may thus influence how farmers seek or process information related to farming organically.

The *oikonomia*-bounded rationality framework helps us describe Amish decision-making with respect to farming in general and specifically concerning the adoption of organic practices. The precise ways in which the bounded rationality framework plays out may vary from one Old Order Amish settlement to the next, even though they face almost identical socio-economic contexts and choices. In discerning oikonomia-bounded rationality behaviors, we wrestle with the ontological challenge of discerning what the farmers know about farming systems given that we too as researchers operate under significant informational constraints about individual farm operations. Perhaps the best we can offer here is to demonstrate the range of bounded rationality behaviors that appears to guide their choices. There are various *oikonomia*-bounded rationality behaviors that describe how acquiring and processing information and attitudes toward risk and uncertainty intersect with fundamental values in decision-making processes within Amish communities and households.

We describe a continuum of behaviors that involve distinctive approaches to both information gathering and the criteria that guide decisions. At the positive extreme may be what is commonly called ambiguity aversion where information that can help to address uncertainty is difficult to obtain (Epstein 1999), and some agents may be averse to change as a result of this ambiguity or uncertainty. At the opposite end, "extremetization" (Brock 2010) can occur when a decision-maker anchors on a particular issue and makes broad generalizations out of extreme cases. Value loss

[6]Since Simon first coined bounded rationality as a concept, there has been much empirical evidence in support of this way of viewing decision making especially as compared to empirical evidence for unbounded rationality (Conlisk 1996; Gabaix et al. 2006).

aversion and anchoring fall in the middle in this continuum and all of these behaviors may lead to principled satisficing, a concept we introduce first.

12.2.1 Principled Satisficing

The concept of principled satisficing helps center the other *oikonomia*-bounded rationality concepts. For our analysis this means that the Amish, for value-based reasons, may be satisfied with less utility or information than they would be under the assumptions of rational utility maximization. "Satisficing" tends to be common when decision-makers face complex choices with limited information, because it assumes that decision makers can only build on their perception of the situation. The concept of satisficing acknowledges that individuals "satisfice," that is, they may select a 'satisfactory' solution to a complex problem, rather than maximizing profits or some other resource as is typical in unbounded rationality (Simon 1955). Instead, decision makers, "satisfice" or "optimize" with the information they know and understand (March and Simon 1958; Gasson and Errington 1993).[7]

Principled satisficing is a type of satisficing that is influenced by values. In this way, it serves as a hybrid of *oikonomia* and bounded rationality. Values influence decision makers and keep them from maximizing, as they would if they were engaged in profit-oriented rational decision-making (Chua and Juurikkala 2008). An emphasis on farming as a family activity illustrates principled satisficing as family farming in this context may not maximize income potential, but it is consistent with Amish contentment theology and the Christian principle of living simply. Here are the four bounded-rationality heuristics we offer along a continuum.

12.2.2 Ambiguity Aversion

This form of bounded rationality demonstrates the most active level of engagement with new information, and is based on the real-world constraints that can limit a decision-maker's access to full information about the uncertainty involved with making a decision (Epstein 1999). This ambiguity is even more likely for a major systems change, such as organic farming, and can mean that farmers have to transform their regular patterns, habits and routines in order to take on the new system. This change may require significant new learning and much continuing uncertainty. As Bell states, "farmers can't spend all day reading farming magazines, cruising the Internet, talking to sales people and having coffee with locals at the

[7] Satisficing may in part explain why decision-makers stop their thinking or information gathering early on, based on an attitude that continued searching "seems like a waste of time" (Mansourian and Nigel 2007, p. 686).

café in town" when they are making decisions about their farm (Bell 2004, p. 138). The tendency of farmers to avoid or be averse to the uncertainties associated with change may be one of the main reasons more do not practice organic agriculture (Fairweather 1998; Clark et al. 1999; de Buck 2001; Bell 2004).

12.2.3 Value Loss Aversion

Value Loss Aversion[8] is another *oikonomia*-bounded rationality term that can lead to principled satisficing and is a variant of the more well-known "loss aversion". Loss aversion dissuades decision makers from making significant changes because they are more concerned about losses on past investments or choices than the potential gains from future decisions, (Kahneman and Tversky 1979; Samuelson and Zeckhauser 1988; Kahneman et al. 1991) and this may lead to information satisficing. When applied to the Amish, the concept needs to be expanded to consider their attention to protecting their common values within the church and the community. At a community level, strong value loss aversion may translate into norms and rules, and hence individual behavior focused on the potential losses over possible or even likely gains, so that decision makers are more likely to maintain the status quo. This is especially relevant to the Amish farmers in our sample, who appear to often display what we call here "value loss aversion", wherein their hesitation about new activities, technologies, or even information is based on the sense that change inherently threatens core Christian family values relative to traditional behaviors in the vein of (Chua and Juurikkala 2008).

12.2.4 Value Anchoring/Extremetization

Anchoring is a way to drastically simplify complex problems so that the decision-maker does not fully consider the advantages and disadvantages of an entire decision; instead, they may focus on one aspect of the decision and weigh that outcome heavily (Tversky and Kahneman 1974; Gigerenzer and Goldstein 1996; Todd and Gigerenzer 2000; Lunneryd and Öhlmér 2006). The issues that decision makers focus on which lead to anchoring and "extremetization" may be closely connected with values, hence the term "value anchoring" which is inspired by the other oikonomia-bounded rationality issues illustrated earlier inspired by Chua and Juurikkala (2008). Some anchors are so extreme that they do not seem to be well grounded economic or ecological indicators of the core decision, and this phenomenon will be denoted as "extremetization" (Brock 2010). For example, the idea that organic farmers may be cheating on their practices may cause

[8]See (Kahneman and Tversky 1979; Bandiera and Rasul 2006).

conventional farmers to not consider organic with an open mind. Anchoring and "extremetization" can both lead to satisficing behaviors.

12.3 Field Methods

To study these complex decision making-processes within a blended *oikonomia* and bounded rationality framework, we conducted semi-structured interviews. The interviews focus on key questions about why the Amish farm and why they farm the way they do particularly focusing on the decision to adopt organic dairy. *Oikonomia*-bounded rationality plays out differently at the community, settlement and household levels. The subtle and complex nature of the factors influencing organic adoption decisions required detailed interviews with church leaders and farmers after descriptive surveys were executed. Bounded rationality concepts emerged from the farmer interviews following a grounded theory approach where the theory is informed by the empirical data. Following the interviews, the data was organized according to *oikonomia* and bounded rationality themes.

Southwest Wisconsin was selected for several reasons. About a quarter of the state's Amish dairies are located in Southwest Wisconsin (Cross 2004). And, as mentioned above, this region is an ideal location to study organic adoption decisions since it has over 30 % of the state's organic dairy farms (Brock and Barham 2009) despite accounting for only roughly 10 % of the state's dairy farms.

Farm decision-making was studied among Amish and other local dairy farmers based on interviews with 25 Amish farmers[9] as part of a larger sample which included organic, conventional and managed grazing farmers outside of the Amish community (Brock 2010). The sample was selected semi-randomly but was modified to ensure diversity in farming approaches. Amish farmers from the two largest Amish settlements in the Kickapoo Valley, Hillsboro and Cashton[10] were interviewed in person, for 40–90 min. There was also a concerted effort to get a mix of organic and non-organic dairy farmers in the sample. Some key informants from the Old Country Cheese Amish Co-op, who were not associated with the survey, were also interviewed.

Many of the farmer interviews were arranged through stopping by the farmstead un-announced and finding a suitable time to come back for an interview. The majority of interviews were conducted between August of 2006 and June of 2008 by

[9]The interview sample was mainly drawn from 100 respondents to a 2004 mail-based survey which was a full sample survey of Amish farmers from several settlements who sell milk to, Old Country Cheese, an Amish cheese cooperative based in Cashton, WI.

[10]Roughly one-half of the Amish sample was selected from the Cashton settlement, and one-half was selected from the Hillsboro settlement.

the primary researcher.[11] Letters were sent to the farmers with response postcards enclosed so that a meeting time could be arranged since the Amish do not have phone access. Notes were typed up shortly after each interview, as they were not recorded. Interview data, Amish newsletters and other literature were used to discern motivations to farm and farm management decisions especially related to organic farming among the Hillsboro and Cashton settlements.

12.4 *Oikonomia*-Bounded Rationality Reasons for Farming Among the Amish

A critical take-off point of our empirical analysis is a discussion of the complex reasons why the Amish farm, and specifically why they favor dairy farming. Their deep commitment to dairy farming connects not only to family and to how one raises children, but also to what holds the church community together and nourishes their faith. In other words, the interplay of family, labor, history, and faith is crucial to understanding the motivations of Amish farming and their farm decision-making.

The Amish in this area are farming because they believe the farm is a good place to raise a family (98 % of respondents); it is a lifestyle consistent with their faith (97 % of respondents), they enjoy working outdoors, and they enjoy the independence (93 % of respondents). Only about one-third of the Amish dairy farmers (32 % of respondents) felt that the ability to earn a good income was the reason that they farm (Brock et al. 2006). These motivations illustrate clearly why an integrated value based approach is critical to study farming and farm decision-making.

In particular, the Amish believe that farming makes family and community more reliant on God. This idea is well expressed by an interviewee who said, "there is a virtue of working with the soil and remembering the Creator." As an Amish elder cites from scripture, you need to rely on "the rain and the snow come down from heaven"[12] when you are farming. Another Amish farmer loosely refers to the scripture teaching us not worry about everyday life because God provides for our needs, and he goes onto to say that the farming operation forces him to rely on God's control.[13] God's role in the production process is seen in very tangible ways as another Amish responds to compliments about their flower bed with the statement, "The Lord makes them grow, we just try to do our part." As an entry in an Amish newsletter states, "[Although] Farming is not a top paying job [. . .] Isn't farming still the most important and best for the family?" (Anonymous 2006). Many of these

[11]However, there were a few interviews with Amish elders and other informants that were co-conducted with Dr. Dail Murray, a professor from UW-Marinette who was studying other socio-anthropological topics related to Amish in the Kickapoo.

[12]Isaiah 55: 10.

[13]Matthew (6:25 and 6:32).

Amish individuals are committed to farming despite low levels of satisfaction with farm income because of the connection they feel with God, family and community.

The Amish illustrate principled satisficing as they are more satisfied with their net farm income than either conventional or grazier farmers (30 % of the Amish were somewhat to very satisfied compared to 5 % of conventional farmers and 6 % of grazier farmers (Brock and Barham 2009). While these comparisons only offer *prima facie* evidence, consider that even though the Amish produce Grade B milk and receive a lower price for milk than non-Amish fluid milk producers, the Amish still have a higher level of satisfaction with milk prices. One Hillsboro Amish expresses this extreme sense of contentment: "We have [. . .] a good living here. We have enough to eat." It would be difficult to imagine a non-Amish farmer expressing that having enough to eat is a good living. This example illustrates how Amish may require less to be "satisfied." Amish contentment theology emphasizes simplicity and consumption of only basic goods, such as food, housing, and clothing, and this may be the key to their survival on the land despite milking a small number of cows and getting much lower prices for their milk (Hostetler 1993).

Rather than focusing on labor that maximizes income, the Amish pursue work activities that bind together the family unit and the larger community, and thus sustain the interaction of the social, spiritual and the economic spheres of the *oikonomia* framework. Indeed, contrary to most labor-leisure household economics model assumptions, farm labor is widely viewed by Amish as a benefit rather than a cost because it binds the family and church community together. Moreover, the Amish see hand labor as a calling described in Genesis as they state they want to do work "by the sweat of our brow". Indeed, the Amish view dairy farming in particular as the most desirable farm enterprise because of its significant and constant labor demands. The fact that 80 % of Amish farmers in Wisconsin have dairy herds, compared to 22 % of farms across the state, exemplifies this preference (Cross 2007). One Amish elder goes so far as to say, "Those who have a farm but don't milk cows scare me [. . .] milking ties you down." Based on the commitment required, this elder seems to consider being "tied down" as a benefit not a cost.

By contrast, vegetable and crop farming only require intensive labor effort during the growing season whereas dairy farming operations have continuous labor demands for milking, feeding, and tending the animals, at least twice a day throughout the year. The Amish exemplify these *oikonomia* dimensions of farming as a lifestyle, although it is important to mention that family labor and the values that a farm upbringing instills in children are also an important focus for many types of non-Amish dairy farmers. Indeed, surveys of non-Amish dairy farmers in the area reveal that family and the associated rural lifestyle ranks as the top reason for farming for them as well (Brock 2010).

Amish dedication to farming may also rest on value loss aversion rationale as evidenced by the argument made in their national newsletters that farming is necessary to maintain a Christian/Amish faith life. In addition to the strong *oikonomia* values that underpin the Cashton and Hillsboro Amish settlements' dedication to farming, there is also uncertainty about the potential negative ramifications of other forms of work on their families, church, and community. According to some members of the

Cashton and Hillsboro Amish, other settlements where day labor instead of farming has become the predominant way of generating income have made sacrifices in terms of their community and spiritual well-being in ways that are not totally known or understood. Some of the Amish fear that disposable income will cause their children to leave the church (Kraybill and Nolt 2004). When asked what is the biggest obstacle for the Amish people, an elder responded by stating, "The devil. He's very busy [. . .] In prosperity; the devil gets a lot of people. When prosperity goes too far, it is dangerous- we should make a living but not more." An avoidance of this uncertainty plays a major role in their choice to dairy farm and to remain dedicated to it despite financial stresses. This uncertainty seems to be rooted in bounded rationality behaviors and may help us understand why farming is slower to change than other sectors (Kraybill and Nolt 2004) and why there are divergent views amongst the Amish about organic farming.

The Amish in Southwest Wisconsin remain dedicated to farming despite economic pressures because of the social/spiritual value of farm labor. Farm input costs have risen, and are more volatile, and milk prices have had several low periods especially for the Grade B milk that the Amish produce. In response, there has been some shift away from relying on dairy income among the Cashton and Hillsboro Amish, but they continue to keep at least part of their household connected to farming. As one Cashton elder states, when they first settled here in the 1960s dairy farming was about 85 % of the Amish family income and 15 % came from non-dairy income, and now it may be the reverse, 85 % non-dairy and 15 % dairy. It seemed evident from interviews and Amish farm newsletters that the prices of farm products were often not enough to cover their costs of production. Many of the Cashton and Hillsboro Amish say that they would rather be full time farming, and would prefer to have more dairy cows if they could make it economically viable. In fact, the Amish maintain very small herd sizes (around 13 cows per family). These herd sizes are generally smaller than they were for the Amish a generation ago, and are far smaller than the average Wisconsin farm (which is now close to 100 cows).

The decrease in herd sizes among the Amish may be due to their inability to make a substantial income from the dairy operation. One farmer reported that he had "been squeaking by for a long time," and these kinds of comments were common amongst Amish farmers. Another farmer stated, "I gotta' go to the saw mill to support my cows." In fact, in both the Cashton and Hillsboro settlements, Amish families operate a variety of cottage industries on their farms which include saw mills, woodworking, bakeries, quilt fabricating, and processing of food products, all of which are used to secure additional income for the household.

In summary, the Amish clearly are dedicated to farming and particularly dairy farming for financial reasons. And unlike most dairy farmers in Wisconsin, their income from the dairy is not their primary source of income. In fact, their continued dedication to farming seems to run against the principles of rational decision making which makes an *oikonomia*-bounded rationality framework more useful for studying decisions around farming than a standard, income and leisure maximizing neoclassical framework.

12.5 Amish Church Structure and Adoption Decisions: Overview

This section explores how the Amish church structure affects management decision-making; this lays the groundwork for reviewing the complex subtleties of the organic dairy decision in the next section. Church rules and the non-hierarchical nature of the Amish church as a whole and the resulting diversity between settlements is an important factor in explaining diverse farm management decision making across Amish communities. Although the Amish are by no means static in terms of technology adoption, "changes are clearly not encouraged, vocalized or rewarded in Amish society" (Hostetler 1993, p. 302). When major changes occur in Amish technology choice, they typically focus on issues of survival. The decrease in the herd size discussed earlier illustrates choices based on survival. Change can also occur because of diversity in how the rules are enforced between different districts in a settlement. Sometimes practices and technologies just get adopted, and they never really become a serious issue because they are not deemed a public threat. Although community norms that shape their lifestyle, such as simplicity in consumption standards, have contributed to the Amish economic viability on the land, community norms on management practices may also make it more difficult for them to compete on the modern farm landscape. It seems that Amish technology and management adoption choices create tensions between maintaining values and the need for economic survival on the land.

The *Ordnung* (Ordinance), or church rules, are the social glue and guiding principle of the Amish community and tend to change very slowly over the years (Kraybill and Nolt 2004). These rules unify the church body and they evoke a value-loss aversion principle when it is not clear how and to what extent the values will be compromised. Some technologies and management practices are avoided as a precautionary measure that combines *oikonomia*-based values and bounded rationality. The *Ordnung* reflects the tension between adopting useful ways of the world (e.g. technologies and practices) in the midst of economic challenges and establishing guidelines for navigating economic sustenance and shared values. One Cashton Amish elder states: "If we don't have an Ordinance to live by, we can't move forward as a community. What we have in common holds us together [. . .] we are trying not to slip like other communities. It is made for the safety of the people like rules of the roads." Any decision that may have a connection to common values in the midst of uncertainty emphasizes safety and caution. It is all about a slippery slope or a sliding board as some Amish may describe it. They fear that giving an inch, will give up a mile.

The *oikonomia*-bounded rationality nature of the *Ordnung* is also evident because it appears that the Amish elders do not discuss the *Ordnung* in a very systematic way during their meetings. Such discussion instead is left for the bi-annual *Ordnung* review and is used as a means to help create unity before taking Communion. Different ideas even if they do not directly challenge the *Ordnung* are suspect. There is much more emphasis on unity *per se* than there is on remaining static or looking forward with respect to technology and management systems. From time to time, the church leaders, with the endorsement of their congregations, will revise an aspect of the *Ordnung* to adapt to changes in the community environment. However, once "an understanding" becomes inscribed in the *Ordnung*, it is very challenging to revise it, at least for several years.

Cultural differences between settlements are helpful for understanding adoption decisions. The decentralized nature of Amish society facilitates subtle change and divergence between different settlements in terms of adoption decisions (Hostetler 1993). There is a different *Ordnung* for each self-governed settlement, and so the tension between adopting useful ways of the world and maintaining Amish values can play out differently in each settlement. Between two Old Order settlements Hillsboro and Cashton, there are key differences even though they share the same core values as one Cashton Amish stated about the Hillsboro Amish, "They are totally different people but not that much different—they have some of the same values about the future of the children."

They both moved to Wisconsin, in part, to maintain a more conservative lifestyle from the pressures of urbanization in Indiana and Ohio. They both milk by hand and do not utilize a bulk tank. However, one Hillsboro Amish states the following about Cashton Amish, "There is a different blood pumped through them, they are more conservative in some ways." One Cashton Amish elder explained that the differences between the settlements would be something that "you [as an outsider] might not understand." For example, the Cashton Amish do not observe daylight savings time and stay with what they call "slow" time whereas the Hillsboro Amish do observe daylight savings time. The Cashton Amish could be considered more conservative and removed from society in this respect but they are also less reliant on farming than the Hillsboro Amish, which could be considered more liberal. Thus, the comparison between these two settlements is not simple and more important, the reasoning behind the differences are subtle and cannot be easily understood by an outsider.

12.6 Organic Management Adoption Decisions

The previous sections discussed key *oikonomia*-bounded rationality principles, or reasons for dairy farming and essential parts of connections between Amish church structure and management and technology adoption decisions, that lay the groundwork for discussing organic dairy adoption decisions. Considering the emphasis on family labor, and the labor-intensive methods of organic farming that could bring the family together, the organic approach might appear to be a straight-forward choice for Amish farm families. However, the *Ordnung* described earlier does not offer a clear stance on organic practices. In fact, the *Ordnung* as a general rule, addresses modern practices that are more visually perceived or mechanized (e.g., the tractor) than less visible and manual management practices and technologies, such as chemical fertilizers and hybrid seeds (Kraybill and Olshan 1994).

While Amish farms may be well suited to organic production, the percent of Amish dairy producers who use certified organic practices is small.[14] Because the decisions governing stewardship and practices like organic do not necessary address overarching Amish values directly, the decisions tend to differ by settlement even for Amish settlements that have a lot of commonalities, such as Cashton and Hillsboro. Overall, a higher percentage of Hillsboro Amish (15 %) produce organic dairy compared to Cashton producers (8 %). The Cashton Amish who do not practice and market organically tend to have more extreme bounded rationality views about organics.

This section begins with a summary of overarching Amish views on organic methods. The core assessment of Amish organic adoption decisions is made in the context of the local community, individual settlements, and individual households and then briefly compared to non-Amish producers. This community – settlement approach among the Amish is distinct from the standard approach to adoption decision-making where individual farm households are frequently the beginning and end of the analysis.

Although there are no explicit edicts about organic agriculture in the *Ordnung* for Amish settlements, there is an overarching sentiment about stewardship and what is natural. One Hillsboro Amish expressed it this way, "I feel that the Christian God made the earth with laws of the land, and if you go against those laws, there are going to be problems." As one Cashton Elder adds, "It's always been our underlying thought that being stewards of the earth is not being destructive or detrimental." Although there is a commitment to follow God's rules in terms of what is "natural," this mainly translates to Amish settlements merely abstaining from the growth hormone, rBST and genetically modified organisms. As a writer in an Amish newsletter stated about genetic engineering, "I could be wrong, but to me that conflicts with the laws of God" (King 2006 [letter to the editor]). However, there may be quite a few Amish farmers who buy conventional seeds not knowing whether the seed is GMO or not (Brock 2010). This idea of what is natural or not, does not translate to a clear stance among Amish communities on organic methods overall.

12.7 Organic Decision-Making: The Community Level

One needs to know something about the management and history of Old Country Cheese co-op if one is going to understand organic adoption decisions of Cashton and Hillsboro Amish farmers. Old Country Cheese was started in 1982 by a group

[14] "In 1997, Ohio had no certified organic dairies but in a few years there were over one hundred, and ninety percent of them were Amish and Old Order Mennonite" (Kraybill et al. 2013). Across the United States, the organic option seems to be playing an important role in Amish agriculture in some parts of the country as noted earlier. Greenfield Farms (http://www.gffarms.com/) in Ohio is one example (Kraybill 2001; Mariola and McConnell 2013). Another example, closer to the Wisconsin settlements in this paper, is Kalona Organics in Iowa (http://www.kalonaorganics.com/who_we_are.html).

of Cashton Amish committed to having a market for canned milk, non-refrigerated milk storage on Amish farms.[15] Their intention was to ensure a future in dairy farming for themselves and future generations of Amish by marketing their own cheese from the canned milk.

Old Country Cheese continues to be the only marketing outlet for canned milk as all other milk buyers in the area require farmers to acquire a bulk tank. Currently, a number of other Amish settlements sell their milk to Old Country Cheese. The Cashton Amish established the board structure, continue to manage it, and have control over major decisions including decisions around establishing and maintaining an organic line. The following discussion focuses on the views of a few Cashton Amish elders in particular because they in effect have the marketing decision making power over all of the other Amish dairy farmers (i.e. in Hillsboro and other settlements) who sell their milk to Old Country cheese.

At the time of the study, Old Country Cheese Co-op was marketing some of its cheese organically, but they had not kept a consistent commitment to the organic line in previous years. This lack of consistency is partly due to the negative attitudes about organic among some of the Cashton Amish elders who have influence on the board. This lack of consistency is a key point because it means that some Amish farmers could go through the expense of converting to organic, a 3 year certification process, but not have a market for their milk since Old Country Cheese is the only marketing outlet for organic canned milk. The Old Country Cheese co-op was involved with the organic market two times in the past 10 years -once in 2004 and once since 2006. It is important to understand the decision making process of key leaders on the board as it impacts the ability of the whole community to market organically.

For some of the key leaders, organic dairy was viewed as a potential threat to some key Amish values so value loss aversion was at play in various ways. For example, the Amish have an idea of unity or "being one body" in a scriptural sense. One of their concerns with organic was that some members of the community may be receiving higher pay prices for organics than non-adopters, and this would then translate to not everyone being on an equal plane. If some of the farmers became organic, and prospered, this could challenge the importance of humility and equality which are core values for the Amish people. Another example of a threat to unity was the idea that organic certification created a system where neighbors were supposed to report others who were cheating.

Another value within many Amish communities is to minimize interactions with the government. This value anchoring can be explained by negative associations with government affiliation, which is in turn related to the Amish history of persecution and desire to remain autonomous from political affairs. Their alternative goal is to focus their allegiance with God and their community. One Amish elder was falsely convinced that the Organic Valley cooperative was able to expand into

[15]Some other settlements in eastern states adopted bulk tanks as early as the 1969 (Kraybill and Olshan 1994).

an expensive large new building because of a government grant. He could have also have been concerned that organic farming was regulated by the government, though that was not stated explicitly. Thus, some Amish leaders made a connection between organic farming with government affiliation that appeared to create an anchoring of behavior towards non-adoption or remaining conventional in their marketing efforts.

12.8 Organic Dairy Decision Making: Cashton Settlement and Households

There are also divergent views about organic among individual households within the Cashton settlement. For example as discussed in the community section, some Cashton household perceptions of organic dairy are related to the idea that organic adoption may be divisive and that all Amish dairies should be equal. Some of the Cashton elders and farmers feel that all milk is the same, reflected in the comment that "milk is milk". If they believe that all milk is good and wholesome, then they may feel threatened when other farmers claim that "organic" is somehow better than conventional milk. The "milk is milk" concept may also be related to the prevalent idea among non-organic farmers (both Amish and non-Amish) that organic milk is a "marketing scam". One Cashton Amish elder who was not a proponent of organic challenged an organic Cashton Amish to a bet that the milk from their family farm would test out the same in terms of quality based on typical milk quality measures. When the results showed that the milk was similar, the Cashton elder felt vindicated in his thinking.

The organic farmer was not dissuaded, and responded that the market demand trumped the test results. So, in other words, why did it matter if the test indicated that the milk was the same if the consumer believed it was better and was willing to pay more for the milk? It was also interesting that the Cashton elder did not know what qualities were being tested. Instead he was anchored on the value notion that all milk is good and wanted to prove that. If individuals are anchored on a certain idea like "milk is milk", they tend to practice information satisficing and thus they stop gathering additional information about organic even if it may be an appropriate and more profitable management system for their farm.

The idea that organic farmers cheat illustrates a more extreme version of anchoring on "milk is milk." One Cashton farmer when asked whether he considered adopting organic, retorted that he knows about truckloads of "hot milk" (i.e. milk with antibiotic residue) that was marketed as organic. Some of the Cashton farmers were convinced that a certain feed mill sold conventional grain labeled as organic grain.[16] This idea about cheating within organics was fairly common among both Amish and non-Amish conventional farmers, and it was sometimes the first response they gave to questions about organic dairy.

[16]A cross-check of this story suggested that it may have been just gossip.

The focus on the possibility of cheating within organics among some of the Cashton Amish may also be related to the value loss aversion concept in the sense that it was connected to cheating and a more subtle temptation idea. Although the farmers who talked about cheating within organics stated that the Amish would not cheat, they seem to have an idea that the organic system created the temptation to cheat. This may be because a lot of the compliance is on the honor code in the sense that there is no real way to know if people are cheating unless it is reported or it is discovered in a spot check. The temptation to cheat on the organic rules creates a problematic situation for the Amish in terms of value loss aversion. The uncertainty involved with the situation is key to the idea of value loss aversion here because the Amish in this area do not have much direct experience with organic rules and the impact of having the temptation to potentially cheat could have on the community. These reactions are counterintuitive in some sense as the organic rules seem similar to the Amish *Ordnung* as there are rules about what is right and wrong and there is not always direct oversight on the individual's behavior (as with organic farming management). There are temptations associated with not following the *Ordnung* rules, and these rules can be only enforced through community.

The diversity about ideas on organic within Cashton also tends to be clustered along family lines. For example, let's consider the organic Cashton farmer who was challenged to the bet about the superiority of organic milk described earlier. A number of the family members of this organic Cashton farmer were also farming using organic methods. The family descendants of the elder with negative attitudes towards organics had similar issues with anchoring and extremetization associated with organic dairy. The importance of family units in adoption decisions may be especially profound in areas where there are divergent views about organic like in the Cashton settlement. As one Cashton Amish farmer states, "It is easier if you grew up with it [. . .]". He had not grown up with it so was intimidated by the unknowns associated with it. If you grew up doing organic farming, you may be confident to try it on your own farm and may be more convinced of the potential benefits.

12.9 Organic Dairy Adoption Patterns: Hillsboro Settlement and Households

More Hillsboro Amish practice organic farming than do Cashton Amish and their views about organic farming choices are less extreme though both wrestle with balancing distinctive *oikonomia* objectives and bounded rationality concerns. Overall, the Hillsboro Amish maintain a more positive view of organic dairy and its effects on relationships between households. There may have been some ambiguity aversion early on with respect to organics among a number of the Hillsboro Amish, but additional information shifted their view towards a positive view of organics. In this case, the tight social network of the Amish community actually led to more rapid changes in management instead of perpetuating negative perceptions and/or divergent perspectives as it did in the Cashton settlement. As one Hillsboro farmer states, "There was a period in the 80's where [we] used some pesticides but that was mostly out of ignorance. Now we have more understanding." Some of the Hillsboro farmers started organic farming while they were in Indiana before moving to Wisconsin. These farmers "were the first ones", they "were on the train" and then the "others who had not grown up that way" got on "the caboose" as one Hillsboro farmer describes. Another Hillsboro Amish farmer who arrived in Wisconsin more recently, stated the "community was more organically minded when he got here so he learned a lot." The community created an environment where knowledge and positive ideas about organic spread within the community. So it seems that the values associated with organics, i.e. the concern about the impact of pesticides on the environment, was attributed to acquiring more information and thus was a blending of *oikonomia* and bounded rationality.

In contrast to the Cashton Amish, some of the Hillsboro Amish connected their social/spiritual *oikonomia* framework to organic farming in a positive way. For example, for one Hillsboro Amish family, there is a connection with organic farming and the Protestant Work Ethic. As one farmer stated "organic farming is well-suited to family labor as the whole family needs to all get out there with a hoe and be together" in order to counter weed pressures. Thus, with their decision to adopt organic practices there is potentially a positive synergy between the agro-ecological and the social realms with the *oikonomia* framework.

The tight social network of the Hillsboro Amish community also facilitated a group certification strategy that significantly reduced the costs for individual farmers to become certified organic. Under this scheme, a group of farmers that agree to use the same management system is treated as one unit by the certification agency. They are doing this group certification through ICO (Indiana Certified Organic). The Cashton Amish formed an ICO group after hearing about it from the Hillsboro Amish, and this created some limited exchange between the two groups.

This more positive environment for organics within the Hillsboro settlement lead to much more subtle and positive forms of *oikonomia*-bounded rationality issues even for those farmers who decided not to market organically. It seems that in contrast to the Cashton Amish, for the Hillsboro Amish, their reasons for not going organic may be more about keeping prices low for consumers and relations with neighbors and this demonstrates principled satisficing or alternatively preferences for helping others. As one of the Hillsboro Amish states, they are not certified organic because they feel pretty well-established and should save "organic for the young struggling farmer". This Hillsboro farmer felt good about continuing to produce milk for the conventional market because this allowed him to "provide quality milk at an affordable price" rather than the higher priced organic market. As another Hillsboro farmer offers, they choose to market their milk conventionally, so they still have the option to buy hay from their Amish neighbors. In these cases, the farmers' motivations for not marketing their milk organically is consistent with the principled satisficing idea as they are not maximizing potential profits with higher priced organic milk for value based reasons.

In addition to the cultural factors and *oikonomia*-bounded rationality issues associated with organic dairy, there was also some risk aversion among Hillsboro Amish towards marketing organically because of the costs of organic feed and hay and the uncertain state of the organic line within Old Country Cheese. The Hillsboro farms tend to be smaller than the Cashton farms and not all of them were feed self-sufficient. (Many non-Amish intensive graziers shared these concerns because sourcing organic feed can be challenging.). Although not many Amish cite marketing uncertainty related to the unpredictable decisions of Old Country Cheese described earlier, it is likely to have played a critical role in discouraging adoption. Their investments in organic certification offered no organic price premium during the periods when the co-op did not sell organic cheese. This uncertainty may have created some risk or ambiguity aversion that limited organic dairy adoption among both the Hillsboro and Cashton Amish. For example, one Hillsboro Amish who had been organic when the co-op was selling organic milk, switched to conventional fertilizer after the co-op decided to leave the organic market. But when the co-op decided to re-enter the organic market, the farmer was understandably nervous about once again incurring the cost of getting re-certified without any assurance about how long Old Country Cheese would continue to purchase his milk.

12.10 Summary

Oikonomia and bounded rationality frameworks help to explain why and how the Amish farm on the landscape. Organic dairy seems particularly well suited to the Amish style of farming and their lifestyle, so the relatively low adoption rates of this practice may be surprising to outside observers. Despite the popular perception that Amish farmers are organic, not all Amish practice organic and/or other sustainable practices. The Amish have reasons for farming and for deciding to adopt organic dairy and some of these reasons are specific to their faith and culture. The Amish emphasis on contentment and simplicity plays an important role in explaining how and why the Amish farm in the midst of economically challenging times.

There can be a tension between maintaining common values and enabling financial survival on the land when Amish elders decide what technology and management systems are allowed. Decisions are made with the goal of attaining unity among the church leaders. The Amish have minimal overarching structure beyond the individual church settlements and their written and unwritten rules. This minimal bureaucracy in combination with intimate social relations within settlements may contribute to the diversity in how *oikonomia* values and bounded rationality themes play out in complex adoption decisions within and across communities.

The intersection between the frameworks of *oikonomia* and bounded rationality described in this chapter at the community, settlement and individual levels is helpful for understanding the complexities of individual factors when studying farm management decisions. This framework may also be helpful for understanding the other complex systems-based decisions outside of the Amish communities. All farmers may have value-based factors shaping farm decisions besides economics, and they make decisions in the face of a significant amount of uncertainty. Many other-non-Amish farmers exhibit principled information satisficing as they seem to value hand labor over efforts to gather information about their farm and alternative farming practices. This can mean many farmers do not really systematically consider alternatives like organic dairy. Although the focus on unity is not as explicit with other non-Amish farmers as it is within Amish society, it is clear that social pressure within the broader farm culture can potentially result in less adoption of emerging management techniques than there might be otherwise.

Adoption decisions of the Cashton and Hillsboro Amish offer insights about farm decision-making on the broader landscape in the sense that adoption decisions are complex and can involve oikonomia/bounded rationality behaviors in addition to full-information profit-maximizing criteria. Although there is not as clear of a leadership structure within non-Amish communities as within Amish communities, there may be opinion leaders in farming communities who set the tone for what farming practices are acceptable. There are also norms within rural communities that make it difficult to be different than other farmers thus making it difficult for some farmers to convert to organic farming (Bell 2004; Brock 2010). For example, some of the non-Amish farmers also had similar anchoring issues with the idea that all milk is milk, as reported here for some of the Cashton Amish. Relatedly, there is also a focus among non-Amish on cheating, but there is not a subtle temptation focus like there is with the Cashton Amish. There are also patterns of organic adoption decisions among families and spatial clustering that probably reflects social networks (Brock 2010; Lewis et al. 2011), but this is not as likely to be pervasive as it is within the tight social confines of the Amish church. This chapter points a path forward for further work on these themes.

References

Anonymous. 2006. Blessings on the farm. *Family life,* 27 (March).

Bandiera, O., and I. Rasul. 2006. Social networks and the adoption of new technology in Northern Mozambique. *The Economic Journal* 116: 869–890.

Barlett, P. 1980. *Agricultural decision making anthropological contributions to rural development*. New York: Academic Press.

Bell, M. 2004. *Farming for us all: Practical agriculture and the cultivation of sustainability*. University Park: Pennsylvania State University.

Bennett, J. 1982. *Of time and enterprise North American family farm management in a context of resource marginality*. Minneapolis: University of Minnesota Press.

Berry, W. 1997. *The unsettling of America culture & agriculture*. San Francisco: Sierra Club Books.

Brock, C. 2010. *An integrated household economics approach to decision-making: Dairy system choice among organic, Amish, Graziers, and conventional farmers in Wisconsin*. Dissertation, University of Wisconsin, Madison.

Brock, C., and B. Barham. 2009. Farm structural change of a different kind: Alternative dairy farms in Wisconsin—graziers, organic and Amish. *Renewable Agriculture and Food Systems* 24(01): 25–37.

Brock, C., B. Barham, and J.D. Foltz. 2006. *Amish dairy farms in southwestern Wisconsin*, PATS research report, vol. 17. Madison: Program on Agricultural Technology Studies.

Chua, J., and O. Juurikkala. 2008. The personal gift in sound business enterprises: Bounded rationality, incommensurable values and economic agency. *Revista Empresa y Humanismo* 11(1): 67–88.

Clark, S., K. Klonsky, P. Livingston, and S. Temple. 1999. Crop-yield and economic comparisons of organic, low-input, and conventional farming systems in California's Sacramento Valley. *American Journal of Alternative Agriculture* 14(3): 109–121.

Conlisk, J. 1996. Why bounded rationality? *Journal of Economic Literature* 34(2): 669–700.

Cross, J.A. 2004. Amish settlements in Wisconsin. *Wisconsin Geographer* 20: 2–9.

Cross, J.A. 2007. The expanding role of the Amish in America's dairy industry. *FOCUS on Geography* 50(3): 7–16.

Daly, H., and J. Cobb. 1989. *For the common good: Redirecting the economy toward community the environment and a sustainable future*. Boston: Beacon.

de Buck, A. 2001. *The role of production risks in the conversion to more sustainable arable farming*. PhD Thesis. Wageningen: Universiteit Wageningen.

Dilly, B. 1994. *Religious resistance to erosion of the soil and the soul among three farming communities in North East Iowa.* Dissertation, University of California Irvine, Irvine, CA.

Epstein, L.G. 1999. A definition of uncertainty aversion. *The Review of Economic Studies* 66(3): 579–608.

Fairweather, J. 1998. Understanding how farmers choose between organic and conventional production: Results from New Zealand and policy implications. *Agriculture and Human Values* 16: 51–63.

Gabaix, X., D. Laibson, G. Moloche, and S. Weinberg. 2006. Costly information acquisition: Experimental analysis of a boundedly rational model. *The American Economic Review* 96(4): 1043–1068.

Gasson, R., and A. Errington. 1993. *The farm family business*. Wallingford: CAB International.

Gigerenzer, G., and D. Goldstein. 1996. Reasoning the fast and frugal way: Models of bounded rationality. *Psychological Review* 101(4): 650.

Gottfried, R. 1995. *Economics, ecology, and the roots of the western faith: Perspectives from the garden*. Lanham: Rowman & Littlefield Publishers.

Goudzwaard, B. 1997. Christianity and economics. In *Signposts of god's liberating kingdom,* vol. 1, 229–240. Potchefstroom: Potchefstroom University/Institute for Reformational Studies.

Goudzwaard, B. 2000. *Capitalism and progress. A diagnosis and western society*. Toronto: Authentic Media.

Heasley, L. 2005. *A thousand pieces of paradise: Landscape and property in the Kickapoo valley*. Madison: The University of Wisconsin Press.

Hostetler, J. 1993. *Amish society*, 4th ed. Baltimore: John Hopkins Press.

Jackson, M. 1998. Amish agriculture and no-till: The hazards of applying the USLE to unusual farms. *Journal of Soil and Water Conservation* 43(6): 483–486.

Kahneman, D., and A. Tversky. 1979. Prospect theory: An analysis of decision under risk. *Econometrica: Journal of the Econometric Society* 47(2): 263–291.

Kahneman, D., J.L. Knetsch, and R.H. Thaler. 1991. Anomalies: The endowment effect, loss aversion, and status quo bias. *Journal of Economic Perspectives* 5(1): 193–206.

King, R. 2006. Letters to the editor – Food laws are not effective. *Family Life,* 5.

Kraybill, D. 2001. *The riddle of Amish culture*. Baltimore: John Hopkins Press.

Kraybill, D., and S. Nolt. 2004. *Amish enterprise from plows to profits*. Baltimore: John Hopkins Press.

Kraybill, D., and M. Olshan. 1994. *The Amish struggle with modernity*. Hanover: University Press of New England.

Kraybill, D., K. Johnson-Weiner, and S. Nolt. 2013. *Amish farming*. Baltimore: John Hopkins University Press.

Lewis, D.J., B.L. Barham, and B. Robinson. 2011. Are there spatial spillovers in the adoption of clean technology? The case of organic dairy farming. *Land Economics* 87(2): 250–267.

Lunneryd, D. 2003. *Unique decision making with focus on information use: The case of converting to organic milk production*. Dissertation, Swedish University of Agricultural Sciences Uppsala, Uppsala.

Lunneryd, D., and B. Öhlmér. 2006. *Problem detection and definition – The case of farmers' choice of organic milk production*. Paper presented at the 99th EAAE Seminar 'Trust and Risk in Business Networks', EAAE, Bonn, Germany.

Luthy, D. 2003. Amish settlements across America. *Farming Magazine*, 17–21.

Mansourian, Y., and F. Nigel. 2007. Search persistence and failure on the web: A "bounded rationality" and "satisficing" analysis. *Journal of Documentation* 63(5): 680–701.

March, J.G., and H.A. Simon. 1958. *Organizations*. Hoboken: Wiley.

Mariola, M.J., and D.L. McConnell. 2013. The shifting landscape of Amish agriculture: Balancing tradition and innovation in an organic farming cooperative. *Human Organization* 72(2): 144–153.

Meeks, M.D. 1985. God and land. *Agriculture and Human Values* 2(4): 16–27.

Padel, S. 2002. *Conversion to organic milk production: The change process and farmers' information needs*. Dissertation, University of Wales, Aberystwyth.

Salamon, S. 1992. *Prairie patrimony: Family, farming, and community in the midwest*. Chapel Hill: University of North Carolina Press.

Samuelson, W., and R.J. Zeckhauser. 1988. Status quo bias in decision making. *Journal of Risk and Uncertainty* 1(1): 7–59.

Simon, H. 1955. A behavioral model of rational choice. *The Quarterly Journal of Economics* 69(1): 99–118.

Smith, A. 1776. *The wealth of nations*. Random House, Inc., Bloomington Indiana.

Sommers, D.G., and T.L. Napier. 1993. Comparison of Amish and non-Amish farmers – A diffusion farm structure perspective. *Rural Sociology* 58(1): 130–145.

Stinner, D.H., M.G. Paoletti, and B.R. Stinner. 1989. In search of traditional farm wisdom for a more sustainable agriculture: A study of Amish farming and society. *Agriculture, Ecosystems & Environment* 27(1/4): 77–90.

Todd, P.M., and G. Gigerenzer. 2000. Précis of simple heuristics that make us smart. *Behavioral and Brain Sciences* 23(05): 727–741.

Tversky, A., and D. Kahneman. 1974. Judgment under uncertainty: Heuristics and biases. *Science* 185(4157): 1124–1131.

Vonk, M. 2011. *Sustainability and quality of life: A study on the religious worldviews, values and environmental impact of Amish, Hutterite, Franciscan and Benedictine communities*. Amsterdam: Buijten & Schipperheijn.

Worster, D. 1994. *Nature's economy*. Cambridge: Cambridge University.

Young, M.D. 1992. *Sustainable investment and resource use: Equity, environmental integrity, and economic efficiency*. Park Ridge: Parthenon Publishing.

Part V
Re-thinking Ethics in the Organic Movement

Chapter 13
Framework for Re-thinking Ethics in the Organic Movement

Bernhard Freyer, Jim Bingen, and Milena Klimek

13.1 Values in the Contemporary Organic Movement

Today, organic actors embody a mix of traditional and new values that emerge from the challenges confronted in a modern/post-modern society (see e.g., Grosglik, Chap. 8; Brock and Barham, Chap. 12). Several phenomena illustrate this blend of modern and post-modern features and values of organic farming and marketing: the diversity among the types of organic farmers and their practices (see e.g., Constance et al., Chap. 9; Hunt et al., Chap. 10); the variety of markets for organic products (see e.g., Grosglik, Chap. 8), and, the range of consumer values about organic (see e.g., Hatanaka, Chap. 3; Getter et al. Chap. 7). The values of the organic pioneers continue, but they are expressed in diverse patterns within the organic movements (see Chap. 2).

In contrast to other reviews of the organic movement that focus on specific organic actors,[1] our discussion applies an ethical framework to compare the ethics of different actors all along the organic chain (Sect. 13.1.1). We argue that our identified value types are consistent across these actor groups. To support these positions we refer to the former chapters in this volume and review and summarize

[1](e.g., Hall and Mogyorody 2002; Yiridoe et al. 2005; Van Huik et al. 2006; Aertsens et al. 2009a).

B. Freyer (✉) • M. Klimek
Department of Sustainable Agriculture Systems, Division of Organic Farming, University of Natural Resources and Life Sciences (BOKU), Vienna, Austria
e-mail: Bernhard.Freyer@boku.ac.at; Milena.Klimek@boku.ac.at

J. Bingen
Department of Community Sustainability, Michigan State University, East Lansing, MI, USA
e-mail: Bingen@msu.edu

B. Freyer, J. Bingen (eds.), *Re-Thinking Organic Food and Farming in a Changing World*, The International Library of Environmental, Agricultural and Food Ethics 22, DOI 10.1007/978-94-017-9190-8_13

the diverse discourses on values in recent years by all organic actors (Sects. 13.1.2, 13.1.3, and 13.1.4). We close with a final comment on how well the typology stands the test of describing current diversification of the organic agrofood chain (Sect. 13.1.5).

13.1.1 Typology of Ethical Values in the Organic Movement

The uniqueness of organic agriculture, in comparison to other agriculture and food system movements, stems from its long history in Europe, beginning in the 1920s, as a value[2] based movement opposed to conventional farming practices (see Chap. 2). Always called a movement, and developed and continually cultivated by non-governmental organizations (Spoor 2002, p. 6), organic farming has invariably focused on creating an agricultural production and food system rooted in close human-nature, value oriented relationships (Padel and Foster 2005, p. 117). Over the years, the movement has maintained its tradition of protest against the industrialization of society (see Chap. 2). Core features of the movement have included: a commitment to self-determination in farming, processing and trade; and openness to a wide range of actors (from gardeners to consumers to intellectuals) who have brought in diverse ideas about nature, farming and e.g., the exclusion of synthetic chemicals (see e.g., Jones, Chap. 11).

As organic production, processing and trade became more widespread, it also became more regulated (Rigby and Cáceres 2001, p. 25). Different actors in the expanding and increasingly global organic market began demanding more controls over organic (see Constance et al., Chap. 9). In order to justify and to distinguish organic products, but also to protect organic values from misuse and fraud (e.g., Hyman 2000; Zorn et al. 2009), control and certification processes were put in place that responded to the interests of different actor groups. But this step has also led to a bureaucratized reality that has pushed many of the initial core values of the movement to the sidelines (see e.g., Hatanaka, Chap. 3).

With this diversification, the question arises, can these different types of practicing organic and discussing values be systemized and described by a typology applicable along the organic agrofood chain? In this chapter, we identify and define such a typology as a means to review systematically the diverse dimensions of values practiced in the organic movement.

Our entry point is derived from the typology used by Alrøe and Noe (2008) (see also Verhoog et al. 2007) that offers a basis for classifying the range of value sets represented by farmers, consumers, processors and supply chain actors. We adapt their three-part typology of organic farmer orientations/motivations and values in order to propose our own that is: sensitive to a range of ethical foundations;

[2]The meaning of value in most sociological research on organic "is almost identical to the motives of different stakeholders in the organic sector" (De Wit and Verhoog 2007, p. 455).

has historically characterized ethics in the organic movement (see Chap. 2)[3]; and continues to be relevant for understanding the more recent differentiation of values with respect to what is called "conventionalization" (Darnhofer et al. 2010) (see Constance et al., Chap. 9). The following describes our typology of ethical values in the organic movement.

Nature and philosophy (Nature/Philo): Practicing organic involves a commitment to participating in the cycles of nature and adapting an ecocentric/holistic and philosophical, bio-dynamic or spiritual understanding of belonging to an organic eco-community (cf. Bookchin 1982, p. 46); this often includes support for small-scale, family-based innovations (see also the collectivist-immaterial in Meeusen et al. 2003). Organic is understood as a logo-poietic system i.e., self-organizing (autopoiesis) that creates its meaning (logo) (Alrøe and Noe 2008) and is held together through principles. This type also includes those actors who are more spiritual (e.g., deep ecology) (Naess 1973, 1986) or those for whom organic is in harmony with religious views (Lockie et al. 2002, p. 26).

Environment and protest (Enviro/Protest). Organic is fundamentally a protest by farmers and environmental activists against industrialized agriculture and seen as a means to support/sustain a healthy environment (Willock et al. 1999; Baker et al. 2004; De Wit and Verhoog 2007). Such actors often oppose the use of inorganic fertilizer, synthetic pesticides, food additives or genetically modified organisms (Alrøe and Noe 2008). The philosophical background of this group ranges from pathocentrism to biocentrism (see Sect. 2.2.1).

Economics and markets (Econo/Market): Organic is primarily for profit and accepted primarily for individualist and materialist values (Meeusen et al. 2003). Industrialized large-scale farmers, consumers, and retailers who reflect an anthropocentric, utilitarian and hedonist orientation illustrate this type. Organic is seen as part of the globalized market system that relies upon organic standards designed to meet this economic niche (Alrøe and Noe 2008).

[3] Alrøe & Noe's typology reveals similarities to that used by Memery and others (see Memery et al. 2005). These typologies use three clusters: food quality and safety; human rights and ethical trading; and, environmental (green) issues (see also Best 2008). Browne et al. (2000) apply also three categories of consumers distinguishing the true ethical consumers described as consuming primarily organic products; semi-ethical who will be sometimes and those who would be ethical consumers if premiums are low and access was easy. This classification is close to that of Alrøe and Noe, however differences exist specifically in defining the semi-ethical type. A further example to classify ethics is that of the Ethical Matrix developed by Ben Mepham and Sandra Tomkins. The matrix was originally applied for students in an educational context, to make an ethical assessment about the impacts of certain choices in relation to food production. This concept is built on three types of ethical reasoning in line with those of Barnett et al. (2005): (1) "Wellbeing" is related to the concept of utilitarianism and consequentialism, which is close to the Econo/Market type; (2) "Autonomy/Rights" refers to responsibilities and duties against others; and (3) Fairness/Justice that appeals to exclude any unfair and hence unjust action. Thus, (2) and (3) are relevant for the definition of Nature/Philo and Enviro/Protest type.

13.1.2 Farmers' Values

Before analyzing farmers' values in detail, it is important to remember that those who have chosen to convert to organic must go through a transition period (Lamine and Bellon 2009). There is evidence that the social environment influences the decision to convert (Rantzau et al. 1990). To change from conventional to organic often means socializing with a new peer group and becoming part of a new culture (Verhoog et al. 2003). However, during this time, some might also decide to continue with their (first) non-organic "value-community". Therefore, the need for such a socio-cultural re-orientation or adaptation into a new social environment is less evident for those who adopt a high input conventionalized organic approach.

In the following discussion of farmers' diverse motivations and values, we clearly recognize the diversity in backgrounds, intentions and motivations to convert to organic (e.g., Fairweather 1999). When studying the factors that influence the decision to convert to organic, we should be careful not to compare today's converters with those who first converted. For example, Lund et al. (2004) identified different values between the pioneers and more contemporary organic farmers. Best (2008, p. 100), as well, in a study of German organic farms, reported that recent converters were more specialized, and found direct marketing of less importance. It is hard to assess however, if these trends were the result of changes in societal values or specific differentiation in the organic movement. Based on country-specific dynamics, the development of organic farms, farmer motives and values differ from country to country (Willer and Kilcher 2011).

Technical, economic, and labor issues have been reported as main barriers to the adoption of organic farming (Midmore et al. 2001; Khaledi et al. 2007). Some studies report that bureaucracy and market access present challenges to adoption (Schneeberger et al. 2002). But it has also been recognized for some time, that those who decide to convert, are confronted with pressure from their farmer friends because organic deviates from dominate social norms or runs the risk of failing (Duram 2000, p. 20; Dimitri and Oberholtzer 2009). Especially in large-scale, cash crop regions (Kölsch 1988), or where there are large numbers of non-organic farmers (Schramek and Schnaut 2004), organic converters may quickly become outsiders. As a result, some simply decide to use high input organic practices (feedstuff and organic fertilizer) that often parallel those of conventional farming.

Kaufmann et al. (2009) argue that subsidies appear to be more influential than social factors in the decision to convert. However, with subsidies and a positive social environment, often the adoption rate often increases. In those regions where organic is mainstream or where direct marketing and agro-tourism strategies dominate (FAO/ITC/CTA 2001), organic farmers become highly respected by consumers (Kaufmann et al. 2009) (see e.g., Getter et al., Chap. 7). Those following a more classical organic approach tend to relate more with engaged or attentive organic consumers than they do with their neighbors who continue intense conventional and non-organic practices.

This suggests that the future of organic is highly dependent on shared values between farmers and consumers (see Hatanaka, Chap. 3). However, when organic and conventional farms are in close proximity, social ties between farmers may often trump differences in practices (see Hunt et al., Chap. 10).

13.1.2.1 Nature/Philo

There appears to be a renaissance in this type of farmer (see Getter et al., Chap. 7). For these, mostly small farmers, the integration of nature with religion, spirituality or community is central. Earnings from the farm are, of course, important, but they often are of secondary concern. These farmers commonly comprise counter-culture opposed to "industrialized" (or conventionalized) organic farming and marketing (Reed 2002; Tovey 2002; Allen et al. 2003).

These farmers orient their activities around a broad set of values that are often in line with the four IFOAM Principles (see Chap. 1). But this does not mean that they know about these principles. Moreover, this group generally finds that "conventional" organic policy and practices contradict their core values. Consequently, many practice organic, but forego official certification (governmental recognition), available government subsidies and the use of the official organic label. These organic farmers tend to criticize the weakness of values in official standards and find many "organic" practices (i.e., the substitution of organic for synthetic mineral fertilizer) in conflict with the IFOAM Principles and with their personal environmental, economic and social values. Moreover, they also tend to be critical of the current economic model that undermines fairness and justice in society. In response, they engage and practice a new (organic) lifestyle that includes many ecological and social oriented activities.

Many of these farmers also engage in what is called "organicPlus", or activities that go beyond the regulations and that specifically address the IFOAM Principles (Padel and Gössinger 2008; Gössinger and Freyer 2009a; Zander et al. 2010). These commonly include adherence to strict animal welfare standards, involvement with school groups or the integration of disabled individuals into production and marketing activities (Neuberger et al. 2006). Many practice on-farm "nature protection" that contributes to the conservation of traditional and local crop species and that exceeds most written guidelines or that is eligible for subsidies (Padel and Gössinger 2008). More concretely, these farmers seek to preserve the ideas of the organic pioneers. They represent the ideal image of farming practices from a former time and thus often serve as reference group for promoting organic.

A second relevant group of the Nature/Philo type includes those who adopt a holistic approach to their search for new farm style patterns outside the mainstream economy, and experiment with new lifestyles beyond agricultural activities. This group commonly includes younger people who might not have a farming background, but tend to be knowledgeable about farming issue. Some are involved in loose or formalized collaborative activities such as eco-villages (Grundmann and Kunze 2012). They often create communities in which organic values are a logical element in, and central to the organization of their social and economic activities.

They practice diverse production and processing activities, do not rely on off-farm inputs, and are primarily engaged in local and regional direct marketing, as well as in smaller organic associations. They seek to develop an innovative food and energy autarky, and to integrate environmental, economic and social innovations into their groups (e.g., Sargisson and Sargent 2004). Those with "ethical" orientations and holistic lifestyles represent a group of "societal transition innovators" seeking a sustainable society.

13.1.2.2 Enviro/Protest

For these farmers, there is no question that organic is the best way to farm. Environmentally sound and healthy practices are central (consistent with the IFOAM Principles of Health and Care). They may also take a pragmatic view of organic practices that allow them to accept some exceptions to organic principles on a case-by-case basis (see e.g., Jones, Chap. 11). Marketing their organic products may involve cooperating with non-organic retailers, but they prefer marketing via farmer groups that operate with their own rules and values, and seek economic independence. These farmers also are engaged in diverse types of collaboration e.g., machinery, processing and storing products or to develop new solutions for energy independence.

This type also includes farmers who are engaged in publically promoting the organic movement. Those farmers often assume positions in international, national or regional organic associations, and develop processing and market structures, are linked with political organizations or work on promoting more subsidies and participating in control and certification activities. They tend to cooperate with others involved in nature protection, climate change, religious communities or educational associations, including those who practice new types of societal and economic collaboration. However, in contrast to the Nature/Philo type, they pursue their more pragmatic approach through more established structures. Their role is to ensure that the organic movement protects and practices its values, but without being extreme or "revolutionary" (see Hunt et al., Chap. 10). In addition, they seek to bring organic into the societal debate without provoking or demanding comprehensive societal change, as is the case with the Nature/Philo type.

13.1.2.3 Econo/Market

For the Econo/Market organic farmer, economic and marketing concerns are the key motivations for practicing organic; ecological and social values, including the IFOAM Principles, are less central to their practices. In short, these farmers practice organic farming only when they see a convincing economic advantage to do so (see Constance et al., Chap. 9). It is widely understood that financial considerations for both farmers and the corporate food industry has become a significant reason for adopting organic (Hall and Mogyorody 2002). This type of large, corporate agribusiness industrializing or "conventionalizing" organic production[4] has been discussed for many years (Buck et al. 1997, p. 4) (see Grosglik, Chap. 8; Constance et al., Chap. 9).

There is also some evidence, by no means overwhelming, that financial support, largely through subsidies, has become a relevant motive for this group to convert to organic, specifically in the EU. Under these conditions, they are usually unfamiliar with the IFOAM Principles and values in contrast to early converters for whom subsidies were not available (Padel et al. 2007a, p. 116).

In some cases, such market-driven production practices directly contradict the IFOAM Principles. For example, Padel et al. (2007a) found significant differences between the IFOAM Principles and industrial-type production practices in pig and poultry (feeding) and horticulture farming, as well as in crop production (organic fertilizer) (ibid., p. 9, 10). De Wit and Verhoog (2007) also observed that high off-farm inputs (not always organic) are applied on organic farms in the Netherlands and are often transported over long distances.

Obviously, some organic practices that are consistent with the IFOAM Standards, but involve an oversupply of inputs, compromise the IFOAM Principle of Ecology. Oversupplying nutrients creates several negative consequences for soil fertility, plant health and food quality, and this raises several questions about the awareness/ignorance of IFOAM Principles on the part of farmers. This intensification of organic is not only a question of production, but it risks jeopardizing continued support from both consumers and policy-makers (Darnhofer et al. 2010). This explains in part, why non-governmental standards are more restrictive than federal regulations in the EU or in the US (Padel and Gössinger 2008, pp. 14–15).

13.1.3 Consumers Values

There is little doubt that organic has become widely accepted by consumers and no longer represents only a status or lifestyle symbol (Torjusen et al. 2004; Hughner et al. 2007). However, there are also those who argue that organic should include the values of "local" as a means of incorporating concerns with social justice and ecological sustainability (Allen et al. 2003 cited in Clarke et al. 2008, p. 220)

[4]In recent decades, the 'conventionalization' of organic has been widely discussed within organic farming movement (De Wit and Verhoog 2007; Guptill 2009; Konstantinidis 2012). The conventionalization of organic farming can be described in many ways, however most definitions are concerned with how organic farming has generally followed trends of conventional production. This most commonly has involved the ways in which agribusiness discovers ways to industrialize organic farming and production, which often leads to the occurrence of some farms that fulfill certification requirements without adhering to the principles of organic (Buck et al. 1997; Hall and Mogyorody 2002; Darnhofer et al. 2010). In short, conventionalization of organic practices often minimizes ethical considerations along the organic agrofood chain addressed by the IFOAM Principles by bringing egocentric and utilitarian views to the foreground.

(see Getter et al., Chap. 7). In contrast, Born and Purcell (2006) argue that eating local food is not more ecologically sustainable and socially just than systems at other scale. Sustainability is more a question of content and not one of scale (Hinrichs 2003). So far, the organic system cannot cover the increasing demand of urban centers through "regional" organic products; to eat locally in general can only serve as an orientation. These introductory thoughts make clear how challenging it is for consumers to understand the broader meaning of local and global, as well as their ecological, social and economic dimension. Consumers are confronted with a complex and seemingly confusing organic world, e.g.:

- high premium prices, inefficient access (Thøgersen 2007; Aertsens et al. 2009b, p. 1157),
- lack of information concerning organic quality and production, intellectual and ethical challenges (Zagata and Lostak 2012),
- taste and aesthetic differences (e.g., Schuldt and Hannahan 2012),
- over-labeling and confusion with different types of labels, or lack of trust in labels (Cliath 2007), or
- lack of trust in the certification process (see Hatanaka, Chap. 3; Adamoli, Chap. 6) (Hamm et al. 2002; Yiridoe et al. 2005; Hughner et al. 2007; Aertsens et al. 2009b; Janssen and Hamm 2011).

The decision making process of consumers for organic products is based on a set of factors—such as, personal attitudes, subjective and social norms, values, and (perceived) behavioral control (see e.g., Fazio 1990; Aertsens et al. 2009b, p. 10). The attitudes and motives of consumers, why they consume organic food or doubt organic creditability, provide insights into how far the IFOAM Principles or similar ethical arguments are of relevance for their food consumption behavior. Depending on many factors such as differing demographics, methods and models of data collection and even consumer ethics and behavior, one finds that consumers are inconsistent in their interpretation of the meaning of organic (Yiridoe et al. 2005), as well as their attitudes and behavior (e.g., Vermeir and Verbeke 2006).

13.1.3.1 Nature/Philo

The ecocentric/holistic-oriented consumers use a range of purchasing criteria that include human values, local production and traditions, animal welfare, political issues (environmentalism), fair producer prices and further ethical orientations (Padel and Gössinger 2008; Hjelmar 2011; Zander et al. 2012). These consumers also express health, and specifically freshness, concerns among the central reasons for buying in farmers markets (Trobe 2001). Schösler et al. (2012) (Netherlands) argue that this reflects a return to a more natural lifestyle, one that involves a distancing from materialistic lifestyles and reverting to a more meaningful moral life. The authors suggest that the orientation toward a nature-oriented lifestyle represents the wishes of a much larger part of Dutch society, not only those who currently shop for organic food (cf. Zander and Hamm 2010).

Consumers are sensitive to local and regional products, and there is considerable interest "in the availability of locally grown products, reliable information and easy comparison with non-organic products" (Gottschalk and Leistner 2012). Local means "to be able to trust", since these consumers commonly doubt the quality of imported organic products. Communicating and sharing values through direct contact to the producers is one of their mains motivations. The opportunity to "know the farmer" is a very strong value in the local market. However, even when this possibility exists, only a minority of consumers have the opportunity for close contact with local farmers.

With the growth in the number of farmers' markets (specifically in US), community supported agriculture, urban farming and food cooperatives, many alternative forms of producing and procuring food have taken a 'stand' against corporate organics and have actively included values associated with such innovations (Tovey 2002). Therefore, this consumer group attitude also carries a political message.

Moreover, the values of these consumers are already close to the principles, and thus they represent an exception to most consumers, not a model for them. There are many similarities between the IFOAM Principles and the values of these consumers.

13.1.3.2 Enviro/Protest

Environmentally and politically oriented consumers can trace their roots to the 1960s and the reaction to the use of organochlorine pesticides (see Sect. 2.3). These consumers act individually or as part of groups engaged in environmental, social and economic issues (e.g., Fair Trade) (e.g., Browne et al. 2000; Raynolds 2002; Lyon 2006). Similar to the Nature/Philo, the Enviro/Protest consumer takes a critical view of environmental issues, but is less directly engaged in alternative economic and social systems. This type is more reflective about organic purchases, is commonly aware of the real differences between local and industrialized organic production and marketing, and respects the higher prices for organic product quality. This type includes what Johnston (2008) describes as the "citizen-consumer hybrid" who "votes with their money" for reasons related to both individual self-interest (consumerism) and a sense of social collective responsibility to a social and ecological commons (citizenship) (see Adamoli, Chap. 6). These value oriented consumers are, however, not specifically aware of the IFOAM Principles in part because products do not carry an "IFOAM label".

13.1.3.3 Econo/Market

In contrast to the early years in the organic movement, many now justify buying organic (Clarke et al. 2008, p. 225) for a wide variety of reasons, including enjoyment, personal health benefits (ibid., p. 228), and food safety (Michaelidou and Hassan 2008). These are "pragmatic consumers" who purchase organic for reasons of convenience (Hjelmar 2011) and health. This type of buyer requires that organic foods be available in local supermarkets, easily identifiable by a label, and with a minimal price difference from conventional products. These conditions are important for buyers who purchase organic for largely instrumental reasons, such as concerns with personal or family health issues. These consumers are also those who purchase local produce (e.g., in farmers markets) and who accept local or regional products even if they are not organic, largely because they are more affordable.

This is also the group that is largely responsible for the increase in organic consumer sales and responds to lower prices and convenient access to organic products[5] as well as easily understandable information about a product's origin rather than ethical issues related to buying organic. Marketing experts identify this group as those who buy organic largely in response to food safety concerns (pesticide and hormone free), environmental protection, taste (Hamm et al. 2002), and animal welfare concerns.[6]

For a majority of consumers, neither ethical concerns nor the IFOAM Principles frame their food purchasing behavior.[7] Instead, 'private good' attributes (freshness, taste and health benefits) and convenience are central (Wier et al. 2008). These consumers commonly show little concern for, or commitment to organic farmers *per se*, or to the importance of assuring the ecological, social and economic robustness of organic farming (Howard and Allen 2006). Clarke et al. (2008, pp. 223, 225) describe this type of consumer ethics as "ordinary ethics". These ethics entail caring about family, taste, and using health concerns to guide everyday choices. This group does not have "strongly held ideological or spiritual blueprints for action" (ibid., p. 224) and instead of representing a counter-culture, purchases organic as the new "yuppie chow" (Guthman 2003).

13.1.4 Supply Chain Values

Today, consumers confront a wide diversity of purchasing opportunities for organic products. In addition to the range of products branded by large, corporate and often multi-national food companies,[8] the number of value-driven organic markets and cooperatives, sometimes established by farmers to serve their local communities are growing. At first glance it might appear as if the different market approaches complement each other, and each plays a specific, but different role in providing organic food.

[5](see also Pollan 2006; Fromartz 2007; Kirchmann et al. 2008; Paarlberg 2009; Hjelmar 2011).

[6](e.g., Bennett and Blaney 2002; Frewer et al. 2005; Hughner et al. 2007).

[7](Magnusson et al. 2001; O'Donovan and McCarthy 2002; Thøgersen 2002; Lea and Worsley 2005; Padel 2005; Yiridoe et al. 2005).

[8]See for example, www.msu.edu/~howardp/organicdistributors.html

However, these markets address organic values differently. The organic market is confronted with the challenge that local and direct market schemes offer the opportunity to express organic values, but are limited in meeting the demand for products (Stagl 2002) and often are mainly available to middle class consumers (Hinrichs 2003). However, to cover the continuing growth in demand for organic products in urban areas, there is need to collaborate with conventional grocery markets. These tendencies raise the issue of whether organic marketing becomes "conventionalized" (Loconto 2010) and no longer follows a direct marketing approach.

13.1.4.1 Nature/Philo

Similar to our discussion of Nature/Philo farmers, the local orientation of the Nature/Philo in the supply chain is a prototype for ethical trade that is expressed in diverse forms of CSAs and similar initiatives where citizens accept more active engagement.

The roots of organic marketing are in diverse forms of direct marketing through local markets or on-farm sales. Consistent with these marketing venues, the increased demand for organic food has given rise to new organic marketing opportunities, including diverse direct marketing possibilities such as food networks, CSAs, box schemes and other initiatives (Renting et al. 2003). This type integrates social criteria into the organic market system. "Knowing the farmer" creates trust (e.g., Hinrichs 2000; Jarosz 2000; Marsden et al. 2000; Pivato et al. 2008) and can lead to solidarity (Principles of Fairness and Care), acceptance of "higher" prices and understanding when not all products are available or are of varying quality and size (i.e. the Principle of Ecology), unlike uniform products on most store shelves.

Direct communication offers the opportunity for sharing values and strengthening trust between farmers and consumers (Wier et al. 2008). But markets are diverse and distinct in their potential to communicate, to create closer social ties, to share values between consumers and farmers, and in the economic risk they pose for farm production. While the roles of consumers and farmer in farmer markets are often rooted in commodity relations, the consumer plays a relatively passive role. On the other hand, there is some evidence regarding the positive impact of direct marketing on local economies and the formation of social networks, where the consumer takes over an active role (Brown and Miller 2008). More specifically, Community Supported Agriculture—CSA (e.g., Guthman 2004; Hole et al. 2005) creates the conditions for a de-commodification of the agrofood system (Hinrichs 2000) and can help to rebuild local communities (Goland 2002). CSAs also help inform consumers about organic and to discuss values.

In general, these local marketing initiatives offer numerous opportunities for creating closer relationships among farmers and consumers. These relationships may be expressed in countless ways, including practical and concrete collaboration such as contracts, common production and labor use planning and agreeing upon the economic benefits for the farmer and the consumer (Cooley and Lass 1998).

Local food systems can contribute to bolstering farm income and promoting rural development (Renting et al. 2003). These forms attract (agro-) tourists and increase the awareness of consumers for local and regional products that often enhance the environment or protect the natural heritage. The income from such value based marketing in rural areas may often be limited (Gale 1997); demand for seasonal and local produced organic food from small-scale farmers does not necessarily strengthen the income of all small-scale farmers (Jarosz 2008). These insights remind us of the importance of looking critically at each type of farmer-consumer value driven collaboration.

Within the market context it is also of interest how the input industry is linked with values. The organic input industry receives little or no attention either in the IFOAM Principles or in discussions of ethics in the organic movement. As illustrated in many of the advertisements in publications such as *Acres, The Voice of Eco-Agriculture,*[9] the firms that sell organic supplies do not use the Principles in their promotional material. It is rare to find animal breeders, for example, who incorporate organic values in their promotional materials (van Bueren et al. 2003; Niggli 2007). Furthermore, there are no systematic studies of this industry.

13.1.4.2 Enviro/Protest

The Enviro/Protest Type operates with mainly regional products, but also uses organic fair trade or convenience products to satisfy the consumer demand and to attract urban middle and upper class consumer. This type includes entrepreneurs who operate organic-only supermarkets or operate as regional organic wholesalers for smaller supermarket chains, individual shops and restaurants. These new innovators in organic retail also represent community values (Lyons 2007). However, those suppliers often have to follow conventional business models (Higgins and Lockie 2001), and concentrate their buying from farmers who meet their standards and specifications.

Alternative entrepreneurs see themselves differently from large corporate big retailers (Cody, 2005 cited in Lyons 2007, p. 158), but they are nevertheless embedded in a capitalist market structure, and increasingly ask for uniform products that fulfill technical standards that can be supplied only by industrial organized farms that produce for the mass market. As a result, they must hold larger farmers to commercial standards and at the same time find it hard to accommodate the lower and uneven production capacity and quality from small farmers (Burch and Rickson 2001). Some large-scale farmers have the option to supply these organic supermarkets. For these farmers, retailers offer a certain financial security.

The economies of scale weaken the position of small and medium-sized farmers and conflicts with the ethical foundations of the IFOAM Principles. That is, the hierarchical and profit-driven relationship even between the Enviro/Protest retailers

[9]http://www.acresusa.com/

and farmers put the issues of fairness under pressure. In response to this type of economic pressure, small farmers diversify their markets (Lyons 2007), and follow a marketing path similar to the Nature/Philo type.

We conclude that these regionally oriented entrepreneurs integrate organic values in a more idealistic and holistic manner than big retailers. They reactivate a regionally oriented food network that is more independent from the global market. However, they are limited in significantly fulfilling these values because business constraints force them to adopt the conventional business model. On the other hand, given their values, small-scale businesses are negatively affected by these organic 'superstore' chains and are unable to compete with their logistics, buying requirements and price structures (Lyons 2007).

13.1.4.3 Econo/Market

Since the 1990s, many large conventional processing and marketing firms have jumped on the "organic bandwagon" and have introduced their own organic products. Most large retailers now offer an organic line of products and organic is now an important section in most retail grocery stores (Willer and Kilcher 2012). Retailers have become aware of the organic business opportunities and respond to Econo/Market consumer who seeks low cost organic food (Lyons 2007). Relatively lower organic prices in discount chains allow lower-income consumers to purchase organic food. This encourages some organic farmers to enter uniform mass production (see Enviro/Protest). In many cases, these farmers also bear all the risk of meeting the buyer specifications, just as in marketing conventional produce (Higgins and Lockie 2001).

These companies clearly influence the distribution of power among different actors in the organic food chain by product, between countries and the types of contracts between retailers, processors and farmers. In many countries, local organic outlets have been acquired, thereby negatively affecting the diversity of small marketing and processing units. The role of retailers in the organic market is controversial (see e.g., Grosglik, Chap. 8). Large, corporate retailers commonly pressure organic farmer for lower prices in order to offer less expensive organic product to consumers. At the same time, the increasing interest of conventional retailers has helped encourage both large and small farmers in some countries to convert to organic production. Richter et al. (2000) argue that the expansion of multiple organic retailers might promote organic market development, by establishing an increasingly secure market and financial incentive for farmers considering conversion to organics.

The large quantity of food processed and sold organically in most supermarket and discount chains reflects the institutionalization of the global agrofood system (Raynolds 2004) (see Constance et al., Chap. 9). This segment continues to grow specifically in California, but also in Europe, Australia and New Zealand. The majority of these retailers rarely pay attention to broader issues of organic principles e.g., food miles, labor conditions or fair prices (cf. Latacz-Lohmann and Foster 1997).

However, there are indications of change, including serious alternative approaches to re-balance these relationships between farmers and corporate marketing. This re-balancing is sensitive to the values of IFOAM Principles, but does not specifically refer to them.

To justify the organic premium, and to create trust, more distributors and retailers are giving serious attention to health, environmental and social issues. For example an Austrian retailer certifies organic products with an additional CO_2-Certificate in producing their organic products (Lindenthal et al. 2009; Hörtenhuber et al. 2010, 2011). Others subscribe to the principles of Corporate Social Responsibility (CSR) and specifically, Corporate Social Performance (CSP). Obviously CSR/CSP influence positively trust of consumers into organic products in the retail market (Pivato et al. 2008; Perrini et al. 2010) and positively influence Corporate Financial Performance (CFP). That means on the one hand that consumers are sensitive to social issues; on the other hand, retailers develop marketing strategies to justify the organic premium price by making more visible the values behind the label. Retailers selling both organic and non-organic products are also aware of the conflict which would arise if consumers realize the circumstances under which the majority of their food is produced (Lyons 2007). That is, bringing organic to the shelf affects the overall marketing strategy of non-organic products.

Additionally, it is important to consider the interests and values of the agricultural input industries with respect to becoming part of organic development. Agricultural supply firms are equally important and often-overlooked players that influence the future development of organic. Organic farms do not offer important sales opportunities for these firms. In Austria, for example in cash crop farms, the average organic farmers' expenditure for inputs is only about 1.4 % of their total income, whereas the average expenditure by conventional farmers represents about 27,4 % of their total income (LBG 2010).[10] Thus, it is not surprising that organic farms are of rather limited interest for the agriculture supply industry. In contrast, the food industry actively integrates organic products in their processing, product lines and shelves.

13.1.5 Reflections on Our Typology of Ethics in the Organic Agrofood System

The analysis of the role and awareness of ethics and the IFOAM Principles in the organic movement today identifies similarities in the different sectors or actor groups that are part of, or linked with the organic agrofood chain, but also express specific actor group characteristics.

The typology that we applied here presents a "central" or dominant tendency for each type. Empirical data on ethics along the organic agrofood chain confirm that

[10]http://www.lbg.at/403_DE.0

these three types are relevant and we applied them to each organic sector. Thus, we suggest that the framework deserves further investigation and application.

Similar to Alrøe and Noe (2008) typology, noted, each type follows its own logic, and there is little room to combine attitudes and practices from different types, largely because of the contradictory orientations between the types, specifically between the Nature/Philo and Econo/Market types. There is also a tendency that the Nature/Philo and Econo/Market types are more represented in practice, while Enviro/Protest type may be less so. Moreover, we accept the importance of not polarizing and oversimplifying the differences between two manifestations of organic (Lockie et al. 2002, p. 25; Lockie and Halpin 2005; Constance et al. 2008; Rosin and Campbell 2009) as expressed by the term "bifurcation" (Constance et al. 2013).

Kaltoft (1999) argues that individual farmers, because of their commonalities with one another, represent both a collection of diverse values and diverse organic practices. Most organic actors may use some reasoning from all three rationales, and there is no doubt that in between these three rationales, we find mixed approaches. Therefore, the three labels do not do justice to the complexity of each type. In short, unlike the position by Alrøe and Noe, we do not think that crystal clear boundaries between the different types are essential to demarcate.[11]

Similarly, consumer studies find a variety of motivations, perceptions, and attitudes (Hughner et al. 2007). Consumer research on "willingness to pay",[12] or investigations on environmental behavior, (e.g., the well-known phenomena of differences between behavioral intention and concrete action), shed light on how challenging the classification into three types could be (Shepherd et al. 2005; Tarkiainen and Sundqvist 2005; Vermeir and Verbeke 2006; Arvola et al. 2008). This means that there might be serious reflections by the organic actors about an ethical or principle based "organic lifestyle." However, there are always enough reasons why consumer practice differs from intentions and values. As a result, we find individuals who hold all three types of values combined. Sometimes values are practiced and sometimes not. Therefore, when theory (individuals intention) and practice (individuals behavior) are not congruent, the boundaries between the three types become blurred. What is measured in behavior or willingness to pay studies is exactly a mixture of behavior and behavior intentions.

These observations do not contradict Alrøe and Noe's typology and reflections in principle. We agree that ethics follow a systemic logic (Alrøe and Kristensen 2002) and must always be considered within a specific ecological, economic and socio-cultural context. All three are ideal types, and at least typologies are an approximation of reality (Johnston 2008). As we know from studies of the conversion to organic, it is most helpful to understand those in the process as "on the way" from one type or stage to the next. The types are steps in a learning process

[11](Meeusen et al. 2003; Darnhofer 2005; Van Huik et al. 2006; De Wit and Verhoog 2007).

[12](Gil et al. 2000; Loureiro and Hine 2002; Krystallis and Chryssohoidis 2005; Didier and Lucie 2008; Zander and Hamm 2010; Janssen and Hamm 2012).

and thus do not represent a pure form, but as Kröger and Schäfer (2013) suggest, some difference between "ideal and reality". Of course, the different histories and development of the organic movement (US, Australia or Europe as well as Asia, Africa and Latin America) and related agro-ecological, cultural and economic environments call for further detailed investigations of how the typology represents a realistic picture of the differentiation of values in the organic agrofood chain.

13.2 The Role of Ethics in Different Sectors of the Organic Agrofood Chain

While there is increased popularity, attention and commitment to organic, the explicit recognition of the IFOAM Principles of Health, Ecology, Fairness and Care is not always visible in practice. If IFOAM claims that organic is the most ethical and sustainable form of agriculture and food production, then bringing these principles back to center stage of the organic agrofood chain is one critical aspect of what is needed in "re-thinking organic."

For that we have to broaden our perspective on ethics beyond the organic key actors discussed in Sect. 13.1, to consider additional factors that influence the ethical practices of the organic agrofood chain. To do so, this section first offers insights on the relevance of ethics in IFOAM Standards, control and certification system. We also look at alternative control and certification procedures and modes of collaboration between farmers, consumers and trade that are sensitive to the IFOAM Principles and specifically to integrating social values (see the Principles of Fairness and Care). The position of consumers as the only "unregulated" actors in the organic agrofood chain is critically discussed with reference to our ideas for creating consumer commitment in the organic agrofood chain. Finally, we offer some insights on the current status of ethics in higher education, advisory services and research, and how these sectors contribute to an ethically founded organic movement.

13.2.1 Ethical Challenges in Organic Standard, Control and Certification Systems

While discussions and reviews of diverse farmer and consumer values have always been part of the organic movement, assessments of ethics in organic standards and regulations (e.g., DeLind 2000; DeLind and Howard 2008; Padel et al. 2009, 2010, p. 84), and specifically the organic certification system, are relatively recent (Neuendorff and Spiller 2011). Organic production, consumption and markets are regulated and influenced by standards, control and certification, and in turn are influenced by these activities (see Hatanaka, Chap. 3; see Loconto and Van der Kamp, Chap. 4). In this section we review the relevance of ethics in the standards, in public and private regulations, and in control and certification procedures. We are especially concerned with the specific ethical challenges in the certification process and with the philosophical perspective on the regulatory ethics.

13.2.1.1 The Status of Ethics in the Standards and Regulations

We start this discussion with a perspective on current organic farmers' sensitivity and awareness of the IFOAM Principles. The current generation of organic farmers, generally shares a value-based commitment to food quality, environmental protection, limited resource use, individual health, and independence and sustainability (Padel 2005). But these values tend to focus on largely material issues and qualities, and less on social issues, such as fairness or care. As our historical review in Chap. 2 shows, social and economic values always played a role in the historical evolution of the organic movement, but they were expressed differently in each country.

As Jones notes in her historical analysis of values in the organic movement in Australia (see Chap. 11), values associated with health and environment have also been dominate in their organic principles since the time of the organic pioneers in the 1940s. Norwegian studies have shown, that contemporary organic farmers, as compared to their conventional counterparts, are more likely to have larger farms, are more educated but also sensitive to the organic health and environmentally oriented values (Koesling et al. 2008). But they are also more business-minded and pay less attention to the organic ideas that were important for the founders of organic farming. The advisory service of *Bioaustria* confirms the dominance of economic and market issues among the most recent converters, (see Econo/Market type in Sect. 13.1), who pay little attention to social and economic justice issues.[13] These farmers focus on meeting the increasing mass-market demand for organic products. As such, they are subject to some concerns in society about the integrity and creditability of organic.

In a comprehensive review of the implementation of the organic principles in the European Regulation for organic food, Padel and others (Padel et al. 2009) in a series of publications between 2000 and 2009 analyzed the values expressed in the IFOAM Principles. They found that while labor rights are addressed, the values of fairness, equity, respect, justice, animal welfare and future generations are not considered in the standards (Padel et al. 2009). They conclude that overall, the standards reflect a utilitarian approach to organic. Perhaps the broader question is: to what extent can the certification of environmental indications be used to facilitate acceptance of a value based agriculture and food system (Niggli 2005)?

The push to write more uniform organic regulations that the economic interests of large corporately controlled markets and "industrialized" organic production creates numerous conflicts with the IFOAM Principles. Mutersbaugh (2005) argues that while the growth of the organic market requires regulations, the larger distributors,

[13] Oral communication with Christa Größ; Lothar Greger, both Bioaustria, Austria, June 2013.

wholesalers and retailers benefit more from uniform, globalized standards than do organic farmers. These corporately oriented control or regulatory policies also constrain efforts to promote production that is adapted to local conditions.

Guthman (2004, p. 307) point out that large corporations both influence the guidelines and seek to sell the most profitable products. As she notes, since the guidelines deal primarily with production inputs, and less with processing, this favors larger, corporate players at the expense of small and economically weaker partners in the organic system. This results in a loss of social and economic values and ecological qualities (Guthman 2004, p. 307). This leads to regulations that create advantages for large corporations in the organic market while suppressing opportunities for small farmers in niche markets (see Hall and Mogyorody 2002) (see Constance et al., Chap. 9).

Rigby and Cáceres (2001, p. 28) also critically review the social dimension of control and certification. They state: "objectives such as the sustainability of farm families, farm workers and rural communities, which are frequently espoused by organic groups, are simply not amenable to this type of regulation. Individual producers may be committed to such goals, but most standards do not include them, and it is difficult to see how they could." The broader, underlying question concerns whether current certification standards and processes encourage farming based on standardized, contract-like approaches to production and more standardized industrially processed products.

We argue that without a clear statement and a broader expression of social and economic core values in the organic standards, organic risks becoming a weak version of sustainability or of being branded as "green washing" (see Freyer et al., Chap. 5; Constance et al., Chap. 9). Such a development would negatively affect consumer trust and would risk undermining the organic movement. Therefore, we suggest that the integration of social and economic justice and fairness into the IFOAM Standards and related instruments is essential for the future development of the IFOAM Norms (Raynolds 2000; Alrøe et al. 2006; Jaffee et al. 2009; IFOAM 2012). In contrast, alternative regulatory systems that adhere to the IFOAM Norms (IFOAM 2012) offer another way to integrate social, economic and traditional values (Padel et al. 2010, p. 66), and could lead to a broader integration of social values (Padel and Gössinger 2008). However, it should be recognized that farmers are often look at ethics as ideological (Kaltoft 1999). Consequently many leave or are uninterested in joining such an "ideological" movement.

13.2.1.2 Ethical Challenges in the Certification Process

Compared with the early history of the organic movement (see Chap. 2), the current discourse on ethics has been replaced by bureaucratic procedures for documenting the technical conformity of a farmer's practices with the organic standards or regulations. It is significant to note that even the certifiers feel overburdened by the inflation of standards, regulations and inspection requirements (Schmid and Lockeretz 2007).

Certifying agents have become judges of a farmer's compliance, that is, policing to assure that organic practices conform to the regulations. Yet, the certification process creates opportunities for some flexibility in interpreting the regulations (see Hatanaka, Chap. 3; see Loconto and Van der Kamp, Chap. 4). In this way, the certifying agent can be seen as exercising a value-based judgment and not simply carrying-out a matter-of-fact application of a set of clear and unequivocal rules.

The organic control and certification system is not free of conflict with organic values. Is this system "fair" and how can it be tested against the IFOAM Principles? Certification assures that organic products are produced or processed according to the guidelines. Certification also protects farmers against the abuse of product standards or fraudulent practices. But how these certification requirements influence an organic farmer's values is controversial (see Hatanaka, Chap. 3). The IFOAM Norms can be easily used for marketing based on notions of healthy and safe food, just as standards, control and certification are key elements of modern marketing practices. Since these control and certification processes measure the largely technical aspects of production and processing regulations, they create the conditions for using a largely bureaucratic approach to certification. In doing so, the underlying IFOAM Principles are not only less apparent, but the process becomes more susceptible to fraud.

There is increasing evidence that some certifying agencies find it more profitable to work with larger organic farms, and not with smaller farms (Jahn et al. 2005; Clarke et al. 2008, p. 220; Zorn et al. 2013). In response, many certifiers have been criticized for protecting or defending the economic interests of larger or more corporate farms that rely on significant quantities of commercial, off-farm inputs. Some suggest that these certifying agents are insensitive to local and cultural realities, including the needs and conditions of smaller farms. Moreover, some argue that certification is relatively more costly for smaller farms than it is for larger ones using large quantities of off-farm inputs.

In principle, certifiers, as members of an IFOAM Accredited Certification Body should adhere to a codex of conduct that covers at least two important requirements: (1) "The body making or ratifying certification decisions shall be free from any commercial, financial and other pressures that might influence decisions"; and, (2) "Fee structures and other issues related to payment shall not compromise objectivity" (IFOAM 2012, p. 91). These requirements oblige certifiers to support all classes of farmers, specifically including more economically vulnerable smallholder for whom access to the organic market for their income is the key to food security.

13.2.1.3 The Philosophical Perspective on a Regulatory Framework for Ethics

The message of the IFOAM Principles is one of sharing responsibility, and (economic) risk as well as the benefits, for an ecologically, socially and economically just approach to food production, independent of regulatory control or certification system. The question to pursue is: what makes the implementation of ethics into standards, as well as in control and certification systems, so challenging?

In order to address this question, it is useful to recall that morals cannot be controlled by law (Pieper 1994, p. 20). Each person is free to act independently. As Jean Piaget (1896–1980), argues, we can only discuss moral understanding and behavior; ethical orientations are not subject to "compulsion from outside, but guarantee the highest degree of freedom for all members of a community. Only a rule, which fulfills this objective, is a moral rule" (ibid.).

Each individual has the freedom and duty to translate the ethical message of the IFOAM Principles into concrete practices (cf. Meadows et al. 2005, p. 289). But this 'free decision' does not guarantee that individuals make use of the IFOAM Principles. Even if the values defined as standards, this would not always influence concrete practices (Busch 2011).

Instead there is need for individuals to be with a supportive community in order engage in "virtue" oriented action (see Chap. 2). "The paradox is that it is extremely difficult for individuals, even the most environmentally committed, to act without leadership, without strong social support, and without a market structure that makes sustainable living feasible" (Isenhour 2010, p. 151).

While farmers and consumers are relatively open to apply the instrumentalist values of organic, participating in the organic system with respect to social and economic values requires a mental conversion (Källander and Rundgren 2008, p. 6). Such a mental conversion similar to what Hay (2010, p. 168) argues, to be sustainable and resilient in our own lives, as well as in our organizations and communities, we need more commitment and effort—a new way of life, derived from a view of identity in context within a network of relationships and of personal development in service to society.

Is there a potential to implement social and economic values specifically in private guidelines and control systems? From a philosophical (value of freedom), ethical but also practical point of view this integration has its limits (Whyte and Thompson 2010). Separate from a partly broader formalization of social and economic ethics into private regulations, there are participatory farmer-consumer collaborations using non-regulatory more process oriented approaches that could strengthen ethics in these food networks (e.g., Allen 2008).[14] Also, Fair Trade offers formalized practices representing "a particular type of relationship between 'ethical consumers' and low- income producer households through international trade" (Hayes 2006, p. 447).

13.2.2 Consumers Ethics in the Organic Agrofood Chain

The average shopper now finds a bewildering array of organic (and so-called "natural") food products from which to choose in most grocery stores. Not unlike

[14]E.g., non profit food networks/Food Alliances – http://foodalliance.org/

the organic farmer, new consumers as well, confront the need for going through some type of "conversion process" to learn about and adapt their food purchases based on information from friends, advertisements, and the internet.

There are a small number of groups such as the Organic Consumers Association (in the US) with websites that focus on educating consumers. Numerous types of grocery stores in both Europe and the US are well known for offering a wide range of organic products. But with the exception of some organic consumer cooperative stores, organic marketing pays little or no attention to organic values. Only the most knowledgeable, or already converted and committed consumer will take the time to ask about more value-laden dimensions of the organic products on offer. With the rare exception of some smaller cooperative consumer purchasing groups, it is useful to remember that the consumer is the one "free actor" in the organic system. Unlike the farmers or processors, consumers do not have to be "certified" to shop organic.

13.2.2.1 The Organic Ethical Message and Challenges for Consumers

The average consumer interested in buying organic in a supermarket must often choose a product without really knowing what the "certified organic" label really means. Those who have access to, and interest in, shopping in city or farmers' markets are confronted with another choice: they must distinguish between products from "local" organic farmers and the often less expensive, but local, products from non-organic farmers. Moreover, instead of seeking out "organic," many consumers may prefer regional products (Siderer et al. 2005, p. 334).

Furthermore, linking the local production with local markets is now a common corporate marketing approach for non-organic. As Adamoli notes, it is little wonder that consumers become confused about the importance of organic or of local, or that price, convenience or a well-known label (see Adamoli, Chap. 6) will trump their ethically based decision.

Janssen and Hamm (2011) found that consumer knowledge of organic certification schemes is generally low (Padel et al. 2010, p. 23, 81). Several quality assurance systems and labels in the agrofood chain seek to help the consumer navigate the organic world. The growing number and variety of organic labels, regulations and certifications (Schulze 2008) is positive because it offers a measure of customer assurance that a product is organic. But, at the same time, labeling often leads to confusion and a refusal by consumers to read the product information (Thøgersen 2002; Krystallis and Chryssohoidis 2005; Leire and Thidell 2005; Howard and Allen 2006; Batte et al. 2007). In contrast to the wide variety of value driven labels in the US (Boström and Klintman 2006), the model in Sweden integrates stakeholder groups along the agrofood chain. This assures an internal chain oriented control mechanism that allows consumers to trust the label.

Most consumers are unfamiliar with the organic standards (Codron et al. 2006, p. 16), nor are they aware of issues related to their food and values of ecological quality. In the absence of direct relationships with producers, consumers must trust labels and the certification of products (Zanoli, 2004 in Aschemann et al. 2007,

p. 141). The IFOAM Principles are not advertised, even in organic markets. Thus, it is not surprising that consumers are unaware of the ethical foundation of organic. In contrast to numerous and ongoing discussions about organic practices and principles in several farmer journals (Schmid and Kilchsperger 2005; Schmid 2007, 2009b), consumer magazines rarely, if ever, discuss the IFOAM Principles.

Given the numerous constraints and limitations that consumers face, they must rely on the integrity of the labels and the standards and certification processes behind them. To avoid consumer fraud. Janssen and Hamm (2011) suggest that stricter production standards and control procedures would improve the position of products in the organic market.

At the same time, it appears that the actual prices for many organic products are less than consumer perceptions of organic prices. It is commonly observed that consumers do not choose organic because they perceive that organic products are more expensive than conventional products. But Hamm et al. (2007) found that consumers often lack knowledge about the real price of organic compared with non-organic products. In reality, Hamm and others found that organic products were only somewhat more expensive than non-organic products.

To conclude these observations on consumer knowledge about labels, standards and organic certification: even when there is a significant amount of information available, most consumers find it difficult to understand organic ethics. Knowing the ethics, standards, labels and certification systems becomes its own field of competence. It is as if the consumer needs to be trained in the skill of identifying cases of mislabeling, symbols and descriptions of regulation and certification processes (Yiridoe et al. 2005). Furthermore Dimitri and Oberholtzer (2009) argue, that the likelihood of buying organic products is related to education, independent of age, race, or ethnic.

13.2.2.2 Diverse Legal Obligations Among Organic Agrofood Chain Actors

This section discusses how different groups of actors along the organic agrofood chain are part of a regulatory system and how this affects a value-based movement. We describe the different commitments of organic actors and the ethical conflicts that that these differences raise.

A farmer's decision to convert the farm to organic requires a long-term acceptance of the organic system, even though the conversion period is fixed by law for a specific number of years. During this period, farmers often must accept lower prices for their "not-yet-organic" product (e.g., Acs et al. 2007; Kerselaers et al. 2007). If for some reason they fail to meet the organic standards and regulations, they may loose their organic certification (Padel et al. 2010, p. 84) or in the EU in some cases be required to return the 5 years of transition subsidy payments received and be ineligible for organic certified status (cf. Zorn et al. 2013). In response, some organic farmers have dropped their effort to be officially certified. They instead argue that organic is less a question of adherence to standards, but more a question of acting responsibly and creating trust with consumers (see Hatanaka, Chap. 3).

Processors must also meet specified organic regulations and rules for processing, labeling and packaging (see Klonsky and Greene 2005; Luttikholt 2007). However, in contrast to farmers, they do not receive processing subsidies, nor are they required to respect a conversion period before offering organic products. To process or sell both organic and non-organic products in the same business as long as they are handled and stored separately. There is no obligation to continue with organic processing, they are free to leave their organic business at any time.

Retailers and specifically wholesalers are relatively free from regulations for selling organic products. They may offer both organic and conventional products on the same shelf. They are free to discontinue selling organic products at any time. Organic fruits and vegetables must be clearly labeled and displayed separately from conventional products. Similar to processors, they are subject to no specific obligations, except of course to fulfill their contracts with farmers or processors.

As noted earlier, consumers are the only actors in the organic agrofood chain free to act independently of public, governmental regulations or standards (autonomy of the individual to act without control and certification) (Fenner 2010, p. 59). With the exception of those who belong to organic purchasing associations, there are no standards or even guidelines for ethical shopping. Consumers are free to choose to pay a premium price for organic products, or simply to not buy organic.

Is this consumer freedom problematic for the relationships among other actors in the organic system? What are the consequences for each of the other organic actors, including the farmers, processors, wholesalers and retailers? Furthermore, does the commitment to "buying organic" trump the commitment by other consumers to "buying local" or "buying seasonal"? (Halberg et al. 2006, p. 286). In other words, is purchasing organic products regardless of their origin, more important or valuable than purchasing products from a specified region? If consumers are not concerned whether smaller organic producers in the global North or South are fairly and justly remunerated for their labor, then is the value of social and economic justice simply to be left aside (see Fotopoulos and Krystallis 2002; Hill and Lynchehaun 2002; Onyango et al. 2007; Bellows et al. 2008; Hjelmar 2011)?

Consumers are individually responsible for the rationale used to purchase organic. For some, it might be a question of health. For others, eating (local) organic (from smaller farmers) may represent a commitment to the social and economic commitment. The role of the consumer as an actor in the organic system raises the question about the forms of, or opportunities for responsible consumer engagement in the organic system.

However, the formalization of consumer commitment through some type of individual control and certification sounds rather preposterous. Yet, considering the ethical responsibilities of the organic consumer does push us beyond purely instrumentalist discussions and analyses of consumer purchasing behavior. If we appropriate the neo-liberal mantra of "all power to the consumer," then how can we begin to explore the concrete and realistic opportunities for consumers to "vote ethically" with their purchases?

13.2.3 Control and Trust in Farmer-Consumer Collaboration

This section reviews different types of collaborative relationships along the organic agrofood chain. Each contributes to more socially driven and economically balanced approaches that specifically respect the IFOAM Principles of Fairness and Care in non-governmental regulations and certification schemes (Padel and Röcklinsberg 2009). Central is this question: how forms of collaboration and standard-setting processes could operate as a social contract relationship among actors along the organic agrofood chain?

Alternative organic guarantee systems (e.g., participatory guarantee system) offer one framework for such approaches (Padel et al. 2010, p. 66). Others include alternative ethical approaches, such as CSR and Fair Trade (see Hatanaka, Chap. 3) or Corporate Moral Responsibility (CoMoRe), that serves as a tool-kit for food companies (Brom et al. 2006). Do these schemes engender engagement and trust-building and thereby help move beyond bureaucratic control (Giovannucci and Ponte 2005), and what can the organic movement learn from these schemes?

13.2.3.1 Alternative Forms of Standard Setting Processes and Collaboration

The creation of site specific and self-reflexive regulatory processes offers the possibility for the emergence of more collaborative trust and guarantees of ethically driven organic practice. Several actor-defined instruments, indicators or checklists for self-assessment (DeLind and Howard 2008, p. 29) could improve transparency in ethical action that is consistent with the IFOAM Principles (cf. De Wit and Verhoog 2007). Such instruments are an option for the creation of a value based mission statement or contracts between partners along the organic agrofood chain.

Giovannucci and Ponte (2005) note that the state should offer a legal guarantee for the implementation of more ethically oriented instruments. In addition, the actors in a specific organic agrofood chain must accept responsibility (with state approval) to develop private regulations that build on the given standards, but specify their concrete value oriented practices (e.g., NGOs, firms and individuals from civil society). To ensure the integration of such values, Padel et al. (2007b) recommend a procedure of participative and deliberative democracy. This process would ensure the participation of representatives of all relevant stakeholder groups, guided by experts who advise with respect to specified rules of ethical dialogue.

There are several advantages to such a procedure: it builds a direct relationship between producers, certifiers, marketers and consumers, and it ensures a better flow on information between all partners (Giovannucci and Ponte 2005, p. 298). Such a system also allows for a democratic decision making process along the value chain (Giovannucci and Ponte 2005). It is important to keep in mind that these types of collaborative relationships are time consuming for all participants. As such, they require broad and deep commitments, e.g., the IFOAM Principles, underlying the process.

13.2.3.2 Old and New Forms of Trust Based Certification Systems

Smallholder Group Certification based on an Internal Control System (ICS) and Participatory Guarantee Systems (PGS) and other trust-based and often regionally oriented approaches, offer some flexibility in addressing and incorporating social, cultural, ecologically and economically specific conditions. PGS are predominantly established in Southern countries—where farm certification is often not affordable (Halberg et al. 2006; Fonseca et al. 2008).

The PGS is an innovative means of bottom up stakeholder participation in the certification process (Fonseca et al. 2008; Källander 2008). In this process, "the organizational structure of the certification body shall ensure that parties significantly affected by the certification system can participate in the development of its principles and policies" (IFOAM 2012, p. 90). Such Participatory Guarantee Systems "... (are) locally focused quality assurance systems that certify producers based on active participation of stakeholders and are built on a foundation of trust, social networks and knowledge exchange" (Padel et al. 2010, p. 69). They allow economically disadvantaged smallholder farmers to become members of the organic movement and to access the organic markets. They are adapted to local conditions and facilitate certification where financial circumstances make it difficult for farmers to pay international control and certification fees. In short, PGS can be adapted to the conditions and economic realities of smallholder farmers (Parrott et al. 2006).

They also help minimize certification costs for European organic farmers, especially those who rely on direct marketing and short supply chains (Padel et al. 2010). The PGS can be seen as a socially and culturally sensitive certification and accreditation process for making the IFOAM Principles of Fairness and Care operational with respect to social and economic justice and access to food for the poor. To do so, PGS requires increased personal responsibility, learning, knowledge and experience and a locally adapted approach of control and certification, involving consumers and other local stakeholders, encouraging improved food access.

Schmid offers a different approach to guaranteeing organic (Schmid 2009a). He suggests that organic guidelines should serve primarily as an orientation for farmers and consumers. In the absence of a third party certification process and procedures, this approach is limited to cooperations between farmers and consumers in a local context. They interact outside of a retailer system, which allows direct communication between the actor groups. In these cases any organic produce is a good of trust. It must stand on its own without any labeling or certification (Darby and Karni 1973; Eichert and Mayer 2008).

Finally, we should keep in mind that there is a consensus that any kind of formalized control and certification is necessary even in trust-based systems following on voluntary agreements (Michelsen 2002a; Kratochvil et al. 2005; Schekahn and Thomas 2008, p. 103; Schüle 2009). But, as Sterba (2003, p. 125) note, "the EU Regulation on organic production should remain a framework and not a listing of detailed prescriptions," to avoid private regulations built on the EU prescription from becoming too complicated.

13.2.3.3 Ethics in Other Standards and Certification Systems

This section briefly discusses some of the ethical dimensions of other standards and certification systems in order to identify their potential as collective frameworks for consumers committed to their role in the organic agrofood chain.

Corporate Social Responsibility. The growing use of corporate social responsibility (CSR) has recently become important for many large, corporate organic food industry players (Pivato et al. 2008). It seeks to be used in some cases for purely profit reasons, with little attention to underlying values (Aupperle et al. 1985; Windsor 2001; Garriga and Melé 2004; Banerjee 2008), thus creating significant controversy among "organic proponents" (Guthman 1998, p. 143). These third-party certifications can be seen as an opportunity to strengthen private governance as well as more socially and environmentally sustainable approaches (see Hatanaka, Chap. 3). But large supermarket chains also use CSR as part of their profit-oriented strategy (Blomqvist and Posner 2004; Shaw 2006; Fliess et al. 2007; Colls and Evans 2008).

Fair Trade. Fair Trade certification embodies many principles similar to those of the IFOAM Principles.[15] Both are based on a holistic concept that refers to environmental, socio-economic, cultural and agricultural values (Cierpka 2000). Both also embody a commitment to social justice (Raynolds 2000), and concerns with food security, food sovereignty, as well as help for farmers in the South to feed their families and to generate income (Gruber and Hauser 2010). The Fair Trade ecological requirements are less detailed than those of the IFOAM Norms, but they provide a comprehensive approach to social, cultural and socio-economic qualities. The regulations and certification practices, with respect to social and economic justice issues in Organic and Fair Trade, already reflect a measure of collaboration between these two movements (EFTA 2001; Moore 2004; Sachs et al. 2007; Nicholls and Opal 2008).

FLO. The FLO (Fairtrade Labelling Organizations International) Charta states: "Fair Trade is, fundamentally, a response to the failure of conventional trade to deliver sustainable livelihoods and development opportunities to people in the poorest countries of the world" (FLO 2009). Groups associated with this alternative trade approach adhere to clearly identify social-economic ethically legitimized criteria to guide their practices.

The FLO groups are split into two wings that are roughly comparable to the bifurcation in the organic movement between the smaller scale and the large, corporate farms (Constance et al. 2008). The first is similar to PGS and specifically related to smallholder farmers.[16] It is "idealistic ... [with respect to] trade justice, structural change and human solidarity, defined by shared understandings of fairness, grassroots development and north south partnerships" (Dolan 2010, p. 9). The second has a more "instrumentalist focus on certification, standardization and market expansion, which certifies and markets fair trade products through

[15] http://www.fairtrade.net/standards.html

[16] http://www.fairtrade.net/fileadmin/user_upload/content/2011-12-27_SPO_EN_FINAL.pdf

mainstream distribution channels, with sales increasingly targeted toward supermarkets and transnational food corporations" (cf. De Schutter 2003; Dine and Shields 2008).

This second group also refers to Corporate Social Responsibility (CSR), but largely with a utilitarian orientation that protects corporate profit maximization (Welford et al. 2003). Raynolds has suggested that: "[Many] corporations are trying to bolster their legitimacy by adopting the rhetoric of environmental and/or social responsibility, though typically this proves to be little more than a corporate face lift" (Raynolds 2000, p. 299). Moore similarly notes: "This presents Fair Trade with a dilemma involving 'the dilution of fair trade ideology by the market' in which the organizations find themselves either remaining 'pure' but probably marginal, or aligning with the mainstream and 'losing their soul'" (2004, p. 83).

Both Moore and Raynolds underline the importance of collaboration between producers and consumers and the need for an increased awareness of the meaning of Fair Trade. However, until now, both the organic and the Fair Trade movements confront similar challenges that stem from operating within a dominant corporate, capitalist market. Moreover, organic is also known for being well established in the conventional trade and market system—following the same rules in pricing and advertising as that of non-organic products that specifically counteracting the social and economic oriented ethics of the Fair Trade approach.

13.2.3.4 Collaboration with Alternative Movements

The organic farming movement incorporates numerous environmental issues and aligns with the commitment of the environmental movement to healthy and diverse food, as well as broader environmentally related health and medical concerns (cf. O'Rourke 2006). These values open organic to collaborations with a range of other value driven groups. Some of these new "partnerships" could include climate alliances, nature protection, health initiatives,[17] alternative energy, and social activists. Collaborative relationships are also possible with established institutions that have easily aligned interests, motivations or values, including health insurance agencies, churches, or initiatives for "Bio-Regions" that involve more than agricultural activities (Polonsky 1994; Peattie and Crane 2005; Santucci 2009). For example, "consumer certificates" that confirm the delivery of an organic food box legitimate the reduction of health insurance bills.[18] This is a realistic approach to honor consumer practices, without compromising the freedom of choice.

From this perspective organic is an issue for a civil society in which farmers are one partner among others in an increasingly value driven citizen-consumer movement. Such broader concerns are not limited to the organic movement.

---

[17] See e.g., www.soilassociation.org/organicstandards

[18] This approach is already established by health insurance companies in US: http://rodaleinstitute.org/2013/from-csa-to-hmo/; http://www.csacoalition.org/our-work/csa-insurance-rebate/

They also include movements such as Slow Food that celebrate traditional food, including the agro-ecological considerations of organic production, as well as the broader value orientation of farmers and consumers (Guthman 2003; Murdoch and Miele 2004). Members of these initiatives bring new ethically oriented social and economic concerns to the attention of the organic movement (Jones et al. 2003).

How relevant are these similarities and relationships in value patterns for broadening the group of committed organic consumers? With respect to the growth of the organic market, it is important to remember that the demand for organic products is not high, and we can not assume that members of environmentally oriented organizations or green party members will consume 100 % organic. The demand for organic products might be much higher if all members of these alternative initiatives would consume organic products. Even well educated environmentally oriented individuals who are sensitive to social and economic justice and interested in organic food, are generally part time organic consumers. Obviously, the complexity of the organic holistic approach asks for a highly reflective actor (cf. Giddens 1991), as those who engage based on ethical principles. Reflectivity however, is not a guarantee that leads to an ethically oriented behavior (Vermeir and Verbeke 2006, p. 170). In addition, several types of practical issues might hinder even convinced organic consumers to purchase predominantly organic products.

13.2.3.5 Collaborative Relationships Between Farmers and Consumers

In a competitive society, characterized by neo-liberalism and individualism, for both individuals and companies, "sharing" is only a preferred choice if there is some added economic value. Independent from public standards, only CSAs and similar collaborative arrangements between farmers and consumers provide a framework for practicing social and economic justice. In contrast Fair Trade is engaged into economically value based collaborations between farmer, processor, trade, retailer and consumers that is largely oriented to an international context, where in most cases no direct contact between farmers and consumers exists.

There are several different types of collaborative, contract and/or trust based relationships between farmers and consumers at farm, local, urban or regional levels (e.g., farmers' markets, box schemes, farm gate sales, CSAs, public kitchens, cooking movements) (La Trobe and Acott 2000; Van Der Ploeg 2000; Rigby and Cáceres 2001; Martinez 2010; Løes and Nölting 2011). Many of these now address concerns with economic justice, including social premiums, advanced payments to prevent smallholder indebtedness, and contracts for farmers that allow for long-term planning (e.g., Renard 2003, p. 90).

These approaches represent ways in which well-informed consumers trust farmer without the need for extensive audits, and in which consumers also assume responsibility for the economic stability of the farm through shared ownership or financial investment (Vos 2000, pp. 246, 251). These relationships have emerged from actor-based interpretations of the IFOAM Principles in various sectors and arise in the practices of local, urban, or regional organic agrofood chains, interpreted and formulated for specific cases, regions or issues of collaboration.

New governance structures (Padel et al. 2010, p. 73), accompanied by participatory learning processes (cf. International Institute for Environment and Development 2009) and based on self-responsibility (e.g., Schmid 2010) could help to develop these local, regional as well as cultural specific approaches. The cultural role of these collaborations is significant because these partnerships provide value based, socially supportive economic features. These include work on farm or in a food co-op, offering lowers prices for low-income customers or for large families, while providing more security for farmers through direct financial support by consumers or the security that consumers will regularly by their products. All these initiatives build on what Alrøe and Kristensen (2004) call "nearness" as a key precondition for building trust between stakeholders.

These forms, not specifically, but generally, are linked with organic farming in which consumers seek a deeper relationship with farming and the land, and that are indicative of new lifestyles and life politics (Ravenscroft and Taylor 2009, p. 215) what is termed "food citizenship." Consumers participate and actively contribute to the agrofood system (ibid., p. 216). It is about de-commodifying food, and the reintegration of people and land (ibid., p. 217). Other approaches include community supported agriculture, farming that shares the risk and rewards of the farming process. These versions of community farming can be classified in terms of three foundations (ibid., p. 222): (1) Co-operations, where non- hierarchical networks are central; (2) New forms of non-profit risk sharing economies; (3) New forms of property ownership. But we have to keep in mind that those collaborations require intellectual and time commitments by the consumer—a commitment that is often difficult to make solely for the purpose of buying food.

There are other wholesaler and retailer systems in between direct marketing formats and corporate market chains. Nationally oriented collaborative relationships between retailer and farmers always raise the issue of how to fulfill consumer demand for products that are grown or processed "outside" the region. As Clarke et al. (2008) describe, some entrepreneurs may bundle and share products, or sell products 'from the region-for-the-region'. When organized on a regional scale, and in order to handle regionally produced and processed food, regionalized businesses can reduce the food miles and follow the principles of nearness (see Byrne et al. 2006). Those models also combine global trade and an eco-localism approach (Curtis 2003; Ritzer 2003). Nationwide and global marketing arrangement of course, does not include direct relationships between farmers and consumers.

In contrast, partnerships between farmers, consumers and large retail firms offer only limited potential for participatory approaches or the significant application of the specific values of the IFOAM Principles related to social and economic justice (Principles of Fairness and Care). That is explained by the unbalanced power distribution between the partners. This raises the question: is the application of the IFOAM Principles limited to specific business models? Currently, we lack innovative ideas on adequate approaches with large retailers that are sensitive to power relationships, or to social and economic justice. The Fair Trade approach opens the door for bridging the gap between the farmer and the consumer. As long as several types of nationally or internationally organized agrofood chains are problematic with respect to transparency, social and economic justice, it is primarily the fair trade approach that enables practicing the Principle of Fairness and Care.

13.2.4 The Role of Ethics in Education and Extension

There is a consensus that organic practice is a learning challenge (Seppänen 2002; Dunn 2005) that requires both knowledge and information (Morgan and Murdoch 2000; Schneeberger et al. 2002; Padel 2005; Zepeda and Deal 2009). For farmers, there is commonly a "lack of information on agro-ecology and necessary skills to manage complex farms [that acts] as a major barrier to the adoption of sustainable agriculture" (Scialabba and Hattaam 2002, p. 144). When we look at the challenge to practice in line with the holistic intention of IFOAM Principles, this obviously goes far beyond the pure technical aspects of organic food production, marketing or consumption.

Therefore, questions arise about opportunities for the stakeholders to learn and get knowledgeable specifically about organic ethics. Further more, does higher education in organic and organic advisory services prepare actors to act ethically? And what could be done to bring in organic ethics into education and extension?

13.2.4.1 Learning Organic Ethics in Higher Education

In recent years, several teaching programs in organic farming largely at the master's level, or in summer schools, have been started at the University level.[19] Often students with idealistic, political and environmental motives, and not their professors or administrators, took the initiative to set up these activities (e.g., Parr and Van Horn 2006). Ethics in organic teaching are offered in some of these programs (e.g., Parr and Van Horn 2006; Parr et al. 2007; Gullino and Pugliese 2008; Francis et al. 2012),[20] however, the relevance of ethics in organic agriculture education system is rather low.

13.2.4.2 The Role of Extension Services and Information to Promote Ethics

Are the IFOAM Principles similar to the standards well known and applied by the organic advisors? Often the advisors of organic organizations provide their

---

[19](Hill and MacRae 1992; Vogl and Hess 1999; Delate and Dewitt 2004; Mittelstrass 2005; Parr and Van Horn 2006; Francis 2009; Francis et al. 2011).

[20]See also https://online.boku.ac.at/BOKUonline/lv.detail?clvnr=267112

own specific values instead of the IFOAM Principles (Schmid 2007; Gottwald and Boergen 2009).[21] Their perspectives overlap somewhat with the IFOAM Principles. But these advisors more often offer technical translations rather than the ethical dimension. To what extent values have any relevance in the advisory process is not well known, but the continuing trend toward economically based decision for converting to organic (Padel 2001; Rigby et al. 2001) suggests that discussions related to values are of little interest.[22]

There are additional institutional challenges when private and public advisory services agencies for non-organic farmer also advise the organic sector (Michelsen et al. 2001; Michelsen 2002b). Even when such services cover a wide range of organic topics from soil management to marketing, they rarely are capable of, or interested in introducing the IFOAM Principles. Value conflicts arise when advisors do both—advise on organic as well non-organic farmers.

When we look for other discussions on values in the organic system, currently organic farmer journals, farmer conferences and on-line organic sources are dominated by debates and information about technical issues, control and certification, and marketing. They rarely discuss the principles or the philosophical foundations of the organic movement, as it is the same on organic farmer and major organic scientific conferences.[23]

13.2.4.3 Creating the Conditions and Environment for Learning Ethics

If ethics, and specifically the IFOAM Principles, are to play a more central role in future development of the organic agrofood chain, information, knowledge and learning opportunities are required to bring these ethics to the center of a reflexive and on-going discourse (cf. Vermeir and Verbeke 2006, p. 170; Stock 2007) in all types of organic learning activities that accompany the organic movement (Geier et al. 2007, p. 271).

These activities should include critical self-reflection on values within the context of everyone's socio-cultural, economic, ecological and political conditions, and a process for sharing widely the various interpretations of values along agrofood chain. While organic advisors currently do not inform producers about values (e.g., Khaledi et al. 2007, p. 39), they could begin to play a constructive role in stimulating such discussions. A precondition for this step is to establish an educational program for advisors to teach ethics in organic farming. Advisors could also serve as ambassadors between the two agricultural worlds (organic,

[21] E.g., http://www.bioland.de/bioland/bioland/die-sieben-bioland-prinzipien.html

[22] Christa Größ (leader of Bioaustria advisory service), personal communication, June 2013.

[23] See e.g., the latest German speaking scientific conference on organic farming 2013: http://www.wissenschaftstagung.de/; Bio Austrian farmer days 2013: www.bio-austria.at/bauerntage – at both conferences there was not much awareness on values and specifically the IFOAM Principles are practically not part of any presentation; in contrast see Millar and West (2009).

non-organic) and different value sets of farming (see Constance et al., Chap. 9 and Hunt et al., Chap. 10), e.g., promoting animal welfare (Lampkin et al. 1999, p. 20), or other sustainable and ethically oriented practices.

To initiate such learning processes, further curricula, including organic ethics, needs to be developed at the university level. In practice (Organic) consumer and organic farmer organizations and journals could be important facilitators for bringing the IFOAM Principles and ethical concerns into the center of a societal debate on food and agriculture.[24]

Morgan and Murdoch (2000) point out that in the non-organic system knowledge tends to be distributed by input suppliers, while in the organic system the farmer is the "knowing agent". That is, farmers have to (re)learn farming, understand the farm as an organism (Paull 2006, 2011), where ecological, disciplinary and systems knowledge is integrated with respect for both human and nature's capacities and demands. The deeper meaning of this notion is also valid for other actors along the organic agrofood chain: the organic agrofood chain should be seen as one organism.

Creating open discussion and exchange "platforms" around the organic agrofood chain and other societal movements could help to bring value oriented discourses of the organic movement from the sidelines into the center of societal debates on food and farming. Such discussions would foster the development of a shared "language" that could allow "foreigners" to gain access to the movement to understand (gain access to) the organic value terminology and thereby consider becoming members of this movement (cf. Baumann 2008). Gössinger and Freyer (2009b, pp. 79, 80) made several suggestions to strengthen the ethical approach through training and advisory services. These are: "(1) a written guide on the development of organicPlus (=ethically based activities that go beyond the IFOAM Standards); (2) specific knowledge of ethical aspects: in particular, a deeper reflection on the IFOAM Principles; (3) written company documentation which describes and reflects their organicPlus approach."

13.2.5 "Organic" Ethics in Science

Over the last decade, research in organic has grown (Watson et al. 2006). But to what extent have ethical question been integrated in organic research? In this sub-section we focus on the following: the relevance of studies of ethics in the organic agrofood chain and studies that specifically deal with IFOAM Principles; the relevance of "organic" scientists personal ethics; and some ideas on re-orienting an ethical driven "organic" research. We conclude with some preliminary thoughts on an ethically oriented organic research agenda.

---

[24]Some examples already exist e.g., http://www.konsumentenverband.ch/; http://www.organicconsumers.org/ (see also Tregear et al. 1994; Browne et al. 2000; Harper and Makatouni 2002; McEachern and Mcclean 2002; Tarkiainen and Sundqvist 2005; de Magistris and Gracia 2008).

13.2.5.1 The Relevance of Ethics and IFOAM Principles—An Issue of Science

There are a series of in-depth investigations about the theoretical foundations of IFOAM Principles (e.g., Kaltoft 1999; Alrøe and Kristensen 2000, 2004; Tybirk et al. 2004; Freyer 2008; Padel et al. 2009). Additionally, there is comprehensive research on animal welfare on organic farms that refers to the standards or EU-Regulations, but with some exceptions, not the IFOAM Principles (Vaarst 2004; Vaarst et al. 2005; Aerts et al. 2006; Sundrum 2007; Vaarst and Alrøe 2012). In addition, there are several studies of organic ethics and values of numerous organic actors from different perspective, as illustrated in this volume. Following the literature, the focus in this volume has been mainly on farmers and consumers, and less on other actors along the organic agrofood chain, e.g., certifiers, advisors, processors or retailers ethical background. Much current research refers to ethical issues, but there is little or no research on the knowledge of organic actors specifically of the IFOAM Principles (Padel 2005; Gössinger et al. 2009c). Questionnaires rarely ask directly about the knowledge, significance, or the role of IFOAM Principles, rather than values and ethics more indirectly. Instead, researchers are more focused on interpreting to what degree the answers of interviewees conform to one of the IFOAM Principles. Do scientists assume *a priori* that organic actors in general do not know the Principles and therefore they do not refer on those in their questionnaires? If so, then there exists a communication gap between the administrators of the Principles, their communication strategy and the organic movement, but also between researchers and the IFOAM strategists.

13.2.5.2 The Value System of Organic Researchers

Several researchers write about ethical issues, such as animal welfare, or connecting ethics and the IFOAM Principles (see also Browne et al. 2000; Darnhofer 2006; Lund 2006; Zollitsch et al. 2007; Vaarst and Alrøe 2012) or with plants (see Willemsen 2009). Many others, some of whom are cited in this volume, do research on ethics and organic agriculture in social science or philosophy (e.g., Alrøe and Noe 2008; Thompson 2010). But what is known about the values of "organic" researchers? In short, little or nothing.

Furthermore, there is no information about the relevance of ethics to the formulation of research programs. Those who do technically oriented research on organic do not specifically identify an ethical approach in conceptualizing their research that could be linked to the IFOAM Principles. They do not address this issue when they formulate their research questions or collaborate with farmers or do science. Moreover, there is no tradition in calls for organic research proposals of asking how the proposed research contributes to ethical discourses in organic or corresponds to the general framework of the IFOAM Principles.

13.2.5.3 Strengthening an Ethically Oriented Organic Research Agenda

From the perspective of science studies, there are at least three steps that could be taken to bring the IFOAM Principles into national and international organic farming research agendas (Geier et al. 2007, p. 272; Gössinger and Freyer 2009b, pp. 80, 86).

- First, the Principles could be applied to help frame organic research. This could be done from two different perspectives. The first is to identify how the research objectives are consistent with the IFOAM Principles, and second how the results shed new light on the Principles, and requires for modifications. With respect to the stakeholders participating in the research, not only the role of the reflexive farmer and consumer as the ones who engage in ethics, but also that of the certifier, advisor, researcher, processors or trader must be studied more in detail.
- Second, there is obviously a gap between organic Principles and practices; the Principles are not well known in daily life. Often it seems that the Principles are not worth more than the paper on which they are written. To better understand these phenomena and to engage in reflective processes supported by transdisciplinary and action research oriented methodologies, these issues should be high on the organic research agenda.
- Third, it is now critical to review and assess how the Principles stand the test in new and diverse practical situations and what kind of organic revision processes could be considered. This is specifically relevant in the context of new developments in the WTO regulations, as well as in European and US agro-policies. More studies are needed to better understand the relevance of the ethical perspective of the IFOAM Principles to solve future challenges in the context of food security, the environment, social and economic justice, climate change and the energy sector.

13.3 The Organic Future Within a Societal Context

In this section, we frame our discussion in terms of the opposing value-based perspectives on organic: the neo-liberal and the social ecological perspective. We argue that the separation between an individualist-materially and a collectivist-immaterial position (Horley 1992; Veenhoven 1999; Meeusen et al. 2005), that is also seen in "organic bifurcation" (see Constance et al., Chap. 9), reflects broader societal mega trends (Beck 2002; Gilg et al. 2005).

Organic agrofood systems are not independent from, but are influenced by societal dynamics. That is, organic is part of societal differentiation toward "multiple modernity" (Eisenstadt 2002) that represents a complex of diverse value sets (Rosin and Campbell 2009) in which consumers specifically make individual decisions on how they deal with the organic issue. The dichotomy in organic is not different from ongoing societal trends toward individualization (see Bachmann-Medick 2006).[25] This is important to recognize when it comes to questions about the future of a value driven organic agrofood system. It explains that the organic pathway is not something independent from societal surroundings. In this context, much more can be done to explore and make more explicit the broader societal relevance of the Principles (Alrøe and Kristensen 2004).

Based on that and with reference to the discussions in this volume, and specifically in this chapter, we review perspectives for a value driven organic approach in the future that also responds to societal realities.

13.3.1 Organic Driven by Industries

From the beginning of the organic movement, and despite the diversity of its origins in different countries and specific societal conditions, it has always been value driven. This value orientation, described with the IFOAM Principles offers an ethical foundation for acting in the whole organic agrofood chain.

Several contributions in this volume demonstrate that these values are often ignored, or at best incompletely translated into practice. They recognize that it is challenging to bridge "theory (value as well as agreed structure) and practice (implementing value and structure respectively)" (Padel et al. 2007b, p. 80). From that perspective, it is not surprising that organic has become an instrumentalized food system subordinated to the economic interests of global food chain actors and the needs of a consumer elite interested in convenience and healthy food.

Within this context, the political and industrial influence on the definition of organic is becoming increasingly significant and with that the power relations in the system are shifting to more hierarchical and institutionalized bureaucratic structures. Ownership and decision-making power in organic has been transferred from the pioneers who acted in a regional context to large corporations (Aschemann et al. 2007, pp. 134, 135). The increasing power of politics and industries in the organic system is a critical development, because the original organic idea of growing toward a social movement already becomes lost (Vogl et al. 2005).

This development, also described as "conventionalization" leads to the critique that organic cannot hold its promise as an alternative to the neo-liberal economies (Campbell and Coombes 1999; Guthman 2008) (see Constance et al., Chap. 9). Furthermore, commercial organic marketing commonly obscures the ethical foundation of organic ("Counter-hegemony or bourgeois piggery") (Johnston 2008). Even more so, organic and related food movements, such as Slow Food, are criticized as being no more than business strategies (Nosi and Zanni 2004; McMichael 2009).

[25] Even in religious movements, we see this dichotomy between organic or non-organic agriculture justified in different arguments. This explains that one common cultural and spiritual identity does not automatically lead to a similar agricultural approach (see Brock and Barham, Chap. 12).

Lyons (2007, p. 12) characterizes the growing number of conventional supermarkets that handle organic as a type of "colonization of … organic …". In doing so, he suggests that this marketing strategy could de-politicize organic. It might be good for promoting organic sales, but participating in conventional or mainstream marketing also distances the critique of the conventional food systems that organic implicitly represents. Thus, the more that organic products are concentrated in supermarkets, the fewer the opportunities for policies or consumers to actively shape relationships with the farmer as they might in regional markets. However, international fair trade marketing represents a hybrid approach since it does appeal to bringing farmers, processors, traders and consumers together around defined values, as well as a political message, even in the absence of close physical proximity.

13.3.2 Organic as a Socio-ecological Movement

With the exception of neo-liberal influences, organic has become a social movement with a well-identified and consistent set of values that includes food safety issues (seasonal and regional food consumption), human health and strict environmental standards (Hess 2004). These values are shared with several other societal movements, and therefore organic has become more than a niche. Organic has become a lifestyle that includes a fundamental and reflexive re-orientation of individual, collaborative, cultural and societal goals and practices (cf. Bachmann-Medick 2006). There is a growing trend in Western society that crosses most political boundaries (conservatives, liberals, and green) that calls into question the current, dominate model of agriculture and seeks a new orientation that often includes organic (UNCTAD 2013). In general, more people are more sensitive to the importance of global solidarity (Hechter 1990), e.g., through Fair Trade (Raynolds 2000), new ideas for a holistic and ecological oriented economy in intentional communities (transition towns and urban farming) (Grundmann et al. 2006; Grundmann and Kunze 2012), and in new types of political consumerism (Jacobsen and Dulsrud 2007). Organic is integrated in these new movements and debates over values are critical to on-going reviews of its goals and practices.

Organic agriculture has become a catalyst for sustainable lifestyles (Gilg et al. 2005) and for some, organic represents a vision of social ecological change for society as a whole that includes much more than food and agricultural production. Organic has become a counter-culture that contradicts elements of the neo-liberal system such as WTO inspired agricultural liberalization (see McMichael 2009, p. 142). However organic is still a minority movement. Therefore, the growth of the organic movement is dependent on the collaboration of other societal movements with similar value orientations.

References

Acs, S., P. Berentsen, and R. Huirne. 2007. Conversion to organic arable farming in The Netherlands: A dynamic linear programming analysis. *Agricultural Systems* 94(2): 405–415.

Aerts, S., D. Lips, S. Spencer, E. Decuypere, and J. De Tavernier. 2006. A new framework for the assessment of animal welfare: Integrating existing knowledge from a practical ethics perspective. *Journal of Agricultural and Environmental Ethics* 19(1): 67–76.

Aertsens, J., K. Mondelaers, and G. Van Huylenbroeck. 2009a. Differences in retail strategies on the emerging organic market. *British Food Journal* 111(2): 138–154.

Aertsens, J., W. Verbeke, K. Mondelaers, and G. Van Huylenbroeck. 2009b. Personal determinants of organic food consumption: A review. *British Food Journal* 111(10): 1140–1167.

Allen, P. 2008. Mining for justice in the food system: Perceptions, practices, and possibilities. *Agriculture and Human Values* 25(2): 157–161.

Allen, P., M. FitzSimmons, M. Goodman, and K. Warner. 2003. Shifting plates in the agrifood landscape: The tectonics of alternative agrifood initiatives in California. *Journal of Rural Studies* 19(1): 61–75.

Alrøe, H.F., and E.S. Kristensen. 2000. *Research, values, and ethics in organic agriculture-examples from sustainability, precaution, nature quality, and animal welfare.* Paper presented at the Eursafe 2000: Two systems-one world, Copenhagen.

Alrøe, H.F., and E.S. Kristensen. 2002. Towards a systemic research methodology in agriculture: Rethinking the role of values in science. *Agriculture and Human Values* 19(1): 3–23.

Alrøe, H.F., and E.S. Kristensen. 2004. Basic principles for organic agriculture: Why? And what kind of principles? *Ecology and Farming* 36: 27–30.

Alrøe, H.F., and E. Noe. 2008. What makes organic agriculture move: Protest, meaning or market? A polyocular approach to the dynamics and governance of organic agriculture. *International Journal of Agricultural Resources, Governance and Ecology* 7(1): 5–22.

Alrøe, H.F., J. Byrne, and L. Glover. 2006. Organic agriculture and ecological justice: Ethics and practice. In *Global development of organic agriculture: Challenges and prospects*, ed. N. Halberg, H.F. Alroe, and M. Trydeman Knudsen, 75–112. Wallingford: CAB International.

Arvola, A., M. Vassallo, M. Dean, P. Lampila, A. Saba, L. Lähteenmäki, and R. Shepherd. 2008. Predicting intentions to purchase organic food: The role of affective and moral attitudes in the theory of planned behaviour. *Appetite* 50(2): 443–454.

Aschemann, J., U. Hamm, S. Naspetti, R. Zanoli, and W. Lockeretz. 2007. The organic market. In *Organic farming: An international history*, ed. W. Lockeretz, 123–151. Wallingford: CAB International.

Aupperle, K.E., A.B. Carroll, and J.D. Hatfield. 1985. An empirical examination of the relationship between corporate social responsibility and profitability. *Academy of Management Journal* 28(2): 446–463.

Bachmann-Medick, D. 2006. *Cultural turns: Neuorientierungen in den Kulturwissenschaften.* Reinbeck: Rowohlt Taschenbuch.

Baker, S., K.E. Thompson, J. Engelken, and K. Huntley. 2004. Mapping the values driving organic food choice: Germany vs the UK. *European Journal of Marketing* 38(8): 995–1012.

Banerjee, S.B. 2008. Corporate social responsibility: The good, the bad and the ugly. *Critical Sociology* 34(1): 51–79.

Barnett, C., P. Cloke, N. Clarke, and A. Malpass. 2005. Consuming ethics: Articulating the subjects and spaces of ethical consumption. *Antipode* 37(1): 23–45.

Batte, M.T., N.H. Hooker, T.C. Haab, and J. Beaverson. 2007. Putting their money where their mouths are: Consumer willingness to pay for multi-ingredient, processed organic food products. *Food Policy* 32(2): 145–159.

Baumann, Z. 2008. *Does ethics have a chance in a world of consumers?* Cambridge, MA: Harvard University Press.

Beck, U. 2002. *Individualization: Institutionalized individualism and its social and political consequences*, vol. 13. London/Thousand Oaks/New Delhi: Sage Publications.

Bellows, A.C., B. Onyango, A. Diamond, and W.K. Hallman. 2008. Understanding consumer interest in organics: Production values vs. purchasing behavior. *Journal of Agriculture & Food Industrial Organization* 6(1): Article 2. On-line: http://www.bepress.com/jafio/vol6/iss1/art2. Accessed 09 Aug 2014.

Bennett, R., and R. Blaney. 2002. Social consensus, moral intensity and willingness to pay to address a farm animal welfare issue. *Journal of Economic Psychology* 23(4): 501–520.

Best, H. 2008. Organic agriculture and the conventionalization hypothesis: A case study from West Germany. *Agriculture and Human Values* 25(1): 95–106.

Blomqvist, K.H., and S. Posner. 2004. Three strategies for integrating CSR with brand marketing. *Market Leader Summer:* 33–36.

Bookchin, M. 1982. *The ecology of freedom*. Palo Alto: Cheshire Books.

Born, B., and M. Purcell. 2006. Avoiding the local trap scale and food systems in planning research. *Journal of Planning Education and Research* 26(2): 195–207.

Boström, M., and M. Klintman. 2006. State-centered versus nonstate-driven organic food standardization: A comparison of the US and Sweden. *Agriculture and Human Values* 23(2): 163–180.

Brom, F., E. de Baker, M. Deblonde, and R. de Graaff. 2006. *Corporate moral responsibility*. The Hague: LEI.

Brown, C., and S. Miller. 2008. The impacts of local markets: A review of research on farmers markets and community supported agriculture (CSA). *American Journal of Agricultural Economics* 90(5): 1298–1302.

Browne, A.W., P.J. Harris, A.H. Hofny-Collins, N. Pasiecznik, and R.R. Wallace. 2000. Organic production and ethical trade: Definition, practice and links. *Food Policy* 25(1): 69–89.

Buck, D., C. Getz, and J. Guthman. 1997. From farm to table: The organic vegetable commodity chain of Northern California. *Sociologia Ruralis* 37(1): 3–20.

Burch, D., and R. Rickson. 2001. Industrialised agriculture: Agribusiness, input-dependency and vertical integration. In *Rurality bites: The social and environmental transformation of rural Australia*, ed. S. Lockie and L. Bourke, 165–177. Sidney: Pluto Press.

Busch, L. 2011. *Standards: Recipes for reality*. Cambridge, MA: MIT Press.

Byrne, J., L. Glover, and H.F. Alrøe. 2006. Globalization and sustainable development: A political ecology strategy to realize ecological justice. In *Global development of organic agriculture*, ed. N. Halberg, H.F. Alrøe, and M. Trydeman Knudsen, 49–74. Wallingford: CAB International.

Campbell, H., and B. Coombes. 1999. Green protectionism and organic food exporting from New Zealand: Crisis experiments in the breakdown of Fordist trade and agricultural policies. *Rural Sociology* 64(2): 302–319.

Cierpka, T. 2000. *Organic agriculture and fair trade two concepts based on the same holistic principal*. Retrieved from: http://www.ifoam.com Accessed.

Clarke, N., P. Cloke, C. Barnett, and A. Malpass. 2008. The spaces and ethics of organic food. *Journal of Rural Studies* 24(3): 219–230.

Cliath, A.G. 2007. Seeing shades – Ecological and socially just labeling. *Organization and Environment* 20(4): 413–439.

Codron, J.-M., L. Siriex, and T. Reardon. 2006. Social and environmental attributes of food products in an emerging mass market: Challenges of signaling and consumer perception, with European illustrations. *Agriculture and Human Values* 23(3): 283–297.

Colls, R., and B. Evans. 2008. Embodying responsibility: Children's health and supermarket initiatives. *Environment and Planning A* 40(3): 615.

Constance, D.H., J.Y. Choi, and H. Lyke-Ho-Gland. 2008. Conventionalization, bifurcation, and quality of life: Certified and non-certified organic farmers in Texas. *Southern Rural Sociology* 23(1): 208–234.

Constance, D. H., Choi, J. Y., and D. Lara. 2013. Social dimensions of organic production and systems research. *Alliance of Crop, Soil and Environmental Sciences, Crop Management* 12(1). https://dl.sciencesocieties.org/publications/cm/articles/12/1/2012-0429-01-RV. Accessed 09 Aug 2014.

Cooley, J.P., and D.A. Lass. 1998. Consumer benefits from community supported agriculture membership. *Review of Agricultural Economics* 20(1): 227–237.

Curtis, F. 2003. Eco-localism and sustainability. *Ecological Economics* 46(1): 83–102.
Darby, M.R., and E. Karni. 1973. Free competition and the optimal amount of fraud. *Journal of Law and Economics* 16(1): 67–88.
Darnhofer, I. 2005. Resilienz und die Attraktivität des Biolandbaus für Landwirte. In *Bio-Landbau in Österreich im internationalen Kontext. Zwischen Professionalisierung und Konventionalisierung, Bd. 2*, ed. M. Groier and M. Schermer, 67–83. Wien: Bundesanstalt für Bergbauernfragen.
Darnhofer, I. 2006. *Organic farming between professionalisation and conventionalisation – The need for a more discerning view of farmer practices*. Paper presented at Joint Organic Congress, Odense, Denmark, May 30–31, 2006. Available on-line at: http://orgprints.org/7390/. Accessed 09 Aug 2014.
Darnhofer, I., T. Lindenthal, R. Bartel-Kratochvil, and W. Zollitsch. 2010. Conventionalisation of organic farming practices: From structural criteria towards an assessment based on organic principles. A review. *Agronomy for Sustainable Development* 30(1): 67–81.
de Magistris, T., and A. Gracia. 2008. The decision to buy organic food products in Southern Italy. *British Food Journal* 110(9): 929–947.
De Schutter, R. 2003. What is at stake in world trade. In *Fair Trade yearbook: Challenges of Fair Trade 2001–2003*, 8–20. Maastricht/Brussels: EFTA.
De Wit, J., and H. Verhoog. 2007. Organic values and the conventionalization of organic agriculture. *NJAS-Wageningen Journal of Life Sciences* 54(4): 449–462.
Delate, K., and J. Dewitt. 2004. Building a farmer-centered land grant university organic agriculture program: A Midwestern partnership. *Renewable Agriculture and Food Systems* 19(2): 80–91.
DeLind, L. 2000. Transforming organic agriculture into industrial organic products: Reconsidering national organic standards. *Human Organization* 59(2): 198–208.
DeLind, L., and P. Howard. 2008. Safe at any scale? Food scares, food regulation and scaled alternatives. *Agriculture and Human Values* 25: 301–317.
Didier, T., and S. Lucie. 2008. Measuring consumer's willingness to pay for organic and Fair Trade products. *International Journal of Consumer Studies* 32(5): 479–490.
Dimitri, C., and L. Oberholtzer. 2009. *Marketing US organic foods*, Economic information bulletin, vol. 58. Washington, DC: US Department of Agriculture, Economic Research Service.
Dine, J., and K. Shields. 2008. Corporate social responsibility: Do corporations have a responsibility to trade fairly? Can the Fairtrade movement deliver the duty? In *Perspectives on corporate social responsibility*, ed. N. Boeger, R. Murray, and C. Villiers, 144–169. Cheltenham/Northampton: Edward Elgar.
Dolan, C.S. 2010. Virtual moralities: The mainstreaming of Fairtrade in Kenyan tea fields. *Geoforum* 41(1): 33–43.
Dunn, E. 2005. Standards and person-making in East Central Europe. In *Global assemblages: Technology, politics and ethics as anthropological problems*, ed. A. Ong and S. Collier, 173–193. Malden/Oxford: Blackwell Publishing.
Duram, L. 2000. Agents' perceptions of structure: How Illinois organic farmers view political, economic, social, and ecological factors. *Agriculture and Human Values* 17(1): 35–48.
EFTA. 2001. Fair trade Jahrbuch 2001. European fair trade association. Available under http://www.european-fair-trade-association.org/efta/Doc/yb01-ge.pdf. Accessed 09 Aug 2014.
Eichert, C., and E. Mayer. 2008. Relevanz wirtschaftsethischer Konzepte für den Ökologischen Landbau. In *18. Jahrestagung, Neue Impulse in der Agrar-und Ernährungswirtschaft?!* ed. H. Peyerl, 131–132. Wien: Facultas.
Eisenstadt, S.N. 2002. *Multiple Modernities (Ppr)*. New Brunswick: Transaction Publishers.
Fairweather, J.R. 1999. Understanding how farmers choose between organic and conventional production: Results from New Zealand and policy implications. *Agriculture and Human Values* 16(1): 51–63.
FAO/ITC/CTA. 2001. World markets for organic fruit and vegetables—opportunities for developing countries in the production and export of organic horticultural products. *FAO Corporate Document Repository*. Retrieved from http://www.fao.org/docrep/004/Y1669E/y1669e00.htm#Contents. Accessed 11 July 2012.

Fazio, R.H. 1990. A practical guide to the use of response latency in social psychological research. In *Research methods in personality and social psychology*, Review of personality and social psychology, vol. 11, ed. C. Hendrick and M. Clark. Thousands Oaks: Sage Publications.
Fenner, D. 2010. *Einführung in die angewandte Ethik*. Tübingen: Francke UTB.
Fliess, B., H.-J. Lee, O.L. Dubreuil, and O. Agatiello. 2007. *CSR and trade: Informing consumers about social and environmental conditions of globalised production: Part I*. Paris: OECD Publishing.
FLO. 2009. *Trade labelling organizations international*. List of all FLO standards April 2009. Retrieved from http://www.fairtrade.net/fileadmin/user_upload/content/2009/standards/documents/05-09_LIST_OF_FLO_Standards.pdf. Accessed 12 Nov 2013.
Fonseca, M.F., J. Wilkinson, H. Egelyng, and G. Mascarenhas. 2008. *The institutionalization of Participatory Guarantee Systems (PGS) in Brazil: Organic and fair trade initiatives*. Paper presented at the 16th IFOAM world congress, Modena, Italy.
Fotopoulos, C., and A. Krystallis. 2002. Purchasing motives and profile of the Greek organic consumer: A countrywide survey. *British Food Journal* 104(9): 730–765.
Francis, C. 2009. *Organic farming: The ecological system*, vol. 54. Madison: American Society of Agronomy.
Francis, C., N. Jordan, P. Porter, T. Breland, G. Lieblein, L. Salomonsson, and I. Braden. 2011. Innovative education in agroecology: Experiential learning for a sustainable agriculture. *Critical Reviews in Plant Sciences* 30(1–2): 226–237.
Francis, C., T.E. Hansen, A.A. Fox, P.J. Hesje, H.E. Nelson, A.E. Lawseth, and A. English. 2012. Farmland conversion to non-agricultural uses in the US and Canada: Current impacts and concerns for the future. *International Journal of Agricultural Sustainability* 10(1): 8–24.
Frewer, L., A. Kole, S. Van De Kroon, and C. De Lauwere. 2005. Consumer attitudes towards the development of animal-friendly husbandry systems. *Journal of Agricultural and Environmental Ethics* 18(4): 345–367.
Freyer, B. 2008. *IFOAM Principles in the light of different ethical concepts*. Paper presented at the cultivating the future based on science: 2nd conference of the International Society of Organic Agriculture Research ISOFAR, Modena.
Fromartz, S. 2007. *Organic, Inc.: Natural foods and how they grew*. Orlando: Houghton Mifflin Harcourt.
Gale, F. 1997. Direct farm marketing as a rural development tool. *Rural Development Perspectives* 12: 19–25.
Garriga, E., and D. Melé. 2004. Corporate social responsibility theories: Mapping the territory. *Journal of Business Ethics* 53(1–2): 51–71.
Geier, B., I. Källander, N. Lampkin, S. Padel, M. Sligh, U. Niggli, and W. Lockeretz. 2007. A look towards the future. In *Organic farming: An international history*, ed. W. Lockeretz and G. Vogt, 264–273. Oxfordshire: CABI International.
Giddens, A. 1991. *Modernity and self-identity: Self and society in the late modern age*. Stanford: Stanford University Press.
Gil, J.M., A. Gracia, and M. Sanchez. 2000. Market segmentation and willingness to pay for organic products in Spain. *The International Food and Agribusiness Management Review* 3(2): 207–226.
Gilg, A., S. Barr, and N. Ford. 2005. Green consumption or sustainable lifestyles? Identifying the sustainable consumer. *Futures* 37(6): 481–504.
Giovannucci, D., and S. Ponte. 2005. Standards as a new form of social contract? Sustainability initiatives in the coffee industry. *Food Policy* 30(3): 284–301.
Goland, C. 2002. Community supported agriculture, food consumption patterns, and member commitment. *Culture and Agriculture* 24(1): 14–25.
Gössinger, K., and B. Freyer. 2009a. Kommunikation von "bioPlus"-Leistungen. *Ökologie and Landbau* 152(4): 44–46.
Gössinger, K., and B. Freyer. 2009b. Activities and communication arguments of organic companies going beyond organic standards. In *Farmer consumer partnerships communicating ethical*

values: A conceptual framework – CORE organic, ed. S. Padel and K. Gössinger, 42–80. Aberystwyth/Vienna: University of Wales, Aberystwyth & University of Natural Resources and Applied Life Sciences.

Gössinger, K., M. Hametter, and B. Freyer. 2009c. *Ethical oriented activities going beyond organic standards in European enterprises/farms from different theoretical perspectives*. Paper presented at the Bioacademy 2009-Proceedings 2nd scientific conference. Organic farming-A response to economic and environmental challenges. Bioacademy 2009-Proceedings Bioakademie 2009-Sbornik, 75.

Gottschalk, I., and T. Leistner. 2012. Consumer reactions to the availability of organic food in discount supermarkets. *International Journal of Consumer Studies* 37: 136–142.

Gottwald, F.-T., and I. Boergen. 2009. *Wachsen mit den Werten: Wertewandel im Ökolandbau–ein Ländervergleich*. Paper presented at the Werte – Wege – Wirkungen: Biolandbau im Spannungsfeld zwischen Ernährungssicherung, Markt und Klimawandel, Zürich.

Gruber, P.C., and M. Hauser. 2010. Ermächtigung und Biologische Landwirtschaft von Ernährungssicherheit zu Ernährungssouveränität. In *Wie wir überleben!: Ernährung und Energie in Zeiten des Klimawandels*, ed. P.C. Gruber, 87–108. Opladen/Farmington Hills: Verlag Babara Budrich.

Grundmann, M., and I. Kunze. 2012. Transnationale Vergemeinschaftungen: Interkulturelle Formen der sozial-ökologischen Gemeinschaftsbildung als Globalisierung von unten? In *Transnationale Vergesellschaftungen*, ed. M. Grundmann, 357–369. Wiesbaden: Springer.

Grundmann, M., T. Dierschke, S. Drucks, and I. Kunze. 2006. *Soziale Gemeinschaften: Experimentierfelder für kollektive Lebensformen*, vol. 3. Münster: LIT Verlag.

Gullino, M.L., and M. Pugliese. 2008. *The development of an international curriculum on organic farming in China*. Paper presented at the XVI Congresso Internazionale IFOAM/ISOFAR. Modena, Italia.

Guptill, A. 2009. Exploring the conventionalization of organic dairy: Trends and counter-trends in upstate New York. *Agriculture and Human Values* 26(1): 29–42.

Guthman, J. 1998. Regulating meaning, appropriating nature: The codification of California organic agriculture. *Antipode* 30(2): 135–154.

Guthman, J. 2003. Fast food/organic food: Reflexive tastes and the making of 'yuppie chow'. *Social and Cultural Geography* 4(1): 45–58.

Guthman, J. 2004. The trouble with 'organic lite' in California: A rejoinder to the 'conventionalisation' debate. *Sociologia Ruralis* 44(3): 301–316.

Guthman, J. 2008. Neoliberalism and the making of food politics in California. *Geoforum* 39(3): 1171–1183.

Halberg, N., T.B. Sulser, H. Høgh-Jensen, M.W. Rosegrant, and M. Trydeman Knudsen. 2006. The impact of organic farming on food security in a regional and global perspective. In *Global development of organic agriculture: Challenges and prospects*, ed. N. Halberg, H.F. Alrøe, M. Trydeman Knudsen, and E.S. Kristensen, 277–322. Wallingford: CAB International.

Hall, A., and V. Mogyorody. 2002. Organic farmers in Ontario: An examination of the conventionalization argument. *Sociologia ruralis* 41(4): 399–422.

Hamm, U., F. Gronefeld, and D. Halpin. 2002. *Analysis of the European market for organic food*, Organic marketing initiatives and rural development, vol. 1. Aberystwyth: University of Wales, School of Management and Business.

Hamm, U., J. Aschemann, and A. Riefer. 2007. Sind die hohen Preise für Öko-Lebensmittel wirklich das zentrale Problem für den Absatz? *Berichte über Landwirtschaft* 85(2): 252–270.

Harper, G.C., and A. Makatouni. 2002. Consumer perception of organic food production and farm animal welfare. *British Food Journal* 104(3/4/5): 287–299.

Hay, R. 2010. The relevance of ecocentrism, personal development and transformational leadership to sustainability and identity. *Sustainable Development* 18(3): 163–171.

Hayes, M. 2006. On the efficiency of fair trade. *Review of Social Economy* 64(4): 447–468.

Hechter, M. 1990. The attainment of solidarity in intentional communities. *Rationality and Society* 2(2): 142–155.

Hess, D.J. 2004. Organic food and agriculture in the US: Object conflicts in a health–environmental social movement. *Science as Culture* 13(4): 493–513.
Higgins, V., and S. Lockie. 2001. Getting big and getting out: Government policy, self-reliance and farm adjustment. In *Rurality bites: The social and environmental transformation of rural Australia*, 178–190. Sydney: Pluto Press.
Hill, H., and F. Lynchehaun. 2002. Organic milk: Attitudes and consumption patterns. *British Food Journal* 104(7): 526–542.
Hill, S.B., and R.J. MacRae. 1992. Organic farming in Canada. *Agriculture, Ecosystems & Environment* 39(1): 71–84.
Hinrichs, C.C. 2000. Embeddedness and local food systems: Notes on two types of direct agricultural market. *Journal of Rural Studies* 16(3): 295–303.
Hinrichs, C.C. 2003. The practice and politics of food system localization. *Journal of Rural Studies* 19(1): 33–45.
Hjelmar, U. 2011. Consumers' purchase of organic food products. A matter of convenience and reflexive practices. *Appetite* 56(2): 336–344.
Hole, D., A. Perkins, J. Wilson, I. Alexander, P. Grice, and A. Evans. 2005. Does organic farming benefit biodiversity? *Biological Conservation* 122(1): 113–130.
Horley, J. 1992. A longitudinal examination of lifestyles. *Social Indicators Research* 26(3): 205–219.
Hörtenhuber, S., T. Lindenthal, B. Amon, T. Markut, L. Kirner, and W. Zollitsch. 2010. Greenhouse gas emissions from selected Austrian dairy production systems–model calculations considering the effects of land use change. *Renewable Agriculture and Food Systems* 25(4): 316–329.
Hörtenhuber, S., T. Lindenthal, and W. Zollitsch. 2011. Reduction of greenhouse gas emissions from feed supply chains by utilizing regionally produced protein sources: The case of Austrian dairy production. *Journal of the Science of Food and Agriculture* 91(6): 1118–1127.
Howard, P.H., and P. Allen. 2006. Beyond organic: Consumer interest in new labelling schemes in the Central Coast of California. *International Journal of Consumer Studies* 30(5): 439–451.
Hughner, R.S., P. McDonagh, A. Prothero, C.J. Shultz, and J. Stanton. 2007. Who are organic food consumers? A compilation and review of why people purchase organic food. *Journal of Consumer Behaviour* 6(2–3): 94–110.
Hyman, C.J. 2000. Food for thought: Defending the Organic Foods Production Act of 1990 against claims of protectionism. *Emory International Law Review* 14: 1719.
IFOAM. 2012. *The IFOAM norms for organic production and processing*. Retrieved from http://www.ifoam.org/en/ifoam-norms. Accessed 1 Jan 2013.
International Institute for Environment and Development. 2009. *A sense of direction; Annual report*. London: International Institute for Environment and Development.
Isenhour, C. 2010. *Building sustainable societies: Exploring sustainability policy and practice in the age of high consumption. American Ethnologist, 37*(3): 511–525.
Jacobsen, E., and A. Dulsrud. 2007. Will consumers save the world? The framing of political consumerism. *Journal of Agricultural and Environmental Ethics* 20(5): 469–482.
Jaffee, D., J.R. Kloppenburg, and M.B. Monroy. 2009. Bringing the, "moral charge," home: Fair trade within the North and within the South. *Rural Sociology* 69(2): 169–196.
Jahn, G., M. Schramm, and A. Spiller. 2005. The reliability of certification: Quality labels as a consumer policy tool. *Journal of Consumer Policy* 28(1): 53–73.
Janssen, M., and U. Hamm. 2011. Consumer perception of different organic certification schemes in five European countries. *Organic Agriculture* 1(1): 31–43.
Janssen, M., and U. Hamm. 2012. Product labelling in the market for organic food: Consumer preferences and willingness-to-pay for different organic certification logos. *Food Quality and Preference* 25(1): 9–22.
Jarosz, L. 2000. Understanding agri-food networks as social relations. *Agriculture and Human Values* 17(3): 279–283.
Jarosz, L. 2008. The city in the country: Growing alternative food networks in Metropolitan areas. *Journal of Rural Studies* 24(3): 231–244.

Johnston, J. 2008. The citizen-consumer hybrid: Ideological tensions and the case of whole foods market. *Theory and Society* 37(3): 229–270.

Jones, P., P. Shears, D. Hillier, D. Comfort, and J. Lowell. 2003. Return to traditional values? A case study of slow food. *British Food Journal* 105(4/5): 297–304.

Källander, I. 2008. Participatory guarantee systems–PGS. *Swedish Society for Nature Conservation.* Available at http://www.ifoam.org/about_ifoam/standards/pgs/All-PGS-Publ.html. Retrieved from http://www.ifoam.rg/about_ifoam/standards/pgs/All-PGS-Publ.html. Accessed 2 May 2010.

Källander, I., and G. Rundgren. 2008. *Building sustainable organic sectors*. Bonn: IFOAM.

Kaltoft, P. 1999. Values about nature in organic farming practice and knowledge. *Sociologia Ruralis* 39(1): 39–53.

Kaufmann, P., S. Stagl, and D.W. Franks. 2009. Simulating the diffusion of organic farming practices in two New EU Member States. *Ecological Economics* 68(10): 2580–2593.

Kerselaers, E., L. De Cock, L. Lauwers, and G. Van Huylenbroeck. 2007. Modelling farm-level economic potential for conversion to organic farming. *Agricultural Systems* 94(3): 671–682.

Khaledi, M., R. Gray, S. Weseen, and E. Sawyer. 2007. *Assessing the barriers to conversion to organic farming: An institutional analysis*. Saskatoon: Department of Agricultural Economics, University of Saskatchewan.

Kirchmann, H., L. Bergström, T. Kätterer, O. Andrén, and R. Andersson. 2008. Can organic crop production feed the world? In *Organic crop production–Ambitions and limitations*, ed. H. Kirchmann and L. Bergström, 39–72. Dordrecht: Springer.

Klonsky, K., and C. Greene. 2005. *Widespread adoption of organic agriculture in the US: Are market-driven policies enough?* Presentation at the American agricultural economics association annual meeting, Providence, Rhode Island. http://sarep.ucdavis.edu/sfr/organic/files-images/2008_Deregistration_factors.pdf. Accessed 09 Aug 2014.

Koesling, M., O. Flaten, and G. Lien. 2008. Factors influencing the conversion to organic farming in Norway. *International Journal of Agricultural Resources, Governance and Ecology* 7(1): 78–95.

Kölsch, O. 1988. Soziale Probleme bei der Umstellung auf ökologischen Landbau. *Für eine bäuerliche Landwirtschaft Schriftenreihe* 14: 149–156.

Konstantinidis, C. 2012. *Organic farming and rural transformations in the European Union: A political economy approach.* Doctoral dissertation, University of Massachusetts Amherst: 190 p.

Kratochvil, R., T. Lindenthal, and C.R. Vogl. 2005. *Prozessqualität im Wandel: Beobachtungen am Beispiel der Bio-Wertschöpfungskette in Österreich.* Paper presented at the Ende der Nische, Beiträge zur 8. Wissenschaftstagung Ökologischer Landbau, Kassel Heß, J. and G. Rahmann. (ed.) 2005. Ende der Nische, Beiträge zur 8. Wissenschaftstagung Ökologischer Landbau in Kassel, 01.03.2005 – 04.03.2005: 415–418.

Kröger, M., and M. Schäfer. 2013. Between ideals and reality: Development and implementation of fairness standards in the organic food sector. *Journal of Agricultural and Environmental Ethics* 27(1): 43–63.

Krystallis, A., and G. Chryssohoidis. 2005. Consumers' willingness to pay for organic food: Factors that affect it and variation per organic product type. *British Food Journal* 107(5): 320–343.

La Trobe, H.L., and T.G. Acott. 2000. Localising the global food system. *International Journal of Sustainable Development and World Ecology* 7(4): 309–320.

Lamine, C., and S. Bellon. 2009. Conversion to organic farming: A multidimensional research object at the crossroads of agricultural and social sciences. A review. *Agronomy for Sustainable Development* 29(1): 97–112.

Lampkin, N., C. Foster, and S. Padel. 1999. The policy and regulatory environment for organic farming in Europe: Country reports. In *Organic farming in Europe: Economics and policy*, vol. 2, ed. S. Dabbert. Stuttgart: Universität Hohenheim.

Latacz-Lohmann, U., and C. Foster. 1997. From "niche" to "mainstream"-strategies for marketing organic food in Germany and the UK. *British Food Journal* 99(8): 275–282.

LBG. 2010. *Buchführungsergebnisse*. Wien: LBG Österreich.

Lea, E., and T. Worsley. 2005. Australians' organic food beliefs, demographics and values. *British Food Journal* 107(11): 855–869.

Leire, C., and Å. Thidell. 2005. Product-related environmental information to guide consumer purchases–A review and analysis of research on perceptions, understanding and use among Nordic consumers. *Journal of Cleaner Production* 13(10): 1061–1070.

Lindenthal, T., T. Markut, S. Hörtenhuber, and G. Rudolph. 2009. *CO2-eq.-emissions of organic and conventional foodstuffs in Austria- results summary of 74 CO2 balanced products – Executive summary*. Frick/Freiburg: FIBL/Öko-Institute Freiburg.

Lockie, S., and D. Halpin. 2005. The 'conventionalisation' thesis reconsidered: Structural and ideological transformation of Australian organic agriculture. *Sociologia Ruralis* 45(4): 284–307.

Lockie, S., K. Lyons, G. Lawrence, and K. Mummery. 2002. Eating 'green': Motivations behind organic food consumption in Australia. *Sociologia Ruralis* 42(1): 23–40.

Loconto, A. 2010. Sustainably performed: Reconciling global value chain governance and performativity. *Journal of Rural Social Science* 25(3): 193–225.

Løes, A.K., and B. Nölting. 2011. Increasing organic consumption through school meals' lessons learned in the iPOPY project. *Organic Agriculture* 1(2): 91–110.

Loureiro, M.L., and S. Hine. 2002. Discovering niche markets: A comparison of consumer willingness to pay for local (Colorado grown), organic, and GMO-free products. *Journal of Agricultural and Applied Economics* 34(3): 477–488.

Lund, V. 2006. Natural living—A precondition for animal welfare in organic farming. *Livestock Science* 100(2): 71–83.

Lund, V., S. Hemlin, and J. White. 2004. Natural behavior, animal rights, or making money–A study of Swedish organic farmers' view of animal issues. *Journal of Agricultural and Environmental Ethics* 17(2): 157–179.

Luttikholt, L. 2007. Principles of organic agriculture as formulated by the International Federation of Organic Agriculture Movements. *NJAS-Wageningen Journal of Life Sciences* 54(4): 347–360.

Lyon, S. 2006. Evaluating fair trade consumption: Politics, defetishization and producer participation. *International Journal of Consumer Studies* 30(5): 452–464.

Lyons, K. 2007. Supermarkets as organic retailers: Impacts for the Australian organic sector. In *Supermarkets and agrifood supply chains*, ed. D. Burch and G. Lawrence, 154–172. Cheltenham/Northampton: Edward Elgar Publishing.

Magnusson, M.K., A. Arvola, U.-K.K. Hursti, L. Åberg, and P.-O. Sjödén. 2001. Attitudes towards organic foods among Swedish consumers. *British Food Journal* 103(3): 209–227.

Marsden, T., J. Banks, and G. Bristow. 2000. Food supply chain approaches: Exploring their role in rural development. *Sociologia Ruralis* 40(4): 424–438.

Martinez, S. 2010. *Local food systems: Concepts, impacts, and issues*, Economic research report, vol. 97. Washington, DC: U.S. Department of Agriculture, Economic Research Service.

McEachern, M.G., and P. Mcclean. 2002. Organic purchasing motivations and attitudes: Are they ethical? *International Journal of Consumer Studies* 26(2): 85–92.

McMichael, P. 2009. A food regime genealogy. *The Journal of Peasant Studies* 36(1): 139–169.

Meadows, D., J. Randers, and D. Meadows. 2005. *Limits to growth: The 30-year update; dt. Grenzen des Wachstums. Das 30-Jahre-Update. Signal zum Kurswechsel*. London/dt. Stuttgart: Earthscan/dt. S. Hirzel 2006.

Meeusen, M.J.G., V. Beekman, R.P.M. De Graaff, and S.M.A. De Kroon. 2003. *Organic values in two-fold*. Vol. no 5.03.05. The Hague: Agriculture Economics Research Institute (LEI), Wageningen University and Research Centre.

Meeusen, M.J.G., A. Wijnands, A. Kijlstra, and M. Boekhoff. 2005. *View on organic animal product chains*. Report no 2.05.01. Wageningen/The Hague: Agricultural Economics Research Institute (LEI)/University and Research Centre.

Memery, J., P. Megicks, and J. Williams. 2005. Ethical and social responsibility issues in grocery shopping: A preliminary typology. *Qualitative Market Research: An International Journal* 8(4): 399–412.

Michaelidou, N., and L.M. Hassan. 2008. The role of health consciousness, food safety concern and ethical identity on attitudes and intentions towards organic food. *International Journal of Consumer Studies* 32(2): 163–170.

Michelsen, J. 2002a. Organic farming development in Europe‘ impacts of regulation and institutional diversity. *Advances in the Economics of Environmental Resources* 4: 101–138.

Michelsen, J. 2002b. Organic farming in a regulatory perspective. The Danish case. *Sociologia Ruralis* 41(1): 62–84.

Michelsen, J., K. Lynggaard, S. Padel, and C. Foster. 2001. Organic farming development and agricultural institutions in Europe: A study of six countries. In *Organic farming in Europe: Economics and policy*, vol. 6, ed. S. Dabbert. Stuttgart: University of Hohenheim.

Midmore, P., S. Padel, H. McCalman, J. Isherwood, S. Fowler, and N. Lamkpin. 2001. *Attitudes towards conversion to organic production systems: A study of farmers in England.* Aberystwyth: The University of Wales, Institute of Rural Studies.

Millar, K., and P.H. West. 2009. *Ethical futures: Bioscience and food horizons: EurSafe 2009, Nottingham, United Kingdom, 2–4 July 2009*. Wageningen: Wageningen Academic Publishers.

Mittelstrass, H. 2005. The emergence of a faculty of organic agricultural sciences. *LEISA-LEUSDEN* 21(4): 26.

Moore, G. 2004. The Fair Trade movement: Parameters, issues and future research. *Journal of Business Ethics* 53: 73–86.

Morgan, K., and J. Murdoch. 2000. Organic vs. conventional agriculture: Knowledge, power and innovation in the food chain. *Geoforum* 31(2): 159–173.

Murdoch, J., and M. Miele. 2004. A new aesthetic of food? Relational reflexivity in the ‘alternative’ food movement. In *Qualities of food*, 156–175. Manchester/New York/Vancouver: Manchester University Press.

Mutersbaugh, T. 2005. Fighting standards with standards: Harmonization, rents, and social accountability in certified agrofood networks. *Environment and Planning A* 37(11): 2033.

Naess, A. 1973. The shallow and the deep, long-range ecology movement. A summary. *Philosophical Inquiry* 16(1–4): 95–100.

Naess, A. 1986. The deep ecological movement: Some philosophical aspects. *Philosophical Inquiry* 8(1–2): 10–31.

Neuberger, K., I. Stephan, R. Hermanowsk, A. Flake, F.-J. Post, and T. Elsen. 2006. Farming for Health: Aspects from Germany. In *Farming for health: Green-care farming across Europe and the United States of America*, vol. 13, ed. J. Hassink and M. Van Dijk, 193–211. Netherlands: Springer.

Neuendorff, J., and A. Spiller. 2011. Die ÖLG-Kontrollstellen-Zulassungsverordnung: Ein Beitrag zum Schutz vor Betrug im Ökolandbau? *AGRA-Europe* 55(124): 35–36.

Nicholls, A., and C. Opal. 2008. *Fair trade. Market-driven ethical consumption.* London: Sage.

Niggli, U. 2005. Folgen des Wachstums. Verliert der Öko-Landbau seine Unschuld? *Ökologie and Landbau* 133(1): 14–16.

Niggli, U. 2007. The evolution of organic practice. In *Organic farming: An international history*, ed. W. Lockeretz. Wallingford: CAB International.

Nosi, C., and L. Zanni. 2004. Moving from “typical products” to “food-related services”: The slow food case as a new business paradigm. *British Food Journal* 106(10/11): 779–792.

O’Donovan, P., and M. McCarthy. 2002. Irish consumer preference for organic meat. *British Food Journal* 104(3/4/5): 353–370.

O’Rourke, D. 2006. Multi-stakeholder regulation: Privatizing or socializing global labor standards? *World Development* 34(5): 899–918.

Onyango, B.M., W.K. Hallman, and A.C. Bellows. 2007. Purchasing organic food in US food systems: A study of attitudes and practice. *British Food Journal* 109(5): 399–411.

Paarlberg, R. 2009. The ethics of modern agriculture. *Society* 46(1): 4–8.

Padel, S. 2001. Conversion to organic farming: A typical example of the diffusion of an innovation? *Sociologia Ruralis* 41(1): 40–61.

Padel, S. 2005. Focus groups of value concepts of organic producers and other stakeholders. *Organic revision*. Aberystwyth: University of Wales.

Padel, S., and C. Foster. 2005. Exploring the gap between attitudes and behaviour: Understanding why consumers buy or do not buy organic food. *British Food Journal* 107(8): 606–625.

Padel, S., and K. Gössinger. 2008. *Farmer consumer partnerships communicating ethical values: A conceptual framework*, Vol. CORE organic project series report, Tjele.

Padel, S., and H. Röcklinsberg. 2009. A conceptual framework for the communication of ethical values going beyond standards in organic farming. In *Ethical futures: Bioscience and food horizons*, ed. M. Kate, P. Hobson West, and B. Nerlich, 394–400. Wageningen: Wageningen Academic Publishers.

Padel, S., N. Lampkin, and W. Lockeretz. 2007a. The development of governmental support for organic farming in Europe. In *Organic farming: An international history*, ed. W. Lockeretz, 93–122. Wallingford: CAB International.

Padel, S., H. Röcklinsberg, H. Verhoog, H. Fjelsted Alrøe, J. De Wit, C. Kjeldsen, and O. Schmid. 2007b. *Balancing and integrating basic values in the development of organic regulations and standards: Proposal for a procedure using case studies of conflicting areas*. Aberystwyth/Tjele: Danish Research Centre for organic Farming (DARCOF).

Padel, S., H. Röcklinsberg, and O. Schmid. 2009. The implementation of organic principles and values in the European regulation for organic food. *Food Policy* 34(3): 245–251.

Padel, S., J. B. Vine, M. Huber, M. Stolze, L.M. Jespersen, E. Rüegg, and S.R. Belliere. 2010. The European regulatory framework and its implementation in influencing organic inspection and certification systems in the EU. Deliverable 11, CERTCOST Project, agreement no. 207727. 7th Framework Programme, University of Kassel. http://www.certcost.org.

Parr, D.M., and M. Van Horn. 2006. Development of organic and sustainable agricultural education at the University of California, Davis: A closer look at practice and theory. *HortTechnology* 16(3): 426–431.

Parr, D.M., C.J. Trexler, N.R. Khanna, and B.T. Battisti. 2007. Designing sustainable agriculture education: Academics' suggestions for an undergraduate curriculum at a land grant university. *Agriculture and Human Values* 24(4): 523–533.

Parrott, N., J.E. Olesen, and H. Hugh-Jensen. 2006. Certified and non-certified organic farming in the developing world. In *Global development of organic agriculture: Challenges and prospects*, ed. N. Halberg, H.F. Alrøe, and M. Trydeman Knudsen, 153–180. Wallingford: CAB International.

Paull, J. 2006. The farm as organism: The foundational idea of organic agriculture. *Elementals: Journal of Bio-Dynamics Tasmania* 80: 14–18.

Paull, J. 2011. Attending the first organic agriculture course: Rudolf Steiner's agriculture course at Koberwitz, 1924. *European Journal of Social Sciences* 21(1): 64–70.

Peattie, K., and A. Crane. 2005. Green marketing: Legend, myth, farce or prophesy? *Qualitative Market Research: An International Journal* 8(4): 357–370.

Perrini, F., S. Castaldo, N. Misani, and A. Tencati. 2010. The impact of corporate social responsibility associations on trust in organic products marketed by mainstream retailers: A study of Italian consumers. *Business Strategy and the Environment* 19(8): 512–526.

Pieper, A. 1994. *Einführung in die Ethik*, 3rd ed. Tübingen/Basel: Francke Verlag.

Pivato, S., N. Misani, and A. Tencati. 2008. The impact of corporate social responsibility on consumer trust: The case of organic food. *Business Ethics: A European Review* 17(1): 3–12.

Pollan, M. 2006. *The omnivore's dilemma: A natural history of four meals*. New York: Penguin.

Polonsky, M.J. 1994. An introduction to green marketing. *Electronic Green Journal* 1(2): 1–10.

Rantzau, R., B. Freyer, and H. Vogtmann. 1990. *Umstellung auf oekologischen Landbau. Betriebliche Erfordernisse und Konsequenzen bei der Durchfuehrung des oekologischen Landbaus. Dokumentation und Analyse von Praxiserfahrungen in unterschiedlichen Entwicklungsstadien und deren Ueberpruefung auf verschiedenen Standorten*. Münster-Hiltrup: Landwirtschaftsverlag GmbH.

Ravenscroft, N., and B. Taylor. 2009. Public engagement in new productivism. In *What is land for? The food, fuel, and climate change debate*, ed. M. Winter and M. Lobley, 213–232. London/Sterling: Earthscan.

Raynolds, L.T. 2000. Re-embedding global agriculture: The international organic and fair trade movements. *Agriculture and Human Values* 17(3): 297–309.

Raynolds, L.T. 2002. Consumer/producer links in fair trade coffee networks. *Sociologia Ruralis* 42(4): 404–424.

Raynolds, L.T. 2004. The globalization of organic agro-food networks. *World Development* 32(5): 725–743.

Reed, M. 2002. Fight the future! How the contemporary campaigns of the UK organic movement have arisen from their composting of the past. *Sociologia Ruralis* 41(1): 131–145.

Renard, M.C. 2003. Fair trade: Quality, market and conventions. *Journal of Rural Studies* 19(1): 87–96.

Renting, H., T.K. Marsden, and J. Banks. 2003. Understanding alternative food networks: Exploring the role of short food supply chains in rural development. *Environment and Planning A* 35(3): 393–412.

Richter, T., O. Schmid, P. Damary, P. van den Berge, U. Meier, and D. Halpin. 2000. *International study of retail chains with regard to their activities in the marketing of organic products*. Retrieved from http://orgprints.org/2945/10/Richter-2000-study-of-retail-chains.pdf. Accessed 1 Jan 2013.

Rigby, D., and D. Cáceres. 2001. Organic farming and the sustainability of agricultural systems. *Agricultural Systems* 68(1): 21–40.

Rigby, D., T. Young, and M. Burton. 2001. The development of and prospects for organic farming in the UK. *Food Policy* 26(6): 599–613.

Ritzer, G. 2003. Rethinking globalization: Glocalization/grobalization and something/nothing. *Sociological Theory* 21(3): 193–209.

Rosin, C., and H. Campbell. 2009. Beyond bifurcation: Examining the conventions of organic agriculture in New Zealand. *Journal of Rural Studies* 25(1): 35–47.

Sachs, W., T. Santarius, and P. Camiller. 2007. *Fair future: Resource conflicts, security and global justice*. London: Zed Books.

Santucci, F.M. 2009. I circuiti commerciali dei prodotti biologici. *Agriregionieuropa* 5:17. Available on line at www.agriregionieuropa.it. Accessed 09 Aug 2014.

Sargisson, L., and L.T. Sargent. 2004. *Living in utopia: New Zealand's intentional communities*. Aldershot/Burlington: Ashgate Publishing.

Schekahn, A., and F. Thomas. 2008. Fairness? Na klar!–Aber wie? In *Der kritische Agrarbericht*, ed. T. Frieder, 100–104. Konstanz: AgrarBündnis e.V.

Schmid, O. 2007. Werte und Richtlinien im Wandel. *Ökologie and Landbau* 144(4): 14–16.

Schmid, O. 2009a. Bio-Richtlinien für die Zukunft. Wegweiser statt Vorschriften. *Ökologie and Landbau* 150(2): 16–19.

Schmid, O. 2009b. Wegweiser statt Vorschriften. *Ökologie & Landbau* 150(2): 16–19.

Schmid, O. 2010. Organic standards for the future – Guidelines/signposts rather than rules. *The Organic Standard* 106: 13–16.

Schmid, O., and R. Kilchsperger. 2005. Warum bin ich eigentlich Biobäuerin? *bioaktuell Das Magazin der Biobewegung* (4/05).

Schmid, O., and W. Lockeretz. 2007. Development of standards for organic farming. In *Organic farming: An international history*, ed. W. Lockeretz, 152–174. Wallingford: CAB International.

Schneeberger, W., I. Darnhofer, and M. Eder. 2002. Barriers to the adoption of organic farming by cash-crop producers in Austria. *American Journal of Alternative Agriculture* 17(01): 24–31.

Schösler, H., J. De Boer, and J.J. Boersema. 2012. The organic food philosophy: A qualitative exploration of the practices, values, and beliefs of Dutch organic consumers within a cultural–historical frame. *Journal of Agricultural and Environmental Ethics* 26(2): 439–460.

Schramek, J., and G. Schnaut. 2004. *Hemmende und fördernde Faktoren einer Umstellung auf ökologischen Landbau aus Sicht landwirtschaftlicher Unternehmer/innen in verschiedenen Regionen Deutschlands (unter Einbeziehung soziologischer Fragestellungen)*. Bonn: Geschäftsstelle Bundesprogramm Ökologischer Landbau in der Bundesanstalt für Landwirtschaft und Ernährung (BLE).

Schuldt, J., and M. Hannahan. 2012. When good deeds leave a bad taste. Negative inferences from ethical food claims. *Appetite* 62: 76–83.

Schüle, E. 2009. BioBOOM. *Ernährungs Umschau* 2: 104–106.

Schulze, H. 2008. *Quality assurance systems in the agri-food chain.* Dissertation, Georg-August-University of Göttingen, Göttingen.

Scialabba, N., and C. Hattaam. 2002. *Organic agriculture, environment and food security*. Rome: Food and Agricultural Organization of the United Nations.

Seppänen, L. 2002. Societal integration in organic vegetable farming: Exploring the learning challenges. *The Journal of Agricultural Education and Extension* 8(2): 87–100.

Shaw, H. 2006. CSR in the community: Redefining the social role of the supermarket giants. *Social Responsibility Journal* 2(2): 216–222.

Shepherd, R., M. Magnusson, and P.-O. Sjödén. 2005. Determinants of consumer behavior related to organic foods. *AMBIO: A Journal of the Human Environment* 34(4): 352–359.

Siderer, Y., A. Maquet, and E. Anklam. 2005. Need for research to support consumer confidence in the growing organic food market. *Trends in Food Science & Technology* 16(8): 332–343.

Spoor, G. 2002. *Organic, fair and social: Cooperation between ethical standards systems in agriculture*. Retrieved from: http://orgprints.org/404/01/SocialAccountabilitySustAgri.doc. Accessed 9 Mar 2009.

Stagl, S. 2002. Local organic food markets: Potentials and limitations for contributing to sustainable development. *Empirica* 29(2): 145–162.

Sterba, J.P. 2003. *Morality in practice*. Independence: Wadsworth Publishing.

Stock, P.V. 2007. 'Good farmers' as reflexive producers: An examination of family organic farmers in the US Midwest. *Sociologia Ruralis* 47(2): 83–102.

Sundrum, A. 2007. *Conflicting areas in the ethical debate on animal health and welfare.* Paper presented at the Sustainable food production and ethics: Preprints of the 7th Congress of the European Society for Agricultural and Food Ethics. Vienna: Wageningen Academic Publishers.

Tarkiainen, A., and S. Sundqvist. 2005. Subjective norms, attitudes and intentions of Finnish consumers in buying organic food. *British Food Journal* 107(11): 808–822.

Thøgersen, J. 2002. Promoting green consumer behavior with eco-labels. In *New tools for environmental protection: Education, information, and voluntary measures*, ed. T. Dietz, 83–104. Washington, DC: National Academy Press.

Thøgersen, J. 2007. Consumer decision making with regard to organic food products. In *Traditional food production facing sustainability: A European challenge*, ed. M. Vaz, P. Vaz, P. Nijkamp, and J.L. Rastoin. Farnham: Ashgate.

Thompson, P. 2010. *The agrarian vision: Sustainability and environmental ethics*. Lexington: University Press of Kentucky.

Torjusen, H., K. Sangstad, J. O'Doherty, and U. Kjærnes. 2004. *European consumers' conceptions of organic food: A review of available research*, Professional report, vol. 4, 147. Oslo: National Institute for Consumer Research.

Tovey, H. 2002. Food, environmentalism and rural sociology: On the organic farming movement in Ireland. *Sociologia Ruralis* 37(1): 21–37.

Tregear, A., J. Dent, and M. McGregor. 1994. The demand for organically grown produce. *British Food Journal* 96(4): 21–25.

Trobe, H.L. 2001. Farmers' markets: Consuming local rural produce. *International Journal of Consumer Studies* 25(3): 181–192.

Tybirk, K., H.F. Alrøe, and P. Frederiksen. 2004. Nature quality in organic farming: A conceptual analysis of considerations and criteria in a European context. *Journal of Agricultural and Environmental Ethics* 17(3): 249–274.

UNCTAD. (eds.). 2013. *Wake up before it is too late: Make agriculture truly sustainable now for food security in a changing climate*. United Nations conference on trade and development. Trade and Environment Review 2013, Geneva, Italia. http://unctad.org/en/PublicationsLibrary/ditcted2012d3_en.pdf. Accessed 09 Aug 2014.

Vaarst, M. 2004. *Animal health and welfare in organic agriculture*. Wallingford: CAB International.

Vaarst, M., and H.F. Alrøe. 2012. Concepts of animal health and welfare in organic livestock systems. *Journal of Agricultural and Environmental Ethics* 25(3): 333–347.

Vaarst, M., S. Padel, M. Hovi, D. Younie, and A. Sundrum. 2005. Sustaining animal health and food safety in European organic livestock farming. *Livestock Production Science* 94(1): 61–69.

van Bueren, E., P. Struik, M. Tiemens-Hulscher, and E. Jacobsen. 2003. Concepts of intrinsic value and integrity of plants in organic plant breeding and propagation. *Crop Science* 43(6): 1922–1929.

Van Der Ploeg, J.D. 2000. Revitalizing agriculture: Farming economically as starting ground for rural development. *Sociologia Ruralis* 40(4): 497–511.

Van Huik, M., B. Bock, M. Kaiser, and M. Lien. 2006. *Attitudes of organic pig farmers towards animal welfare*. Paper presented at the preprints of the 6th congress of the European Society for Agricultural and Food Ethics (EurSAFE 2006), 22–24 June 2006, Oslo, Norway.

Veenhoven, R. 1999. Quality-of-life in individualistic society. *Social Indicators Research* 48(2): 159–188.

Verhoog, H., M. Matze, E. Lammerts Van Bueren, and T. Baars. 2003. The role of the concept of the natural (naturalness) in organic farming. *Journal of Agricultural and Environmental Ethics* 16(1): 29–49.

Verhoog, H., E. Lammerts Van Bueren, M. Matze, and T. Baars. 2007. The value of 'naturalness' in organic agriculture. *NJAS-Wageningen Journal of Life Sciences* 54(4): 333–345.

Vermeir, I., and W. Verbeke. 2006. Sustainable food consumption: Exploring the consumer "attitude–behavioral intention" gap. *Journal of Agricultural and Environmental Ethics* 19(2): 169–194.

Vogl, C.R., and J. Hess. 1999. Organic farming in Austria. *American Journal of Alternative Agriculture* 14(3): 137–143.

Vogl, C.R., L. Kilcher, and H. Schmidt. 2005. Are standards and regulations of organic farming moving away from small farmers' knowledge? *Journal of Sustainable Agriculture* 26(1): 5–26.

Vos, T. 2000. Visions of the middle landscape: Organic farming and the politics of nature. *Agriculture and Human Values* 17(3): 245–256.

Watson, C., H. Alrøe, and E.S. Kristensen. 2006. Research to support the development of organic food and farming. In *Organic agriculture: A global perspective*, ed. P. Kristiansen, A. Taji, and J. Reganold, 361–383. Wallingford: CAB International.

Welford, R., J. Meaton, and W. Young. 2003. Fair trade as a strategy for international competitiveness. *The International Journal of Sustainable Development and World Ecology* 10(1): 1–13.

Whyte, K.P., and P. Thompson. 2010. A role for ethical analysis in social research on agrifood and environmental standards. *Journal of Rural Social Sciences* 25(3): 79–98.

Wier, M., K. O'Doherty Jensen, L.M. Andersen, and K. Millock. 2008. The character of demand in mature organic food markets: Great Britain and Denmark compared. *Food Policy* 33(5): 406–421.

Willemsen, A. 2009. Moral consideration of plants for their own sake. In *Ethical futures: Bioscience and food horizons: EurSafe 2009*, ed. K. Millar, P.H. West, and B. Nerlich, 434–439. Nottingham: Wageningen Academic Pub.

Willer, H., and L. Kilcher. 2011. *The world of organic agriculture: Statistics and emerging trends 2011*. Bonn/Frick: IFOAM, FIBL.

Willer, H., and L. Kilcher. 2012. *The world of organic agriculture – Statistics and emerging trends 2012*. Bonn/Frick: IFOAM, FIBL.

Willock, J., I.J. Deary, G. Edwards-Jones, G.J. Gibson, M.J. McGregor, A. Sutherland, J.B. Dent, O. Morgan, and R. Grieve. 1999. The role of attitudes and objectives in farmer decision making: Business and environmentally-oriented behaviour in Scotland. *Journal of Agricultural Economics* 50(2): 286–303.

Windsor, D. 2001. The future of corporate social responsibility. *International Journal of Organizational Analysis* 9(3): 225–256.

Yiridoe, E.K., S. Bonti-Ankomah, and R.C. Martin. 2005. Comparison of consumer perceptions and preference toward organic versus conventionally produced foods: A review and update of the literature. *Renewable Agriculture and Food Systems* 20(4): 193–205.

Zagata, L., and M. Lostak. 2012. In goodness we trust. The role of trust and institutions underpinning trust in the organic food market. *Sociologia Ruralis* 52(4): 470–487.
Zander, K., and U. Hamm. 2010. Consumer preferences for additional ethical attributes of organic food. *Food Quality and Preference* 21(5): 495–503.
Zander, K., U. Hamm, B. Freyer, K. Gössinger, M. Hametter, S. Naspetti, and R. Zanoli. 2010. *Farmer consumer partnerships-How to successfully communicate the values of organic food – A handbook*. Kassel: University of Kassel, Department of Agricultural and Food Marketing.
Zander, K., H. Stolz, and U. Hamm. 2012. Promising ethical arguments for product differentiation in the organic food sector. A mixed methods research approach. *Appetite* 62: 133–142.
Zepeda, L., and D. Deal. 2009. Organic and local food consumer behaviour: Alphabet theory. *International Journal of Consumer Studies* 33(6): 697–705.
Zollitsch, W., C. Winckler, S. Waiblinger, and A. Haslberger (eds.). 2007. *Sustainable food production and ethics: Preprints of the 7th congress of the European Society for Agricultural and Food Ethics*. Wageningen: Wageningen Academic Publishers.
Zorn, A., C. Lippert, and S. Dabbert. 2009. Economic concepts of organic certification. In *Economic analysis of certification systems in organic food and farming,* vol. 5. Stuttgart: University of Hohenheim, CERTCOST.
Zorn, A., C. Lippert, and S. Dabbert. 2013. *Zur Nichteinhaltung Von Vorschriften Des Ökologischen Landbaus In Deutschland Und In Der Schweiz–Statistische Analyse Einzelbetrieblicher Daten*. In Bahrs, E. et al. (ed.). 2013. Herausforderungen des globalen Wandels für Agrarentwicklung und Welternährung. Schriften der Gesellschaft für Wirtschafts- und Sozialwissenschaften des Landbaus e.V. Bd. 48, Münster: 73–84.

Chapter 14
Positioning Organic Ethics

Bernhard Freyer and Jim Bingen

14.1 Summary Overview: Ethics in the Organic Agrofood Chain

14.1.1 Our Entry Point into the Debate

Olivier De Schutters provides the starting point for our final reflections on ethics in the organic agrofood chain. In his report for the Human Rights Council (De Schutter 2010, p. 1) he writes: "The reinvestment in agriculture, [. . .], is essential to the concrete realization of the right to food. However, in a context of ecological, food and energy crises, the most pressing issue regarding reinvestment is not how much, but how. This how should contribute to the progressive realization of the human right to adequate food." The report itself highlights the significance of agro-ecology and similar agricultural approaches such as organic farming to fulfill the human right to food.

For us, "the how" is also central, but not primarily with respect to the technological, economic or socio-cultural arrangements for carrying out organic farming. Our intention has been to understand better the underlying ethical dimensions that would lead to ecological, social and economic justice and sustainable organic agrofood practices. To study this issue we took into account insightful discourses

B. Freyer (✉)
Department of Sustainable Agriculture Systems, Division of Organic Farming, University of Natural Resources and Life Sciences (BOKU), Vienna, Austria
e-mail: Bernhard.Freyer@boku.ac.at

J. Bingen
Department of Community Sustainability, Michigan State University, East Lansing, MI, USA
e-mail: Bingen@msu.edu

B. Freyer, J. Bingen (eds.), *Re-Thinking Organic Food and Farming in a Changing World*, The International Library of Environmental, Agricultural and Food Ethics 22, DOI 10.1007/978-94-017-9190-8_14

on agrarian (e.g., Thompson 1988, 2008; Carruthers 2009) and specifically, organic ethics (Alrøe and Kristensen 2000, 2004; Alrøe et al. 2006; Padel et al. 2007a, 2009). We built on this understanding and offered an additional contribution by reviewing the IFOAM Principles and related ethics and their implementation all along the organic agrofood chain as a means to focus on the relevance of ethics in the organic approach. The underlying idea was to stimulate consideration of the potential of the ethical principles for assessing the diverse patterns in which these ethics arise along the organic agrofood chain today. These findings set the stage for our reflections on the future of an ethically driven organic agrofood chain.

14.1.2 The Ethical Foundation of Organic

From the beginning, organic was an ethically driven movement. This, however, changed over time (see Chap. 2). A driving force of the organic movement has been to act as a response or alternative to mainstream social, economic, ecological or ethical issues. Movement activists and leaders have represented a wide variety of societal interests and groups well beyond farming. The ethical concepts (worldviews) and the normative ethics (deontology, consequentialism, virtue ethics) represented by key figures who have been historically important in the organic movement (Rudolf Steiner, the Müllers, Aldo Leopold, Lady Balfour or Rachel Carson) do not present a uniform ethical foundation for organic. At best, we suggest that the movement is founded and unified around an ecocentric/holistic worldview, expressed in the IFOAM Principles. In addition, with few exceptions (popular agrarian writers in the US, or Vandana Shiva from India (Shiva 2000)) the movement today has become "mainstream".

IFOAM, founded in 1972, decided to identify and label an ethical approach for the worldwide organic movement. With significant contributions from research scientists and members, these first IFOAM Principles have served as the ethical foundation for IFOAM since the 1980s. The most recent version of the Principles (Health, Ecology, Fairness and Care (IFOAM 2009)) provide the ethical framework for all organic agrofood actors (see Chap. 2). No other food and agricultural movement is grounded on such a clearly articulated set of ethical principles. These IFOAM Principles illustrate an ecocentric/holistic moderately deontological ethical position.

14.1.3 Organic Differentiation

The chapters in this volume clearly illustrate the diversity throughout the organic movement. Contemporary organic actors continue to reflect a wide variety of discourses, practices and values. While we refer to the "organic movement", it is by no means uniform, but continues to be full of contradictions, tensions and polarized positions. Actors all along the organic agrofood chain reflect varying relationships with the IFOAM Principles. In short, the contemporary organic movement includes considerable differentiation, including multiple approaches to, or interpretations of the original organic values.

14.1.4 Structuring Ethics Through Typologies

In order to frame a new, value-centered discussion that is sensitive to the positions of, and dynamics among all actors along the organic agrofood chain, we describe and discuss our three part typology: *Nature and Philosophy (Nature/Philo),* the *Protest and Environment (Enviro/Protest)* and *Economics and Markets (Econo/Market).* In contrast to other discussions that use more limited typologies, such as that used by Alrøe and Noe (2008), we apply ours to identify and review the positions of all major actors—farmers, processors, consumers, as well as wholesalers, suppliers, traders and grocery chains. We also identify where this typology is reflected in several of the contributed chapters.

Without question, the Econo/Market type currently dominates all along the organic agrofood chain. Nevertheless, we want to emphasize the importance of considering the diversity in the organic movement. We understand the popularity of looking at the juxtaposition of the Econo/Market and the Nature/Philo type. While such a perspective can be easily popularized, we feel that our intellectual and programmatic challenge resides in thinking more systematically about the multiple dimensions of the Enviro/Protest type and other diverse hybrid formulations. Bringing the holistic ethical approach of the IFOAM Principles back to the center of our discussions represents a first step forward.

14.1.5 The Relevance of Ethics in IFOAM Norms

The IFOAM Norms for production and processing offer a basis for reviewing the current status and relevance of ethics and the role of the IFOAM Principles in the organic institutional environment.

Ethical concerns are clearly expressed in the IFOAM Standards and private regulations with respect to ecological and health issues. But, with the exception of some private regulations, social and economically relevant ethics lack expression. Certification practices also reflect similar patterns of differentiation. Some certification firms or organizations work primarily with larger farms or units whose size generates more income from the certification process. Other forms of certification (the PGS-approach) take a shared farmer-consumer approach that not only is more affordable, but more open and democratic. Of course, in these diverse certification approaches ethics play a different role. An issue to pursue involves whether there could be a way of implementing more elements of the IFOAM Principles into standards and certification in a legally binding way beyond, or in addition to the PGS?

14.1.6 The Consumer and their Ethics

Despite the growing popularity of organic, very few consumers are aware of the organic standards or the process of certification. Moreover, the IFOAM Principles are absent from most consumer organic marketing strategies. For many, “local”, “regional” or “natural” take pride of place and are often thought to be the same as organic.

In addition, consumer behavior is “protected,” but it is not regulated, unlike that of all other actors in the organic agrofood chain. Consumers are “free” to choose what, how and whether they purchase organic. Therefore consumers and others acting without any standards but addressed via the IFOAM Principles are invited to ideally follow the IFOAM Principles directly. Collective consumer arrangements, like PGS, CSA, food coops, etc., offer ways to foster consumer engagement with other actors in the organic agrofood chain, bringing the Principles more to the center of the debate.

14.1.7 Ethics in Organic Education, Extension Services, and Research

Without question, ethical issues are often more obvious in organic research than in non-organic food and agriculture studies. With some rare exceptions, the IFOAM Principles are not generally discussed in university level courses nor are they prominent in the mission statements of advisory services, organic conferences or academic journals. Indeed, the influence on, and contributions of researchers to identifying the ethical foundations of the IFOAM Principles is significant. Studies of ethics concerning organic farmers, or regarding animal husbandry, and consumer behavior are common. However, unlike several calls for research in developing countries that may ask how the proposed research contributes to socio-political or ethical goals such as the Millennium Development Goals, organic research calls do not require assessments of how the applied research contributes to the IFOAM Principles or other ethical dimensions. Furthermore, there is little or no research available on ethics and the activities of advisors, certifiers, processors or traders.

Despite the weak empirical evidence, it seems that much more could be done to raise the awareness for, to provide knowledge of, and to intensify the discourse on the IFOAM Principles in specific agrofood research.

14.1.8 Corporate Industrial Power

Today, the political and corporate industrial influences in organic have become increasingly significant, if not defining. Power in the organic system has shifted from farmers to largely corporately and business oriented hierarchical and institutionalized bureaucratic structures, and globalized markets. Moreover, the dominant large retailers do not contribute seriously to a shift toward a more socially and economically balanced relationship between organic farmers and commercial businesses. The IFOAM Principles do not appear to be of concern to these corporate actors and in many ways, the current neo-liberal system fundamentally contradicts the IFOAM Principles (James et al. 2012).

It is increasingly obvious that the social, economic, and some of the ecological values of organic are under pressure and are at risk of disappearing among various actors in the organic agrofood chain. Many farmers and consumers disagree with the influence of corporate power in organic, but at the same time find few political opportunities to bring about change.

Realistically, while organic may attempt to espouse ethical conventions and to delineate a moral economy, the organic systems cannot be separated from market conventions and, indeed the political economy (Brown and Getz 2008, p. 20). In other words, organic is both 'in and against the market' (see Raynolds 2000, p. 298; Alrøe et al. 2006; Nicholls and Opal 2008, p. 229).

14.2 Bringing Ethics into Organic Practice

Based on our review in this volume, IFOAM needs to rethink how to promote its values to all the actors in the organic system. In this way, IFOAM can distinguish organic from other value-oriented approaches and offer a convincing model that stands apart from other so called sustainable approaches. This final section discusses some of the activities that could contribute to creating a more value driven organic agrofood chain.

14.2.1 Need for a Social Agenda and a New Agrofood Contract with Society

One of the central observations throughout this volume is, that in those cases where social and economic values in the organic agrofood chain do play a minor role, organic is losing its ethical integrity. In contrast, numerous alternative guarantee systems, internal control systems, PGS, CSR, and food coops, but also closely associated movements like fair trade, slow food or transition towns in which "nearness" and shared responsibility are relevant, organic needs to become more authentic and in ways that embrace the IFOAM Principles of Fairness and Care. Consequently, IFOAM should invest in a social agenda that makes its social features more visible and explicit in concrete practice.

If the organic movement is to play a serious role in the future of world food and agriculture, and seeks to embrace and uphold the IFOAM Principles, coalitions with other social groups that share core values, as well as the diffusion of organic into diverse societal institutions, must be put squarely into the center of the organic debate. IFOAM and other organic organizations must also respond to the growing bifurcation of society between the largely unreflective food consumer who ignores ecological and social issues, and those who are more and more sensitized to new types of societal contracts for sharing food, resources, land and labor and to assuming responsibility for addressing food issues (Ravenscroft and Taylor 2009).

A clear and explicitly ethically oriented organic could lead such a broader movement that incorporates several alternative movements to establish "value" communities in mutually supportive relationships. This could include several types of social and ecologically oriented organizations at local and regional levels, new forms of national fair trade,[1] and new types of collaboration among rural and urban areas, including farmers and consumers. Such collaborative arrangements could also involve networks with non-profit organizations, such as Climate Alliances (Collier and Löfstedt 1997; Behringer et al. 2000), activists for human rights (see De Schutter 2010), food security (Power 1999) or food sovereignty (Windfuhr and Jonsén 2005) and others.

In such coalitions, the consumer should no longer take a "passive" position. The new consumer should become a "producer" and share the risk and responsibility for local production. These alliances are not limited to rural-urban relationships, but also include community level collaborative arrangements between local farmers and others. To establish this type of farmer-public collaboration, several types of new institutional forms for co-operation, economies and ownership types need to be created (Ravenscroft and Taylor 2009). Some of these will embody the Principles of what Dahlberg called, "regenerative food systems" (Dahlberg 1993). Such collaboration has consequences for the IFOAM structure. While IFOAM is largely farmer oriented, in the future it must become a more farmer-public organization. The Principles express the potential contribution that organic farming offers and they present a vision for improving agrofood systems globally (IFOAM 2009). They are formulated in general terms, and thereby lend themselves to being applied and followed outside the organic movement as well as within it (Alrøe and Kristensen 2004). Such a change in orientation might also offer a promising point of departure for bringing the IFOAM Principles into a more open public debate that could even go beyond organic. More generally, organic agriculture and specifically the incorporation of ethics into daily practices needs to be seen as a process for social learning (Röling and Wagemakers 1997) in a societal context.

14.2.2 Implementing Ethics in Organic Marketing

What is the potential to bring a more value centered approach into the corporate food industry? Large retailers are, and will continue to be vital for many organic farmers.

[1] http://www.wfto.com

Both urban and rural consumers hold the unique option of purchasing organic or conventionally produced food. While collaborating with (organic) industries may be viewed as "dancing with the devil," not investing in new forms of contracts, based on the IFOAM Principles, would be reckless. This will be, of course, one of the most challenging jobs in the future agenda of a value driven organic agrofood chain.

IFOAM must insist that the principles become part of organic labeling and thus become more visible commercially. While not all actors would agree 100 % with the IFOAM Principles, such a move could stimulate a much-needed critical debate on the role of values in the market.

Classic strategies for promoting the "organic idea" have involved bringing organic into public schools and in public kitchens. It will be important to expand the scope of these programs in order to incorporate discussions of, and sensitivity to the foundations of organic ethics and how organic is related to other food concerns including food costs, as well as regional and seasonal production. Much more is needed in order to bring the vision of a value driven organic agrofood chain into the mainstream.

14.2.3 Bringing Organic Ethics into Research and the Sustainability Debate

In this chapter we have emphasized our position that the IFOAM Principles provide an ethical foundation for more than food and agriculture. In research, the Principles could serve as an ethical framework that also addresses concerns beyond food and farming, such as those challenges related to the Millennium Development Goals. Organic research must not only study markets and consumers, soils, plants and animals. It is also critically important to promote research that addresses global security concerns, including climate change, free access to seed, migration, food security, food sovereignty, urbanization and rural development, gender and employment, land tenure and land grabbing, and obesity as well as malnutrition. Similar to calls that demand identifying the contribution of the research to sustainable development, we think that it is time that calls on organic include the question: how do the IFOAM Principles or ethical concerns contribute to addressing these global issues?

With respect to the Millennium Development Goals and in the light of the IFOAM Principles, organic ethically driven research should examine ways to strengthen and protect the weakest in the society. This should include more systematic assessments of low cost certification systems with minimum administrative requirements that ease the entry of smallholder farmers into the organic market. Moreover, further research is needed to understand how to assure both an income for organic farmers and affordable organic food for low-income customers.

Investigating the contributions of the IFOAM Principles to these broader concerns will require a clear commitment to transdisciplinary research. Simply, this means that we begin to require and foster collaborative development and implementation of research by all concerned citizens and scientists (Davies 2006). This also will require more serious commitments to, and funding for research that fully integrates natural and social science scientific approaches. In other words, citizens become much more than informants. They are actors alongside researchers. To do so will require fundamental changes in how research protocols are defined and implemented. Scientists will need to accept that citizens may have different, but equally valid, ways of knowing.

That does not mean that the science of the future should limit the freedom of researchers or that research must depend solely on its acceptance by citizens. The question is: how to invite open discussions for identifying and defining research problems that include the ethical dimensions. Perhaps this means that we must begin rethinking and redefining "scientific careers" that move us beyond the current pre-occupation with "sustainability," in order to bring transdisciplinary and ethically oriented research values, skills and perspectives into the mainstream.

But ethical concerns, alone, may be insufficient in providing a base for moving toward a sustainable future. This ethical perspective must be accompanied by more analyses that make the currently internalized, or hidden, social, ecological and economic costs of different agrofood chains, transparent. Social, economic and ecological externalities are central for assessing the sustainability of any agricultural approach, that also includes a serious calculation of food balances, food losses and inefficiencies along the agrofood chain, including several other critical social, economic and ecologic impacts of GMO application. Implementing this perspective into the research agenda and political agenda is central for the future of a competitive organic agrofood system.

14.3 Outlook

For the future, we must directly confront our current options. It appears that we are faced with a choice of "investing in the soil and community," or the stock market and so-called cheap and convenient food. If so, are there paths open for an ethically grounded reconciliation of these choices?

Given what we see as the overwhelming politico-economic power of the corporate, multi-national food and farming industry, taking a constructively optimistic perspective on the future may appear Pollyanna-like. As Günther Anders reminded us more than 50 years ago, seem to faced with a situation vis-à-vis corporate power in which "Whether we play the game or not, it is being played [on] us. Whatever we [act or not], our withdrawal will change nothing" (Günther Anders in Baumann 2008, p. 110). Nevertheless, knowing this, the challenge (and responsibility) is to take the organic principles and forge new value driven farmer consumer partnerships as the foundation for the development of a value driven future for the organic movement (Storstad and Bjørkhaug 2003; Hinrichs and Lyson 2007; Conner et al. 2008; Padel and Gössinger 2008; Zander et al. 2010).

14.3.1 Steps Forward

To secure a value driven organic movement, we see three areas in which organic actors could influence the future development of the movement.

Bringing values into the standards: The ethical foundations of organic agriculture should be promoted as a key orientation for (organic) agriculture and food legislation in national and international trade and governmental organizations (Padel et al. 2007b, 2010; Stolze and Lampkin 2009). The latest invitation of the EU for a "Consultation for the review of the European policy on organic agriculture" is a serious step forward and an option to bring ethical concerns more into the broader political and societal debate.[2] Also the latest activity on "Best Practice Reference for Agriculture and its Value Chains" developed by the Sustainable Organic Agriculture Action Network (SOAAN)[3] and approved for the global organic movement by IFOAM, provide a guide with a more detailed translation of the Principles into practices. The aim should be to better translate the values into organic standards specifically in leading organic Western countrys' governmental policies and regulations (Padel et al. 2007b, p. 29) and in international negotiations concerning food standards (Friedmann 2005).

Societal collaborations: For the diffusion and extension of an organic approach, the intensification of alliances with other societal movements with similar value patterns is crucial (Holt-Gimenez and Shattuck 2011). Thus, up to this time the future of organic lies in its ability to serve as an umbrella or a co-creator for diverse individual and societal movements which share similar sets of values. Alrøe and Kristensen (2004) argue that the Principles should be applied outside the organic movement as well. That is, the organic movement should no longer be separated from other value driven societal movements. The future of a value driven organic movement therefore will depend on a successful mutual political acting of diverse value communities.

Communicating and reflecting the values: If values are to play a more central role in the organic movement, they must be continuously reviewed and studied in all sectors of the organic agrofood chain. Specifically, those who are less subject than producers to the organic standards, such as retailers, advisors, researchers, certifiers, politicians and consumers, should be required to engage in discussions of organic values and how they apply to their professional work (see the recommendations in Sect. 13.2). These activities also include the development of common platforms/"value" communities with non-organic farmers who share interests and

---

[2]IFOAM is currently (2012–2013) in an intense process of implementing new activities,—e.g. the establishment of a global certification data base or the legislative process by consultation of "Best Practice Reference for Agriculture and Its Value Chains"; see http://www.ifoam.org/; http://ec.europa.eu/agriculture/organic/home_en

[3]http://www.ifoam.org/en/value-chain/ifoam-best-practice-program

support a common body of knowledge and initiate mutual learning processes (see Hunt et al., Chap. 10) (cf. Læssøe et al. 2012), that would help to reduce social barriers between organic and non-organic.

14.3.2 The Role of the Value Types in Promoting Organic

The debate on the "right way" is, of course, controversial (see Constance et al., Chap. 9, see Hunt et al., Chap. 10). Even the question of whether organic should expand or not is answered differently and depends on the region and specific interests (see Hunt et al., Chap. 10). Furthermore, there is a debate over the cost of the growth in organic. Given that the Econo/Market type already dominates the organic system, Lockie (2009) recommends a pragmatic and flexible interpretation of organic values to prevent organic from becoming a niche. Instead of applying a strict black and white approach, or criticizing the increase of organic convenience products and food (Klonsky 2000; Arvola et al. 2008) that might conflict with quality issues of the IFOAM Principles, Gottschalk and Leistner (2012) argue that large-scale supermarkets could become 'icebreakers' for organic products and induce further purchase. We accept this rather realistic idea. However, we suggest that such a strategy could also end up as a "laissez faire" approach that undermines the holistic intention of organic values. To clarify: Convenience products that are often non-seasonal and non-regional should comprise only a niche of the organic market not a dominant feature of it. In the long run, IFOAM and the organic movement respectively must insist that practices fit into its value framework, and use this to justify the difference from other food products, whether they are labeled or not as sustainable.

Expressed for the whole organic movement, we assume that following irregular distribution curve, the Enviro/Protest should assume a leading position that fits well with the majority of the organic values, but also those from other value driven societal movements, as Lyson (2004) formulated as "Civic Agriculture" that relinks agriculture and community (see Getter et al., Chap. 4; see Constance et al., Chap. 9).

The Econo/Market approach invites those who enter the organic approach without knowing many of the details, yet it builds bridges for those who intellectually and emotionally are not involved into the broader meaning of the organic values. If organic is to avoid ending up as a niche, the Econo/Market type must also play a significant role in the future development of organic agriculture. However, the rules for production, processing and trade need to be more closely aligned with the IFOAM Principles of Fairness and Care (see Freyer et al., Chap. 5; Constance et al., Chap. 9).

The Nature/Philo type is seen as the frontrunner, the innovator and initiator for new forms of holistic lifestyles and new societal collaborations. This type is central for the organic movement as a model for remembering and rethinking rules, and continually readjusting the system toward its values. It also provides the perspective for a more radical and consequent change for a value driven organic or even lifestyle approach, including several other sustainable practices.

14.3.3 Key Risks in the Organic Future

There are several critical developments in the international political and industrial arena that fundamentally frame the future of the organic movement, and that could paralyze the organic system. These call for attention:

- Laws, specifically when it comes to agricultural subsidies that do not honor an agriculture that is internalizing the environmental and social costs in their product price calculations
- The continuous world wide distribution of GMO, that put a GMO free organic production at risk; and related laws that do not protect organic farmers against GMO contamination (Hanson et al. 2004)
- The increase of bureaucratic and technical regulations that are economically unacceptable for small and medium scale farmers, excludes them from the market, weakening food security and hindering food sovereignty
- WTO or other international regulations threaten to undermine social and ecological principles of the organic system (Giovannucci 2003; Blay-Palmer 2005; González and Nigh 2005; Swinbank 2005)

These issues are becoming dominant in political norms and laws, and industrial practices. They fundamentally conflict with the IFOAM Principles, with regard to food quality and health, ecology, self-determination, justice and care, food security and food sovereignty (see Freyer et al., Chap. 5). They also threaten the practice of organic, and lead to a form of agriculture and food quality that would lose its foundation in the IFOAM Principles and Standards. This development is also critical for non-organic farmers. All non-GMO and small and medium size non-organic farms would be affected by such bureaucratization and industrialization of the agricultural sector.

14.3.4 Toward an Ethically Based Path

Thus, in our minds there is no question that an agricultural movement based on the ethical concerns expressed by the IFOAM Principles must fight for its place in world society. In order to offer a different and transparent approach that represents a convincing contrast to other so-called sustainable approaches, IFOAM and its associated member organizations must rethink how to promote and communicate their values to all actors in the organic system and in society. The future of the organic idea depends upon articulating commonly reflected and practiced values by those who produce process, trade and consume food and those who formulate policies and those who research, advise and certify. A broad range of institutions in the environment of the organic movement—education, advisory service, science, political, industries and a consumers who are able to decide between consumerism and citizenship (see Adamoli, Chap. 6) are the surrounding factors and actors intervening into the organic system. The IFOAM Principles must stand the test before today's challenging conditions—in developing a value driven organic agro-food chain, and to fulfill broader responsibilities for society (ecosystems services, food security, etc.), and their formal integration into political regulatory instruments (Dabbert and Eichert 2007). National and international debates on IFOAM Norms, the Principles, Standards and orientations for Accreditation, express an awareness of the continuous need for adaptations and reactions to changes inside of the organic movement and new societal challenges.[4] In this reality, organic must chart its own ethically based path.

References

Alrøe, H.F., and E.S. Kristensen. 2000. *Research, values, and ethics in organic agriculture-examples from sustainability, precaution, nature quality, and animal welfare*. Paper presented at the Eursafe 2000: Two systems-one world, Copenhagen.

Alrøe, H.F., and E.S. Kristensen. 2004. Basic principles for organic agriculture: Why? And what kind of principles? *Ecology and Farming* 36: 27–30.

Alrøe, H.F., and E. Noe. 2008. What makes organic agriculture move: protest, meaning or market? A polyocular approach to the dynamics and governance of organic agriculture. *International Journal of Agricultural Resources, Governance and Ecology* 7(1): 5–22.

Alrøe, H.F., J. Byrne, and L. Glover. 2006. Organic agriculture and ecological justice: Ethics and practice. In *Global development of organic agriculture: Challenges and prospects*, ed. N. Halberg, H.F. Alroe, and M. Trydeman Knudsen, 75–112. Wallingford: CAB International.

Arvola, A., M. Vassallo, M. Dean, P. Lampila, A. Saba, L. Lähteenmäki, and R. Shepherd. 2008. Predicting intentions to purchase organic food: The role of affective and moral attitudes in the theory of planned behaviour. *Appetite* 50(2): 443–454.

Baumann, Z. 2008. *Does ethics have a chance in a world of consumers?* Cambridge, MA: Harvard University Press.

Behringer, J., R. Buerki, and J. Fuhrer. 2000. Participatory integrated assessment of adaptation to climate change in Alpine tourism and mountain agriculture. *Integrated Assessment* 1(4): 331–338.

Blay-Palmer, A. 2005. Growing innovation policy: The case of organic agriculture in Ontario, Canada. *Environment and Planning C: Government and Policy* 23(4): 557–581.

Brown, S., and C. Getz. 2008. Towards domestic fair trade? Farm labor, food localism, and the family scale farm. *GeoJournal* 73(1): 11–22.

Carruthers, P. 2009. The land debate–'doing the right thing': Ethical approaches to land-use decision-making. In *What is land for?: The food, fuel and climate change debate*, ed. M. Winter and M. Lobley, 293–318. London: Earthscan.

Collier, U., and R.E. Löfstedt. 1997. Think globally, act locally?: Local climate change and energy policies in Sweden and the UK. *Global Environmental Change* 7(1): 25–40.

[4]e.g. the diverse activities documented at the IFOAM homepage: http://www.ifoam.org/

Conner, D.S., V. Campbell-Arvai, and M.W. Hamm. 2008. Value in the values: Pasture-raised livestock products offer opportunities for reconnecting producers and consumers. *Renewable Agriculture and Food Systems* 23(01): 62–69.

Dabbert, S., and C. Eichert. 2007. Public synthesis report on the scope of National Action Plans, their evaluation procedures and the operability and appropriateness of the developed evaluation concept at national level, as well as the impact of conflict/synergies and policy proposals for implementing the EU Action Plan in member states *Report D,* vol. 8.

Dahlberg, K. 1993. Regenerative food systems: Broadening the scope and agenda of sustainability. In *Food for the future: Conditions and contradictions of sustainability*, ed. P. Allen, 75–102. New York: Wiley.

Davies, G. 2006. Organic principles and research: What implications the new IFOAM principles of organic agriculture? *Aspects of Applied Biology* 79: 79–82.

De Schutter, O. 2010. Report submitted by the special rapporteur on the right to food. In *United Nations Human Right Council: New York*, ed. O. De Schutter. New York: United Nations Human Right Council.

Friedmann, H. 2005. From colonialism to green capitalism: Social movements and emergence of food regimes. In *Research in rural sociology and development*, vol. 11, ed. F.H. Buttel and A.J. McMichael, 227–264. Oxford: Elsevier.

Giovannucci, D. 2003. Emerging issues in the marketing and trade of organic products. In *Organic agriculture: Sustainability, markets, and policies*, ed. OECD, 187–198. Wallingford: CAB International.

González, A.A., and R. Nigh. 2005. Smallholder participation and certification of organic farm products in Mexico. *Journal of Rural Studies* 21(4): 449–460.

Gottschalk, I., and T. Leistner. 2012. Consumer reactions to the availability of organic food in discount supermarkets. *International Journal of Consumer Studies* 37: 136–142.

Hanson, J., R. Dismukes, W. Chambers, C. Greene, and A. Kremen. 2004. Risk and risk management in organic agriculture: Views of organic farmers. *Renewable Agriculture and Food Systems* 19(04): 218–227.

Hinrichs, C.C., and T.A. Lyson. 2007. *Remaking the North American food system: Strategies for sustainability*. Lincoln: University of Nebraska Press.

Holt-Gimenez, E., and A. Shattuck. 2011. Food crises, food regimes and food movements: Rumblings of reform or tides of transformation? *The Journal of Peasant Studies* 38(1): 109–144.

IFOAM. 2009. *The principles of organic agriculture*. Retrieved from http://www.ifoam.org/about_ifoam/principles/index.html. Accessed 23 Jan 2009.

James, H., M. Hendrickson, and P. Howard. 2012. Networks, power and dependency in the agrifood industry. In *The ethics and economics of agrifood competition*, ed. H. James, 99–126. Dordrecht/Heidelberg/New York/London: Springer.

Klonsky, K. 2000. Forces impacting the production of organic foods. *Agriculture and Human Values* 17(3): 233–243.

Læssøe, J., A. Ljungdalh, P. Kastberg, E. Noe, H.F. Alrøe, T. Christensen, A. Dubgaard, S.B. Olsen, and N. Kærgård. 2012. *Theories on motivation and their implication for supporting communication, learning and decision-making in relation to organic food systems*. Paper presented at the 10th European IFSA symposium: Producing and reproducing farming systems: New modes for sustainable food systems of tomorrow, Aarhus, Denmark. http://orgprints.org/21898/2/21898.pdf. Accessed 9 Aug 2014.

Lockie, S. 2009. Responsibility and agency within alternative food networks: Assembling the "citizen consumer". *Agriculture and Human Values* 26(3): 193–201.

Lyson, T.A. 2004. *Civic agriculture: Reconnecting farm, food, and community*. Medford: Tufts University. University Press of New England, Lebanon, NH, US.

Nicholls, A., and C. Opal. 2008. *Fair trade. Market-driven ethical consumption*. London: Sage.

Padel, S., and K. Gössinger. 2008. *Farmer consumer partnerships communicating ethical values: A conceptual framework*. CORE Organic Project Series Report, Tjele, Denmark.

Padel, S., N. Lampkin, and W. Lockeretz. 2007a. The development of governmental support for organic farming in Europe. In *Organic farming: An international history*, ed. W. Lockeretz, 93–122. Wallingford: CAB International.

Padel, S., H. Röcklinsberg, H. Verhoog, H. Fjelsted Alrøe, J. De Wit, C. Kjeldsen, and O. Schmid. 2007b. *Balancing and integrating basic values in the development of organic regulations and standards: Proposal for a procedure using case studies of conflicting areas*. Aberystwyth/Tjele: Danish Research Centre for organic Farming (DARCOF).

Padel, S., H. Röcklinsberg, and O. Schmid. 2009. The implementation of organic principles and values in the European regulation for organic food. *Food Policy* 34(3): 245–251.

Padel, S., J. B. Vine, M. Huber, M. Stolze, L.M. Jespersen, E. Rüegg, and S.R. Belliere. 2010. The European regulatory framework and its implementation in influencing organic inspection and certification systems in the EU. Deliverable 11, CERTCOST Project, agreement no. 207727. 7th Framework Programme, University of Kassel. http://www.certcost.org

Power, E.M. 1999. *Combining social justice and sustainability for food security for hunger-proof cities: Sustainable urban food systems*. Ottawa: International Development Research Centre.

Ravenscroft, N., and B. Taylor. 2009. Public engagement in new productivism. In *What is land for? The food, fuel, and climate change debate*, ed. M. Winter and M. Lobley, 213–232. London/Sterling: Earthscan.

Raynolds, L.T. 2000. Re-embedding global agriculture: The international organic and fair trade movements. *Agriculture and Human Values* 17(3): 297–309.

Röling, N., and M. Wagemakers. 1997. *Facilitating sustainable agriculture*. Cambridge: Cambridge University Press.

Shiva, V. 2000. *Stolen harvest: The hijacking of the global food supply*. London: Zed Books.

Stolze, M., and N. Lampkin. 2009. Policy for organic farming: Rationale and concepts. *Food Policy* 34(3): 237–244.

Storstad, O., and H. Bjørkhaug. 2003. Foundations of production and consumption of organic food in Norway: Common attitudes among farmers and consumers? *Agriculture and Human Values* 20(2): 151–163.

Swinbank, A. 2005. Developments in the Doha Round and WTO dispute settlement: Some implications for EU agricultural policy. *European Review of Agricultural Economics* 32(4): 551–561.

Thompson, P. 1988. Ethical dilemmas in agriculture: The need for recognition and resolution. *Agriculture and Human Values* 5(4): 4–15.

Thompson, P. 2008. Agrarian philosophy and ecological ethics. *Science and Engineering Ethics* 14(4): 527–544.

Windfuhr, M., and J. Jonsen. 2005. *Food sovereignty: Towards democracy in localized food systems*. Warwickshire: FIAN-International, ITDG Publishing.

Zander, K., U. Hamm, B. Freyer, K. Gössinger, M. Hametter, S. Naspetti, S. Padel, H. Stolz, M. Stolze, and R. Zanoli. 2010. *Farmer consumer partnerships-How to successfully communicate the values of organic food—A handbook*. Kassel: University of Kassel, Department of Agricultural and Food Marketing.

Index

B. Freyer, J. Bingen (eds.), *Re-Thinking Organic Food and Farming in a Changing World*, The International Library of Environmental, Agricultural and Food Ethics 22, DOI 10.1007/978-94-017-9190-8

Printed by Printforce, the Netherlands